Statehood under Water

Legal Aspects of Sustainable Development

Series Editor

David Freestone

This series publishes work on all aspects of the international legal dimensions of the concept of sustainable development. Its aim is to publish important works of scholarship on a range of relevant issues including conservation of natural resources, climate change, biodiversity loss and the role of international agreements, international organizations and state practice.

VOLUME 22

The titles published in this series are listed at *brill.com/lasd*

Statehood under Water

Challenges of Sea-Level Rise to the Continuity of Pacific Island States

By

Alejandra Torres Camprubí

TUTA SUB AEGIDE PALLAS 1683

BRILL
NIJHOFF

LEIDEN | BOSTON

Library of Congress Cataloging-in-Publication Data

Names: Torres Camprubí, Alejandra, author.
Title: Statehood under water : challenges of sea-level rise to the continuity of Pacific island states / by Alejandra Torres Camprubí.
Description: Leiden ; Boston : Brill Nijhoff, 2016. | Series: Legal aspects of sustainable development ; volume 22 | Based on author's thesis (doctoral—Universidad Autónoma de Madrid, 2014) issued under title: Climate change and international security:—Revealing new challenges to the continuation of Pacific islands' statehood. | Includes bibliographical references and index.
Identifiers: LCCN 2016013018 (print) | LCCN 2016013371 (ebook) | ISBN 9789004321601 (hardback : alk. paper) | ISBN 9789004321618 (Hardback)
Subjects: LCSH: Islands of the Pacific—International status. | Global warming—Law and legislation. | Sea level—Social aspects—Islands of the Pacific | Coast changes—Social aspects—Islands of the Pacific
Classification: LCC KZ4730 .T67 2016 (print) | LCC KZ4730 (ebook) | DDC 341.26—dc23
LC record available at http://lccn.loc.gov/2016013018

ISSN 1875-0923
ISBN 978-90-04-32160-1 (hardback)
ISBN 978-90-04-32161-8 (e-book)

This book is printed on acid-free paper and produced in a sustainable manner.

Para mis queridos padres, Charo y Santiago.
Quienes me iniciaron en el significado y las complejidades de 'la condición de Estado' desde mi más tierna infancia.

Contents

Series Editor's Preface

I am very pleased to be able to include *Statehood under Water* by Dr Alejandra Torres Camprubí as the twenty-second volume in the Brill Nijhoff series on *Legal Aspects of Sustainable Development* published under my General Editorship. The aim of this series is to publish works at the cutting edge of legal scholarship that address both the practical and the theoretical aspects of this important concept.

There is no doubt that this thoughtful and highly topical study fits very well into the objectives of the series. In its 2014 Fifth Assessment Report the Intergovernmental Panel on Climate Change (IPCC) predicted that global sea levels could rise nearly a metre by 2100, and continue to rise for several centuries thereafter, notwithstanding any action that may be taken to mitigate greenhouse gas emissions. The existential threat that this represent to low lying and small island nations is increasingly well known, but it also poses a challenge to our understanding of the legal requirements of statehood or of international legal personality – concept that we may need to rethink in this changing world. A world that, geologists tell us, may be entering a new epoch as a result of the impacts of *Homo sapiens* – the Anthropocene.

In this volume, a revised version of her doctoral thesis at the *Universidad Autónoma de Madrid*, Dr Torres Camprubí examines in detail the three key challenges that the small island states of the Pacific seem likely to face: deterritorialization, depopulation and government failure. She re-examines the origins of established concepts and explores a number of fascinating historical and political scenarios, contributing new thought to the ways in which these small countries might respond. She challenges what she calls "the classical view of the State as a monolithic structure with a vocation to exist over time."

Much has been, and will doubtless continue to be, written about what Professor Espósito Massicci in his Prologue calls the "sinking island paradigm." Dr Torres Camprubí provides a fresh new voice in this discourse and her work contains a number of innovative and arguably prescient approaches. Her work is a very thoughtful and constructive contribution to the discussion. This is not only an important academic study, but it also addresses an issue of real political and legal importance that in many ways symbolizes the challenge that climate change poses to current generations. I am delighted to include it in the series and I am sure it will find an extensive readership.

David Freestone
Washington DC

Prologue

The compelling title of this book reminds me of the story once told by David Foster Wallace about those two young fishes greeted by an older fish with a "Morning, boys, how's the water?" The young fishes continued swimming and, after a while, one of them asked, "What the hell is water?" The question urges us to reflect on how we think about things omnipresent, to pay attention to the elements and conditions of our very existence that are taken for granted most of the time and therefore somehow invisible, as it happens with the absence of camels in the *Arabian Nights*, as reported wittily by Jorge Luis Borges.

The allegory chosen by Alejandra Torres Camprubí belongs to such kind of endeavour: 'Statehood under water' is not just another international law interpretation of the sinking island image; the author's sophisticated research defies the ways in which we think about statehood in the Anthropocene – a powerful word that has encountered her at a late stage of her passionate intellectual voyage to provide a new persuasive contextual meaning to the whole argument. Indeed, her study of the effects of climate change and sea-level rise on the continuity of Pacific Island States reaches beyond the mere rigorous analysis of the crucial challenges facing these particular islands. The book certainly achieves such goal, most significant for the Pacific Island States, but also offers a mature and complex revision of the exclusionary Westphalian theory of statehood and its elements, which I predict will frame quite a few discussions on possible new forms of legal personality in international law, including entities with provocative names such as 'water states'.

With this timely book, Alejandra Torres Camprubí joins the many contemporary scholarly debates on the concerns about the adequacy of international law to face the challenges of the Anthropocene. Her careful analysis of different scenarios of partial and total de-territorialization and de-population, together with the concept of governments in exile are exemplary, but do not lead the author to an incautious conclusion. On the contrary, Alejandra Torres Camprubí avoids categorical answers about the continuity of the Pacific Island States, and therefore gives proper consideration and space to the normative evolution of the 'sinking island paradigm' and the future legal scenarios of the international law of the Anthropocene.

I have been fortunate to witness the birth and progress of this book, which is based on the doctoral thesis that Alejandra Torres Camprubí successfully defended at the University *Autónoma* of Madrid in 2014. Now, I am impressed by the balanced structure of the book, its refined analyses and learned discussions; most importantly, I am pleased to know that everyone will have the chance to learn from this book and enjoy this momentous scholarly contribution to some fundamental questions of international law.

Carlos Espósito Massicci
Madrid, 7 October 2015

Acknowledgements

This book would have not seen the light of day had it not been for a long list of people and institutions that believed in both the importance of conducting this project, and in my ability to navigate uncharted seas while keeping my head above water.

I feel indebted, first of all, to the *Universidad Autónoma de Madrid* (UAM), for granting me a very generous full-time doctoral scholarship at a time when the rule of austerity was already predominant in southern Europe. During more than four years, I was able to dedicate my time exclusively to diving into the deepest waters of a risky topic, and eventually establish the foundations of this book. Professor Carlos Espósito Massicci, who supervised my PhD, was and continues to be present every mile of the way, as a lighthouse I can always find above the horizon in stormy or foggy days.

His support also enabled me to develop my research as part of the wider project on 'Global Legal Goods', financed by the Spanish Ministry of Science and Innovation, and to share it with a vibrant scientific community. At the Department of Public Law and Legal Philosophy of the UAM, I had the chance to defend key aspects of this work before well-established scholars with different approaches to international law and politics, who made me understand the importance of addressing the intricacies of the concept of statehood with the right measure of both precaution and openness. I am therefore very thankful to Professors Javier Díez-Hochleitner, Soledad Torrecuadrada, Alfonso Iglesias Velasco, Carmen Martínez Capdevilla, and Irene Blázquez Navarro for participating in my seminars; and to Rosa María Fernández Egea for our discussions on international environmental law and the climate change regime, which encouraged my early publications. Jessica Almqvist and Laura Beck Varela influenced me greatly, as they persistently walked me through the fascinating hallways of international legal theory, critical legal history, and beyond. Nicolás Carrillo Santarelli and Javier Frutos Miranda were exceptional 'doctoral brothers' to share an office with – our dose of good humour, mutual support and *laisser-faire*, undoubtedly made my doctoral journey a better experience. Likewise, I am very grateful to my first students, for refusing to take my word for granted, and repeatedly confronting me with critical questions during my lectures on statehood and international legal subjectivity.

The *Universidad Autónoma de Madrid* also gave me the opportunity to enrich my research by spending two terms as a visiting scholar abroad. Professor Laurence Boisson de Chazournes hosted me in spring 2011 with great warmth at the Department of Public International Law and International Organisations of the *Université de Genève*. Her remarks greatly influenced the final structure of my doctoral thesis, which remain reflected in this book. As I enjoyed bathing in the old times and solemn atmosphere of the *Palais des Nations*' library, Makane Mbengue, Mara Tignino, Antonella Angellini, Eva Fernández and Natalia Delgado filled my stay there with joyfulness and companionship. A year later, I was able to come back to the Faculty of Laws of University College London where I had previously studied, thanks to the invitation of Professor Catherine Redgwell. Our discussions on the role of territory in international law and its place in the concept of statehood gave me the confidence I needed at the time to pursue this project in the way it looks now.

The first stage of this 'transoceanic exploration' culminated on 26 March 2014 with the defence of my doctoral thesis before an international jury, presided by Professor Antonio Remiro Brotóns (*Universidad Autónoma de Madrid*). My gratitude goes to him and to Professors Davor Vidas (Fridtjof Nansen Institute), Luis Hinojosa (*Universidad de Granada*), Jorge Viñuales (Cambridge University), and Catarina García Segura (*Universidad Pompeu Fabra*) for their insightful comments on history, the law on statehood, international environmental law, and international security studies. On that day, the seeds for the second round of reflection on the issue of State extinction for climate change impacts were planted, the fruits of which are now reproduced in this book.

From there on, Professor Davor Vidas guided the second stage of this exploration in many crucial ways. Firstly, thanks to him I had the opportunity to continue my post-doctoral research as a member of an international and interdisciplinary project. The support of the Research Council of Norway, within its KLIMAFORSK Programme, to the research project (no. 235638) on 'Climate Change and Sea Level Rise in the Anthropocene: Challenges for International Law in the 21st Century' (2014–2018), hosted by the Fridtjof Nansen Institute in Oslo, and under which the research for this article was undertaken, is gratefully acknowledged. Through this project, as well as, through the ILA Committee on International Law and Sea-Level Rise chaired by Professor Vidas, I was able to hear more on the main issues raised in my doctoral thesis by the very visionaries who had started to deal with them years before the impacts of climate change and sea-level rise became so acute. Amongst those visionaries is Professor David Freestone, editor-in-chief of the Brill/Nijhoff series *Legal Aspects on Sustainable Development* to which this book now belongs. I am deeply grateful to Professor Freestone and the rest of the editorial team for their trust and the opportunity to publish this reviewed version of my doctoral thesis is such a renowned publishing firm.

Besides, in the context of the inter-sessional meetings of the ILA Committee, Professor Clive Schofield (University of Wollongong) volunteered to help me incorporate to this book a cartographic image of the region. I am therefore very thankful to him and, in particular, to Professor I Made Andi Arsana (*Universitas Gadjah Mada Gedung*), for producing the map illustrating the maritime claims and boundaries in the South Pacific. In addition, thanks to my collaboration with the FNI project, I was introduced into the uncharted waters of the Anthropocene and had the privilege to learn about this new geological epoch from scientific world experts such as Jan Zalasievicz and Mark Williams. As reflected in this book, learning about the Anthropocene was a conceptual discovery that made me put all previous work on State extinction for climate change impacts into a wider perspective. Encouraged by Professor Vidas, in November 2014 I attended the Anthropocene Campus, a scholarly project hosted by the *Haus der Kulturen der Welt* in Berlin which intended to explore new forms of knowledge production, in and about the Anthropocene. I am very grateful to all the organisers and participants of the Campus for the very inspiring discussions on the philosophical, historical, scientific, and artistic underpinnings of the Anthropocene.

Since I drafted the first lines of this research project from a top-roof flat in central London 7 years ago, conversations with Mario Prost and Jean d'Aspremont have been very inspiring. Founding the Interest Group on International Environmental Law of the European Society of International Law with Mario, with an idea to open a door to non-mainstream approaches to international environmental law, enriched me greatly. I am grateful to Professor Hélène Ruiz-Fabri for encouraging this project at the time when she was the President of the Society, and to all the members of the Interest Group for keeping this platform active through their continuous participation. My gratitude also goes to Jean for his patient listening and stimulating brainstorming at different stages of the evolution of this work, as well as for inviting me to share my first full draft with the members of the Department of Public International Law of the University of Manchester. I am also very grateful to Ray Griffiths for his precious critique and proofreading of different versions of this work.

What began as a research project makes more sense every day since I joined the International Litigation and Arbitration Department of Foley Hoag. I am particularly thankful to Paul Reichler for giving me the opportunity to approach statehood and the conduct of international relations from the daily and concrete reality that one finds in international legal practice. My gratitude also extends to all the members of the Department, and especially to my *consoeurs* and *confrères* at the Paris office for their trust and daily support. Needless to say, all opinions expressed in this book remain of my sole responsibility, and do not represent the view of any other person, State, or institution.

Before, during and after the doctoral research, I have been navigating side by side with Alfonso Rengifo, Aurora Mateos, Luciano Donadio Linares, Raquel Miranda, Flore González and Elena Fernández Frontán. They have always been the greatest *compañeros de camino* and I am grateful for their genuine love and friendship. Thanks to Aurora, I was able to participate in the 2009 Copenhagen Climate Change Summit. On the last night of the negotiations, I directly witnessed the dramatic moment when former President of the Maldives, Mohammed Nasheed, begged in tears for the Copenhagen Accord to be adopted as a decision of the Conference of the Parties. He fully embodied the struggle of low-lying island States, and the memory of that night has remained at the back of my mind during all stages of development of this work.

Stefan Ćirić is the *terra firme* I found when this journey was getting to its end. He helped me reach the shore, influenced my understanding of what this long exploration had represented, and inspired the title of this book from one of our favorite songs. My parents, Charo and Santiago, have always helped me find my inner compass when travelling on unmarked roads in all aspects of life. Their kindness, love and support have made this book, and everything else, possible. For this reason, this book is dedicated to them two.

Paris, 24 September 2015

List of Relevant International Instruments and National Legislation

International Instruments

Multilateral Treaties and Declarations (in chronological order of signature or adoption)

Convention on the Rights and Duties of States (Montevideo Convention), adopted on 26 December 1933, entered into force on 26 December 1934, 165 L.N.T.S 19. Text reprinted in *American Journal of International Law*, vol. 28 (Supp) (1974), p. 74.

United Nations Declaration on Human Rights, G.A. Res. 217A (III), 10 December 1948, U.N. Doc.: A/810, p. 71.

Convention Relating to the Status of Refugees, adopted on 22 July 1951, entered into force on 22 April 1954, 189 U.N.T.S., p. 137.

Convention Relating to the Status of Stateless Persons, adopted on 28 September 1954, entered into force on 6 June 1960, 360 U.N.T.S., p. 117.

Convention on the Reduction of Statelessness, adopted on 30 August 1961, entered into force on 13 December 1975, 189 U.N.T.S., p. 175.

International Covenant on Civil and Political Rights, adopted on 16 December 1966, entered into force 23 March 1976, 999 U.N.T.S., p. 171.

International Covenant on Economic, Social and Cultural Rights, adopted on 16 December 1966, entered into force on 3 January 1976, 933 U.N.T.S., p. 3.

Protocol Relating to the Status of Refugees, adopted on 31 January 1967, entered into force on 4 October 1967, 606 U.N.T.S., p. 267.

Vienna Convention on the Law of Treaties, adopted on 22 May 1969, entered into force on 27 January 1980, 1155 U.N.T.S., p. 331.

Declaration of the United Nations Conference on Human Development, adopted in Stockholm on 16 June 1972.

Convention on the Elimination of all Forms of Discrimination against Women, adopted on 18 December 1989, entered into force on 3 September 1981, 1249 U.N.T.S., p. 13.

United Nations Convention on the Law of the Sea, adopted on 10 December 1982, entered into force on 16 November 1994, 1833 U.N.T.S., p. 3.

Convention on the Rights of the Child, adopted on 20 November 1989, entered into force on 2 September 1990, 1577 U.N.T.S., p. 3.

International Convention on the Protection of the Rights of All Migrant Workers and Members of Their Families, adopted on 18 December 1990, entered into force on 1 July 2003, 2220 U.N.T.S., p. 3.

Convention on Biological Diversity, adopted on 5 June 1992, entered into force on 29 December 1993, 1760 U.N.T.S., p. 79.

Convention to Combat Desertification in Those Countries Experiencing Serious Drought and/or Desertification, Particularly in Africa, adopted on 17 June 1994; entered into force on 26 December 1996, 1954 U.N.T.S., p. 3.

Maritime Delimitation Agreements in the Pacific (in chronological order of signature)

Australia–Papua New Guinea, Treaty between Australia and the Independent State of Papua New Guinea Concerning Sovereignty and Maritime Boundaries in the Area between the Two Countries, Including the Area Known as Torres Strait, and Related Matters, signed on 18 December 1978, entered into force on 15 February 1985, 18 I.L.M. 291 (1979).

France (Wallis and Futuna)–Tonga, Convention between the Government of the French Republic and the Government of the Kingdom of Tonga on the Delimitation of Economic Zones, signed on 11 January 1980, entered into force upon signature, *Maritime Boundary Agreements* (1970–84); 273 (1987).

Cook Islands–United States (American Samoa), Treaty between the United States of America and the Cook Islands on Friendship and Delimitation of the Maritime Boundary between the United States of America and the Cook Islands, signed on 11 June 1980, entered into force on 8 September 1983, *Maritime Boundary Agreements* (1970–84); 286 (1987).

Fiji-France (New Caledonia, Wallis and Futuna), Agreement between the Government of the Republic of France and the Government of Fiji Relating to the Delimitation of their Economic Zone, signed 19 January 1983, not yet in force, Maritime Boundary Agreements (1970–84); 276 (1987).

France (French Polynesia)–United Kingdom (Pitcairn, Henderson, Ducie and Oeno Islands), Convention on Maritime Boundaries between the Government of the French Republic and the Government of the United Kingdom of Great Britain and Northern Ireland, signed on 25 October 1983, entered into force on 12 April 1984, *Maritime Boundary Agreements* (1970–84), 280 (1987).

France (Wallis and Futuna)–Tuvalu, Exchange of Notes between France and Tuvalu Constituting an Agreement Concerning Provisional Maritime Delimitation between the Two Countries, signed on 6 August 1985 and on 5 November 1985, entered into force on 5 November 1985, 1506 U.N.T.S. 35 (N° I-25964).

Australia–Solomon Islands, Agreement between the Government of Solomon Islands and the Government of Australia Establishing Certain Sea and Sea-Bed Boundaries, signed on 13 September 1988, not yet in force, *Law of the Sea Bulletin*, vol. 12 (1988), p. 19.

Papua New Guinea–Solomon Islands, Treaty between the Independent States of Papua New Guinea and Solomon Islands Concerning Sovereignty, Maritime and Seabed Boundaries between the Two Countries, and Co-operation on Related Matters, signed on 25 January 1989, but not yet in force, *International Maritime Boundaries* (1993), pp. 1162–1165.

Niue–United States (American Samoa), Treaty between the Government of the United States of America and the Government of Niue on the Delimitation of a Maritime Boundary, signed on 13 May 1997, but not yet in force, *Law of the Sea Bulletin*, vol. 35 (1997), p. 104.

France (Wallis and Futuna)–New Zealand (Tokelau), Agreement between the Government of New Zealand and the Government of the French Republic Concerning the Delimitation of the Maritime Boundaries between Wallis and Futuna and Tokelau, signed on 30 June 2003, entered into force on 12 November 2003, published on the Tokelau information page (www.nzlii.org).

Federated States of Micronesia–Marshall Islands, Treaty between the Federated States of Micronesia and the Republic of the Marshall Islands Concerning Maritime Boundaries and Co-operation on Related Matters, signed on 5 July 2006, but not yet in force, published on the Website of the Office of the President, Republic of the Marshall Islands.

Federated States of Micronesia–Palau, Treaty between the Federated States of Micronesia and the Republic of Palau concerning Maritime Boundaries and Co-operation on Related Matters, signed on 5 July 2006, in force since 2008, unpublished.

Cook Islads–Kiribati, Agreement between the Government of the Cook Islands and the Government of the Republic of Kiribati concerning the Delimitation of the Maritime Boundaries between the Cook Islands and the Republic of Kiribati, signed on 19 August 2012, but not yet in force, unpublished.

Kiribati–Marshall Islands, Agreement between the Republic of Kiribati and the Republic of the Marshall Islands concerning Maritime Boundaries, signed on 29 August 2012, but not yet in force, unpublished.

Kiribati-Marshall Islands–Nauru, Agreement between the Republic of Kiribati, the Republic of the Marshall Islands and the Republic of Nauru concerning the Determination of the Tri-junction Point between the Three Countries, signed on 29 August 2012, but not yet in force, unpublished.

Kiribati–Nauru, Agreement between the Republic of Kiribati and the Republic of Nauru concerning Maritime Boundaries, signed on 29 August 2012, but not yet in force, unpublished.

Kiribati–New Zealand, Agreement between the Government of New Zealand and the Government of the Republic of Kiribati concerning the Delimitation of the Maritime Boundaries between Tokelau and Kiribati, signed on 29 August, but not yet in force, unpublished.

Marshall Islands–Nauru, Agreement between the Republic of the Marshall Islands and the Republic of Nauru concerning Maritime Boundaries, signed on 29 August 2012, but not yet in force, unpublished.

Kiribati–United States, Treaty between the Government of the United States of America and the Government of the Republic of Kiribati on the Delimitation of Maritime Boundaries, signed on 6 September 2013, but not yet in force, unpublished.

Fiji–Tuvalu–France (Wallis And Futuna), Agreed Minutes of the Meeting between the Delegations of the Republic of Fiji, the French Republic, and Tuvalu concerning the Determination of the Tri-Junction Point between Fiji, the French Republic (Wallis and Futuna) and Tuvalu, signed on 9 December 2014, unpublished.

National Instruments

Unilateral Declarations of Maritime Rights by Pacific Island States (in alphabetical order)

Cook Islands: Territorial Sea and Exclusive Economic Zone Act n° 16/1977; Continental Shelf Act, 1964.

Federated States of Micronesia: Territory, Economic Zones and Port Entry, FSM Consolidated Legislation Title 18/Chap. 1.

Fiji: Marine Spaces Act [CAP 158A], Act n° 18/1977, Amended by n° 15/1978; Continental Shelf Act [CAP 149], n° 9/1970.

Kiribati: Marine Zones (Declaration) Act, 1983.

Marshall Islands: Marine Zones (Declaration) Act, 1984 [33MIRC Chap. 1].

Nauru: Sea Boundaries Act, 1997.

New Zealand (Niue): Territorial Sea and Exclusive Economic Zone Act n° 220/1996.

New Zealand (Tokelau): Territorial Sea and Exclusive Economic Zone Act, 1977.

Palau: Constitution of the Republic of Palau, 1979, Article 1, as amended on 15 July 2005.

Papua New Guinea: National Seas Act, 1997; Continental Shelf (Living Resources) Act, Chap. 210 of Consolidated Legislation; Oil and Gas Act, 1998.

Samoa: Maritime Zones Act n° 18/1999; Amendment Act n° 13/2004.

Solomon Islands: Delimitation of Marine Waters Act No. 32, 1978, Chapter 95 of Consolidated Legislation; Declaration of Archipelagic Lines, 1979; Continental Shelf Act 1970, Chapter 94 of the Consolidated Legislation.

Tonga: Territorial Sea and Exclusive Economic Zone Act nº 30, 1978.

Tuvalu: Marine Zones (Declaration) Act, 1983 [CAP24A of Consolidated Legislation].

Vanuatu: Maritime Zones Act (nº 6/2010).

Other Legislation and Policy Instruments of Pacific Island States

Australia, Australian Labor Party, 2007 Migration (Climate Refugees) Amendment Act of the 1958 Australian Migration Act. Amendment proposal available at: http://www.austlii.edu.au/au/legis/cth/bill/mrab2007342/ (5 May 2015).

Australia, Australian Strategic Policy Institute, 'Our Near Abroad: Australia and Pacific Islands Regionalism', November 2011, pp. 1–84.

Fiji, Human Rights Commission, Bill of Rights: International Legal Analysis Updated 2009, available at the website of Fiji's Human Rights Commission at: http://www.fhrc.org.fj/index.php?option=com_docman&Itemid=12. (5 May 2015).

Finland, Finnish Aliens Act (301/2004), with amendments up to 973/2007 included, Section 109(1).

Kiribati, *National Adaptation Programme for Action* (Ministry of Environment and Social Development, 2007).

Kiribati, *Kiribati Land Acquisition and Resettlement Policy Framework* (Tarawa, 2005 – amended 2011).

Maldives, *National Adaptation Programme of Action* (Ministry of Environment, Energy and Water, December 2006).

Solomon Islands, *National Adaptation Programme for Action* (Ministry of Environment, Conservation and Meteorology, November 2008).

Samoa, *National Adaptation Programme for Action* (Ministry of Natural Resources, Environment and Meteorology, 2005).

Sweden, Swedish Aliens Act (2005/716).

Tonga, *Joint National Action Plan on Climate Change Adaptation and Disaster Risk Reduction (2011–2015)*, Second National Communication Project (Ministry of Environment and Climate Change and National Emergency Management Office, July 2010).

Tuvalu, *National Adaptation Programme for Action* (Ministry of Resources, Environment, Agriculture and Lands, Department of Environment, May 2007).

Vanuatu, *National Adaptation Programme for Action* (National Advisory Committee on Climate Change, 2007).

List of Cases

Permanent Court of International Justice/International Court of Justice

Settlers of German Origin in Poland, Advisory Opinion, 1923 P.C.I.J. (Ser. B) 6.
Conditions of Admission of a State to Membership in the United Nations, Advisory Opinion, 1948 ICJ Reports 62.
Competence of the General Assembly for the Admission of a State to the U.N., Advisory Opinion, 1950 ICJ Reports 4.
North Sea Continental Shelf (Federal Republic of Germany v. Denmark), 1969 ICJ Reports 3.
Legality of the Threat or Use of Nuclear Weapons, Advisory Opinion, 1996 ICJ Reports 226.
Gabcikovo-Nagymaros Project (Hungary/Slovakia), Judgment, 1997 ICJ Reports 7.
Continental Shelf (Lybia Arab Yamahiriya/Malta), Judgment, 1985 ICJ Reports 13.
Delimitation of the Maritime Boundary in the Gulf of Maine Area (Canada/United States of America), Judgment, 1984 ICJ Reports 246.
Maritime Delimitation in the Area between Greenland and Jan Mayen (Denmark v. Norway), Judgment, 1993 ICJ Reports 38.
Maritime Delimitation and Territorial Questions between Qatar and Bahrain (Qatar v. Bahrain), Merits, Judgment, 2001 ICJ Reports 40.
Land and Maritime Boundary between Cameroon and Nigeria (Cameroon v. Nigeria: Equatorial Guinea intervening), Judgment, 2002 ICJ Reports 303.
Territorial and Maritime Dispute between Nicaragua and Honduras in the Caribbean Sea (Nicaragua v. Honduras), Judgment, 2007 ICJ Reports 659.
Delimitation in the Black Sea (Romania v. Ukraine), Judgment, 2009 ICJ Reports 61.

Arbitral Tribunals

Deutsche Continental Gas-Gesellschaft vs. Poland, Mixed Arbitral Tribunal, Annual Digest of Public International Law, vol. 11 (1929).
Award of the Arbitral Tribunal in the Matter of an Arbitration between Barbados and the Republic of the Trinidad and Tobago, The Hague, 11 April 2006.

National Courts

New Zealand

Refugee Appeal 72719/2001, RSAA (17 September 2001) – (Tuvaluan claimant).
Refugee Appeal 72313/2000, RSAA (19 October 2000) – (Tuvaluan claimant).
Refugee Appeal 72314/2000, RSAA (19 October 2000) – (Tuvaluan claimant).

Australia

10004726[2010] RRTA 845 (30 September 2010) – (Tongan claimant).
0907346 [2009] RRTA 1168 (10 December 2009) – (I-Kiribati claimant).
N00/34089 [2000] RRTA 1052 (17 November 2000) – (Tuvaluan claimant).

List of Acronyms

ANZUS	Australia, New Zealand, United States Security Treaty
AOSIS	Association of Small Island States
APB	Autonomous Province of Bougainville
AusAid	Australian Aid Agency
BPC	British Phosphate Commission
BRICS	Brazil, Russia, India, China and South Africa
CEO	Chief Executive Officer
CERD	Convention for the Elimination of All Forms of Racial Discrimination
CESCR	Commission on Economic Social and Cultural Rights
CLCS	Commission on the Limits of the Continental Shelf
COP	Conference of the Parties (UNFCCC)
CRC	Convention on the Rights of the Child
DNV	Det Norske Veritas
EEZ	Exclusive Economic Zones
EU	European Union
FNI	Fridtjof Nansen Institute
FSM	Federated States of Micronesia
GEF	Global Environment Facility
GHG	Greenhouse gas
HFA	Hyogo Framework for Action
HLP	Housing, land and property (rights)
IASC	Inter-Agency Standing Committee
ICCPR	International Covenant for Civil and Political Rights
ICESCR	International Covenant on Economic, Social and Cultural Rights
ICJ	International Court of Justice
ILA	International Law Association
ILC	International Law Commission
IPCC	Intergovernmental Panel on Climate Change
ISDR	International Strategy for Disaster Reduction
JNAP	Joint National Action Plans
LDC	Least Developed Countries
M	Nautical Miles
NAPA	National Adaptation Programmes of Action
NATO	North Atlantic Treaty Organisation
NGO	Non-Governmental Organisation

NHRI	National Human Rights Institutions
NPT	Non-Proliferation Treaty
OPEC	Organisation of Petroleum Exporting Countries
OSCE	Organisation for Co-operation and Security in Europe
PAC	Pacific Access Category
PACC	Pacific Adaptation to Climate Change Project
PHRD	Policy and Human Resources Development (Japan)
PIF	Pacific Islands Forum
RMI	Republic of the Marshall Islands
SASC	Special Advisory Services Commission
SOPAC	South Pacific Applied Geoscience Commission
SPC	South Pacific Commission
SPREP	South Pacific Region Environment Programme
TWAIL	Third World Approaches to International Law
UN	United Nations
UNCLOS	UN Convention on the Law of the Sea
UNDP	United Nations Development Programme
UNEP	United Nations Environment Programme
UNFCCC	United Nations Framework Convention on Climate Change
UNHCR	United Nations High Commissioner for Refugees
UNITAR	United Nations Institute for Training and Research
USAID	U.S. Agency for International Development
USD	United States dollar
VCLT	Vienna Convention on the Law of Treaties

List of Figures and Tables

Figure

Tables

Introduction: The Law on Statehood in the Anthropocene

> *All else will be immaterial if statehood is lost.*
>
> M. VUNIBOBO, Fiji's delegate, UN General Assembly, 9 June 2009.

0.1 'Statehood under Water': An Allegory of the Law on Statehood in the Anthropocene

In recent years, the 'sinking island paradigm' has emerged as one of the latest and perhaps most paradigmatic images of today's post-modern global environmental crisis. This expression essentially refers to the present situation of low-lying island States, which are currently experiencing how the long-established promise of State continuity progressively fades away in the face of climate change and sea-level rise. What could be seen as yet another exaggerated representation of the impacts of climate change on legal institutions – such as statehood – is in fact an accurate reflection of the reality of a handful of States. This new form of existential threat is indeed particularly sensed among some of the twelve independent Pacific Island States. Virtually at the mercy of the whims of the southwest Pacific Ocean, Tuvalu, Kiribati and the Marshall Islands stand out as being particularly endangered, as they are all three mostly composed of low coral islands and atolls of only a few meters of altitude above sea-level.[1] The Republic of the Maldives, in the Indian Ocean, is the fourth State in the world to embody this most extreme scenario.

Captured by the media in recent years, this image and its underlying issue have eventually been put on the international political agenda. The origins of the media's attention can be specifically traced back to the relocation in 2005 of the inhabitants of the Carteret Islands, located in the Autonomous Province of Bougainville of Papua New Guinea, as well as that of the Lateu settlement in Vanuatu. In addition to competing for the title of 'first climate refugees' in history, these two events also anticipated the beginning of a political move, headed by Pacific Island States before the main organs of the United

1 According to the *CIA World Factbook*, the highest point in Tuvalu is at 5 m above sea-level. The Marshall Islands' highest altitude is of 10 m (on Likiep), while in Kiribati it is of 81 m (Banaba). See: https://www.cia.gov/library/publications/the-world-factbook/ (14 September 2015).

 | DOI 10.1163/9789004321618_002

Nations since 2007. Through the internationalisation of the issue, the threat to extinction due to climate change impacts has been progressively framed by the international community as an existential threat with the potential to constitute an international security problem.[2] Setting aside the different stages of the evolution of this movement,[3] its very existence has already had important effects. First of all, through expressing their concern for their own continuity in the context of an unprecedented scenario before the UN General Assembly and the UN Security Council, Pacific Island States have recalled how the creation of post-colonial States came to colour the monolithic approach to statehood upon which traditional doctrinal law depended significantly.[4] This early predominant view, which focused on one single State form – the 'territorial State' as a power-based and firmly enshrined institution – essentially established one model of "full international legal personality" difficult to replicate.[5] The existence of such a political move yet revealed that Pacific Island States, as vulnerable and weak players in international affairs as they might be, are, nevertheless, not voiceless.

Secondly, the securitisation narrative serves to highlight the fact that the case of Pacific Island States is extreme not only with regard to the possible extinction of their international legal personality, but also to the fundamentals of today's international law. Located in the *terra incognita* of the peripheries, at the boundary between international law, international politics, and international legal theory, the issue of State extinction due to climate change impacts reveals the limitations of the international legal order itself to adapt to the consequences of the Anthropocene, but also opens the door

2 For an overview of the three main stages of the 'securitisation' of climate change, which led to the recognition of sea-level rise as an existential threat to low-lying island States, see the Verbatim Record of Security Council debate on 'Energy, Security and Climate Change', 17 April 2007, UN doc.: S/PV.5663; followed two years later by the adoption of General Assembly Resolution 63/281 on Climate Change and Its Possible Security Implications, 3 June 2009, 85th plenary meeting, UN doc.A/RES/63/281; and developed further, again before the Security Council, in its Presidential Statement on Maintenance of International Peace and Security and Impacts of Climate Change, 20 July 2011, UN doc. S/PRST/2011/15. As of July 2015, the Security Council has continued to address the question of the specific vulnerabilities of Small Island Developing States in relation to international peace and security, including climate change; see Security Council Report, Monthly Forecast, July 2015, 19.

3 This will be referred to as the Climate Change and International Security Discourse. The details of its launch are developed in chapter 3.2.1.B.

4 Hillary Chalesworth and Christine Chinkin, *The Boundaries of International Law* (Manchester: Manchester University Press, 2000), 124.

5 Ibid.

to new opportunities for development.[6] For indeed, as much as the 'sinking island paradigm' reflects the concrete, present condition of low-lying Pacific Island States, the unprecedented challenge they face necessarily calls for a more transcendental analysis. It is perhaps because of this special impulse that, rather unsurprisingly, the appeal of this issue often begins at least among western scholars with the reminiscence of the Atlantis legend as narrated by Plato in his *Critias' Dialogue*.[7] Mythological references are not to blame, as the possibility for Pacific islands' statehood to succumb 'underwater' is, from an ontological perspective, twofold. As much as it constitutes a concrete reality, it equally embodies a timeless allegory, and represents another chapter in the sustained and long-lasting fascination of international law for islands that Antonio Remiro Brotóns masterly depicted.[8] The study of Pacific islands' jeopardised statehood should thus enlighten us as much about the limits of international law , as about the excesses of our civilisation.

0.2 The Oscillation of the Law on Statehood: Between Stability and Flexibility

Much has been said about the role that the State plays in international society. "International law's main device for representing the world";[9] the

6 A presentation of the main characteristics of the Anthropocene, including the meaning of this term is developed in section 0.3 below.

7 Plato, *Timeos and Critias* (Oxford: Oxford University Press, 2009): "[S]o it was that Poseidon received as one of his domains the island of Atlantis and he established dwelling places for the children he had fathered of a mortal woman in a certain place of the island that I shall describe. [N]ow, seaward, but running along the middle of the entire island, was a plain which is said to have been the loveliest of all plains and quite fertile. Near this plain in the middle of the island and at about fifty stats distance was a uniformly low and flat hill. [T]o make the hill on which she lived a strong enclosure he broke it to form a circle and he created alternating rings of sea and land around it. Some he made wider and some he made more narrow. He made two rings of land and three of sea as round as if he had laid them out with compass and lathe. They were perfectly equidistant from one another. And so the hill became inaccessible to humans".

8 Antonio Remiro Brotóns, 'About Islands', in Lilian del Castillo, ed., *Law of the Sea, From Grotius to the International Tribunal for the Law of the Sea: Liber Amicorum Judge Hugo Caminos* (Leiden and Boston: Brill Nijhoff, 2015), 318–336, at 318.

9 Karen Knop, 'Statehood: Territory, People, Government', in James Crawford and Martti Koskenniemi, eds., *The Cambridge Companion to International Law* (Cambridge: Cambridge University Press, 2012), 95–117, 95.

origin of "international law's value";[10] the "most complete expression of international legal personality";[11] that is, the entity with the "greatest capacity to claim rights and duties under international law".[12] Whichever is the expression chosen by the legal scholar – some more metaphorical, apologetic or poetic than others – they all tend to reiterate one same idea: that the State is a structural and a defining concept of the international legal order. Even contemporary moves refusing to acknowledge the State's ontological monopoly, and seeking to challenge it by upholding, in contrast, the importance of the individual or of the peoples as the ultimate recipients of any legal rule,[13] necessarily do so by referring to the State. Whether one philosophically and ethically chooses either to uphold or to discourage the State's central role in the international order, one thing is clear, that applies to all – the State 'is'. *Qua* socio-political reality, it cannot be denied. *Qua* legal concept, it cannot be by-passed.[14]

The burden of attributing legal consequences to the complex range of facts that contribute to the creation or to the extinction of a State lies in what is often referred to as 'the law on statehood'. A timely debate remains open about the truly 'legal' nature of this highly controversial area of international law, inherently dependent on the regulatory and explanatory agendas of international lawyers themselves, and constrained by their professional dynamics.[15] This "scholarly hunger for doctrinal domestication", as Jean d'Aspremont puts it, has fed the progressive grasp of the phenomena of State creation and State

10 Hugo Krabbe, 'L'Idée Moderne de l'État', *Recueil des Cours de l'Académie de Droit International*, vol. 13 (1926): III, 576: "le droit international doit également emprunter sa valeur au pouvoir de l'État".

11 Chalesworth and Chinkin, 124. As the authors point out, "other entities may assert some degree of international personhood *for particular purposes*", such as individuals and NGOs. [Emphasis added].

12 Ibid.

13 Charles Rousseau, *Droit international public*, (Paris: Sirey, 1970), vol. II (Les Sujets de droit), 9. The author distinguished between the 'classical theory' – where States are considered the only subjects of international law – and the 'realist theory', developed within the French School by Politis and Scelle, who claimed that the individuals are the only real subjects of international law.

14 Rousseau, 15: "[I]l est difficile de s'entendre sur une définition unique de l'Etat. Si l'on veut sortir de cette équivoque, il faut dans l'Etat distinguer deux choses: car celui-ci est a la fois un phénomène politico-social et un phénomène juridique".

15 Jean d'Aspremont, 'The International Law on Statehood: Craftsmanship for the Elucidation and Regulation of Births and Deaths in the International Society', *Connecticut Journal of International Law*, vol. 29 (2014): 2, 201–224, 224.

extinction from a legal standpoint.[16] But the controversial nature of the law on statehood is also partly due to the fact that it is affected by the central role of its primary object, the State. The law on statehood thus finds itself torn between the need to provide for the *stability* inherently ascribed to such a solidly grounded institution as the State, and the need to offer a sufficient level of *flexibility* to adapt to the constantly changing political realities in which States emerge, and are terminated.

One of the ways found by the law on statehood to satisfy the quest for stability is through the presumption of continuity of statehood. Duration is indeed an essential element of the State,[17] to the extent that once a political entity joins the community of States, the continuity of its existence is presumed regardless of the level of unsteadiness of the political context in which it is called to operate.[18] If a State 'is', it is presumed to 'persist'. To be sure, a presumption is nothing but an assumption of fact required by law, rebuttable if evidence to the contrary is introduced.[19] Yet its weight explains to a certain extent why the vast array of studies on the creation of States in international law strikingly contrasts with the poor level of attention attracted by the question of State extinction.[20]

There is a second reason explaining such disproportion. Although it is generally accepted that there can be an end to the State,[21] the idea of a "pure extinction of the State" seems to have so far been refused. Mariano Aznar points out that because of a firmly rooted "terror of sovereignty vacuum," State extinction has mostly been addressed in connection with the law on State succession.[22] The extinction of one State thus tends to be associated with the emergence of a new State, following a *political process* of redistribution of power – such as merger with another State, annexation, absorption

16 Ibid.

17 Rousseau, 17: « la durée est un élément essentiel de l'État ».

18 Krystina Marek, *Identity and Continuity of States in International Law* (Geneva: Librairie Doz, 2nd ed. 1968), 5.

19 Davor Vidas, 'Sea-Level Rise and International Law: At the Convergence of Two Epochs', *Climate Law*, vol. 4 (2014), 70–84, 82.

20 An illustration of this disproportion may for instance be found in James Crawford's 870 pages landmark treatise *The Creation of States in International Law* (Oxford: Clarendon Press, 2nd ed. 2006), which devotes one last chapter, out of 17 in total and 18 pages long, to 'The Extinction of States'.

21 Marek, 5.

22 Mariano Aznar, 'The Extinction of States', in Eva Rieter and Henri de Waele, eds., *Evolving Principles of International Law: Studies in Honour of Karel C. Wellens* (Leiden: Nijhoff, 2012), 25–51, 25.

or dismemberment.[23] This association correlatively implies that, from a conceptual perspective, State creation and State extinction could be understood as parts of one single circular process. The present book challenges such a conception, for it deals with situations when, in a perspective, a State may cease to exist due to the disappearance of its *material substratum* – as opposed to a process of redistribution of power – without being followed by the creation of another State. Considering how international law is deeply embedded in geography, and how it rests heavily on the notion of State-based territory,[24] these cases represent an extreme ontological challenge.

Since the possibility that the pure extinction of a State may actually materialise is unprecedented,[25] international law is sparse in clues – let alone definitive answers – on how to deal with this particular scenario. On the rare occasions in which the scenario of material disintegration of a State was even considered from a purely hypothetical perspective, international legal scholars – rather hastily – concluded that the State disappears. This is due to the prevailing view that for long considered that extinction of States must be determined by reference to the same criteria by which the emergence of a State is determined.[26] For instance, in 1987, Ulrich Fastenrath considered it clear that:

> [A]ccording to international law, a State *becomes extinct* with the disappearance of *one* of the criteria of statehood (territory, people and government), *either because it has physically ceased to exist* or has merged into a larger unit or split up into smaller units, thereby removing the social foundation of the former State.[27]

---

23 Some historical examples of these political processes leading to State extinction include, for instance, the voluntary absorption of the German Democratic Republic; the extinction by merger of Yemen; the dissolution of the Czechoslovak Republic, and that of the Socialist Federal Republic of Yugoslavia. See Crawford, *The Creation of States in International Law*, 700–723.

24 Daniel Bethelhem, 'The End of Geography: the Changing Nature of the International System and the Challenge to International Law', *European Journal of International Law*, vol. 25 (2014), 9–24, 14.

25 Aznar, 25.

26 Ineta Zimele, 'States, extinction of', in Rüdiger Wolfrum ed., *Max Planck Encyclopedia of Public International Law* (Oxford University Press, 2013), online edition, para. 1; entry updated as of May 2007.

27 Ulrich Fasternrath, 'States, Extinction', *Encyclopedia of Public International Law*, vol. 10 (1987), 465–467, 465. [Emphasis added].

Such cases, which for long merely constituted hypothetical scenarios, are increasingly turning into a reality. As the voices of Pacific Island States get more heard – whether in diplomatic circles, civil society or public media, their active political move has inspired a growing number of international legal scholars to embrace this unprecedented and challenging issue.[28] The matter has more recently captured the attention of the International Law Association, and a new Committee on International Law and Sea Level Rise was established in November 2012 upon approval of the ILA Executive Council.[29] This increased scholarly and institutional attention is thus gradually introducing complexity and nuance into the debate, an illustration of which can be found in the cautious language used by Matthew Craven when writing on the same topic in the second half of the 1990s:

> [W]hile the territory of a State becomes submerged by the sea, or where the population of a State evacuates en masse to other territories [...] *it should be possible to conclude* that the State has ceased to exist.[30]

The debate as it currently stands has led to the differentiation of at least two streams of opinion in the literature on State extinction due to climate change impacts.[31] On the one hand, the pioneering studies of Rosemary Rayfuse, for instance, consider that when a State's territory becomes uninhabitable and the

28 See, for instance, the collectively authored publication edited by M.B. Gerrard and G.E. Wannier, *Threatened Island Nations: Legal Implications of Rising Seas in a Changing Climate* (Cambridge: Cambridge University Press, 2013), as well as the book by Lilian Yamamoto and Miguel Esteban, *Atoll Island States and International Law: Climate Change Displacement and Sovereignty* (Berlin: Springer, 2013). Articles dealing with the consequences of climate change for statehood include Emily Crawford and Rosemary Rayfuse, 'Climate Change and Statehood', in Rosemary Rayfuse and Shirley V. Scott, eds., *International Law in the Era of Climate Change* (Cheltenham, Northampton: Edward Elgar, 2012), 243–253; and Jane McAdam, *Climate Change, Forced Migration and International Law* (Oxford: Oxford University Press, 2012), Chapter 4, 119–158.

29 See Davor Vidas, David Freestone and Jane McAdam, "International Law and Sea Level Rise: The New ILA Committee", *ILSA Journal of International and Comparative Law*, vol. 21 (2015): 2, 397–408; and Davor Vidas, "International Law and Sea Level Rise: The Role of the International Law Association", *MEPEILAN eBulletin*, vol. 4 (2014): 1, available at: http://www.mepielanebulletin.gr/default.aspx?pid=18&CategoryId=4&ArticleId=174&Article=International-Law-and-Sea-Level-Rise-The-Role-of-the-International-Law-Association. (9 August 2015).

30 Matthew Craven, 'The Problem of State Succession and the Identity of States under International Law', *European Journal of International Law*, vol. 9 (1998), 142–162, 159.

31 Vidas, 'Sea-Level Rise and International Law', 79.

population is forced to evacuate, or when complete submergence of the State ensues, "the claim of statehood will fail".[32] In contrast, other scholars, such as Derek Wong and Jane McAdam argue that the idea of extinction of the international legal personality of a State following the loss of its material elements cannot be automatically inferred;[33] that, in matters of State extinction, the law is not so clear-cut.[34] This dichotomy is nothing but a contemporary manifestation of the 'open nature' of the concept of statehood – that is, of its inherent predisposition, as a legal concept with paramount political ramifications, to be subject to constant evolution and reinterpretation.[35] The oscillation of the law on statehood from stability to flexibility is therefore fundamentally informed by the operation of two factors: the evolution of normative standards, and the adaptation to events that bring about innovative challenges. Both 'normativity' and 'topicality' feed one another, for, as new factual situations come up, new normative positions on how they *should* be dealt with are raised along.[36]

Flowing from this background, the purpose of this book unfolds in two main stages. First of all, it seeks to identify how climate change impacts, and sea-level rise in particular, affect *each* of the dimensions of the State by degrading them from a physical perspective (coastal erosion, population relocation, governments in exile); and it examines in parallel the effects that the transformation of the State as a physical entity may correlatively have on its continuity as a political entity with legal attributes. Secondly, the book seeks to pinpoint the elements of 'normativity' and 'topicality' that are likely to guide this new stage in the regulatory and explanatory agenda informing the law on statehood. Ultimately, the aim of this work is to assess the concept of statehood

---

32 Rosemary Rayfuse, 'International Law and Disappearing States: Maritime Zones and Criteria for Statehood', *Environmental Policy and Law*, vol. 41 (2011), 281–287, 384.

33 Maxine Burkett, 'The Nation Ex-Situ: On Climate Change, Deterritorialised Nationhood and the Post-Climate Era', *Climate Law*, vol. 2 (2011): 1, 345–374.

34 McAdam, *Climate Change, Forced Migration and International Law*, 127; and Derek Wong, 'Sovereignty Sunk? The Position of 'Sinking States' at International Law', *Melbourne Journal of International Law*, vol. 14 (2013), 346–391, 360.

35 Conveying an awareness of how approximations to the concept of State tend to be incomplete, Charles Rousseau borrowed the words of the French poet Paul Valéry "[L]es mots de grande importance, ceux qui traduisent les notions fondamentales de la vie sociale, sont en général des symboles vagues, imprécis et indéterminés". In Rousseau, 14–15. See also Crawford, *The Creation of States*, 718, who concludes that "while statehood is a legal concept with a determinate, though flexible, content it is probably the only such concept in the field of legal personality".

36 Wong, 360.

when confronted with extreme change in environmental conditions. It is yet important to stress here that such conditions as discussed in this book are not simply an issue of climate change. There is a larger framework to the changes currently enveloping: the Anthropocene.

0.3 Changing Scales: From Climate Change to the Anthropocene

The focus of the international community on the climate and carbon cycles – and the recognition of human influence in altering them – is just one amongst other very significant alterations of equally determinant cycles in the functioning of the Earth system – such as biochemical cycles, terrestrial ecosystems, ocean acidification, and effects on the state of biodiversity conservation and prevention of the extinction of species – to mention only some prominent aspects of human-induced change to the Earth system processes.[37] With awareness of the scale of human impacts on a range of different natural processes in terrestrial, atmospheric and marine components of the Earth system, the term 'Anthropocene' began to be used by some scientists in the early 2000s to refer to the understanding that human activities have in fact become a geological force in their own right.[38] This means that the Earth may be leaving its most recent known, and currently 'official' geological epoch, the Holocene, which began around 11,700 years ago and has been characterised by a relative environmental stability that has proved to be crucial for the development of human civilisation as we know it today.[39] The Anthropocene, in contrast, is seen as being "characterised by change, uncertainty and, probably,

37 On this point, see for instance, Will Steffen and Eric Lambin, 'Earth System Functioning in the Anthropocene: Human Impacts on the Global Environment', in *Interactions between Global Change and Human Health* (Vatican City: Pontificiae Academiae Scientiarum *Scripta Varia*, 2006), 112–144.

38 The first publication specifically focused on the 'Anthropocene' can be found in Paul J. Crutzen and Eugene F. Stoermer, 'The Anthropocene', *IGBP Global Change News*, vol. (2000), 17–18; followed by Stephen R. Palumbi, 'Humans as the World's Greatest Evolutionary Force', Science, vol. 293 (2001), 1786–1790; and Paul Crutzen, 'Geology of Mankind', *Nature*, vol. 415 (2002), 23.

39 For a thorough overview of the current status of the Anthropocene in geology and its bearing on international law, see Davor Vidas, Jan Zalasiewicz, and Mark Williams, 'What is the Anthropocene – and Why is it Relevant for International Law?', *Yearbook of International Environmental Law* forthcoming (2015).

considerable instability in the behavior of the Earth system".[40] This is therefore another point of connection with climate change and, in particular, with sea-level rise, whose fluctuations have proved to be irregular, and with potentially dramatic effects on human livelihoods. The risks associated with climate change impacts can therefore be understood as an illustration of the wider spectrum of environmental instability that characterises the Anthropocene.

Such idea began to be taken into consideration by the International Commission on Stratigraphy as a possible new geological unit to be eventually part of the formal Geological Time Scale. The Anthropocene Working Group of the Subcomission on Quarternary Stratigraphy, specifically created in 2009 for that purpose, is currently examining the 'environmental signatures' that can be uniquely attributed to the Anthropocene, the epoch that 'humanity has created'. So far, the term Anthropo-'cene' indicates a disposition to give it the status of an epoch, which will require finding changes in key criteria (biostratigraphical, sedimentological and geochemical) that are of comparable magnitude as evidence of earlier epoch boundaries – between the Pleistocene and the Holocene.[41]

The key outstanding issue regarding the formalisation of the Anthropocene regards its starting date.[42] In addition to whether the 'Anthropocene' will at all gain the required recognition as a new geological time-unit in stratigraphy, this aspect of its lower boundary is perhaps the most controversial point, as several possible dates and events have been proposed. Some scholars suggested the beginning of agriculture as a possible starting date for the Anthropocene: the clearing of forests, land conversion and irrigated rice

40 Ibid. See also, Jan Zalaziewicz, Paul Crutzen and Will Stefffen, 'The Anthropocene', in Felix M. Gradstein, James Ogg, Mark D. Schmitz, Gabi Ogg, eds., *The Geologic Time Scale 2012* (Leiden: Elsevier, 2012), 1033–1040.

41 See the official website of the Anthropocene Working Group available at: http://quaternary.stratigraphy.org/workinggroups/anthropocene/ (8 August 2015). See Jan Zalasiewicz et al., 'Are We Now Living in the Anthropocene?', *GSA Today*, vol. 18 (2008), 4–8. Following up directly on this publication, see also Jan Zalasiewicz, Mark Williams, Alan Haywood and Michael Ellis, 'The Anthropocene: a New Epoch of Geological Time?' *Philosophical Transactions of the Royal Society of London- A*, vol. 369 (2011), 835–841. On the other outstanding question of whether the Anthropocene should be considered a new Epoch or a new Age, see Colin N. Waters, Jan A. Zalasiewicz, Mark Williams, Michael A. Ellis and Andrea M. Snelling, 'A Stratigraphical Basis for the Anthropocene?', *Geological Society of London* (*Special Publications Online First*), 24 March 2014, available at: http://sp.lyellcollection.org/content/early/2014/03/21/SP395.18 (20 March 2015).

42 See Steven W. Running, 'A Measurable Planetary Boundary for the Biosphere', *Science*, vol. 337 (2012), 1458–1459.

cultivation between 8000 and 5000 years ago induced an additional release of CO_2 and methane into the atmosphere that possibly prevented the coming of the new Ice Age.[43] Yet the majority of the scientific scholarship abide to the suggestion that the beginning of the Anthropocene ought to be established at a much later time. The Industrial Revolution, which is considered to have originated in either Great Britain in the early 18th century or in the thermo-industrial revolution of the 19th century Western civilisation, was initially, drawing on Paul Crutzen's proposal in 2000, suggested as the period marking the transition from the Holocene to the Anthropocene.[44] This period essentially marked the shift from the use of primary energy sources (such as water or wind), highly reduced in magnitude and location, to the discovery of a much bigger storage of energy in fossil fuels. Their exploitation allowed industrial societies to use far more energy as during the agrarian period, which in turn brought about great population growth in a very short time.[45]

The period that followed the Second World War and was characterised by an exponential global economic growth, known as the 'Great Acceleration', has thereafter been increasingly taken into consideration.[46] One of the main (geological) consequences of this phenomenon is the shift of more than half of global population from rural to urban areas, and the correlative development of big cities, with their associated deposits.[47] Accordingly, the latest proposals, supported also by the large majority of the Anthropocene Working Group members, are focusing on the mid-20th century, also characterised as the start of the 'Nuclear Age'. The specific event pointed to as stratigraphically relevant is the first nuclear bomb explosion on 16 July 1945 in Alamogordo (New Mexico), which was then followed by additional bombs detonated at an average rate of one every 9.6 days until 1988.[48] From a scientific perspective, this proposal is

---

43 William F. Rudiman, 'The Anthropogenic Greenhouse Gas Era Began Thousands of Years Ago', *Climate Change*, vol. 61 (2003), 261–293. For an updated view from the same author, see 'The Anthropocene', *Annual Review of Earth and Planetary Sciences*, vol. 41 (2013), 45–68.

44 Waters et al., 5–6.

45 Often referred to as 'the Malthusian Revolution', from the scenario anticipated by Thomas R. Malthus *An Essay on the Principle of Population: A View of Its Past and Present Effects on Human Happiness; with an Inquiry into Our Prospects Respecting the Future Removal or Mitigation of the Evils which It Occasions* (London: John Murray, 1826).

46 See, in particular, Will Steffen, Paul Crutzen and John R. McNeill, 'The Anthropocene: Are Humans Now Overwhelming the Great Forces of Nature?' *Ambio*, vol. 36 (2007), 614–621.

47 Waters et al., 6.

48 A recent paper published in the *Quaternary International* and authored by 26 pioneer experts on the Anthropocene (all of whom are members of the Working Group) indicates

based on the fact that the explosions left an easily identifiable marker in the chemostratigraphic record – that is, a trace of nuclear particles in different geological layers.[49] If the Anthropocene were to be formally included in the Geologic Time Scale, and the onset of the nuclear age accepted as its lower boundary, this geological unit created by humans would begin with the consciously acquired capacity of humankind to destroy itself as a species.[50]

Societal consequences of the recognition of this new geological unit have thus begun to be considered by scholars in different disciplines within the social sciences.[51] While historical reflections on the Anthropocene tend to focus on the forms and causes of the evolution of human behavior and the organisation of society through time;[52] the realm of philosophy is faced with the challenge of rethinking ontological patterns which were constituted in a stable material world.[53] The idea conveyed by the Anthropocene is that the Earth system characteristics are neither 'human' nor 'natural', but are in fact highly integrated composites of both.[54] Cartesian reductionisms – such

---

that the 'Nuclear Age Boundary' is progressively gathering the support of the majority of the scientific community. See Jan Zalasiewicz et al., 'When Did the Anthropocene Begin? A Mid-Twentieth Century Boundary Level is Optimal', *Quaternary International*, (online 12 January 2015), 1–8.

49 Ibid.

50 In addition to humanity 'destroying itself', nuclear force – though composed by humans – can be considered at the level of major 'natural' forces, what has been described by Jan Zalasiewicz as being a "geologically instantaneous planetary reshaping", in 'Our Brave New World', *New Scientist*, 8 November 2014, 26–27.

51 For initial writings on those aspects, see two articles included in *Philosophical Transactions of the Royal Society A*, vol. 369 (2011): Crispin Tickell, 'Societal Responses to the Anthropocene', 926–932; and Davor Vidas, 'The Anthropocene and the International Law of the Sea', 909–925.

52 Philosophical and historical reflections include, for instance, Will Steffen, Jacques Grinevald, Paul Crutzen and John McNeill, 'The Anthropocene: Conceptual and Historical Perspectives', *Philosophical Transactions of the Royal Society A*, vol. 369 (2011), 842–867, 843.

53 Observation made by Bernd Scherer, director of the Haus der Kulturen der Welt (Berlin) during the opening of the Anthropocene Campus, a two-year project financed by the German Parliament and seeking to gather experts from all fields with a view to exploring new ways to create inter-disciplinary and innovative knowledge-production on the Anthropocene. Website of the Anthropocene Campus is available at: http://www.hkw.de/en/programm/projekte/2014/anthropozaen_curriculum/anthropozaen_curriculum_1.php (20 March 2015).

54 This idea has been consistently presented in all scholarly writings on the Anthropocene, coming from different disciplines. Amongst legal scholars, it has been referred to, for

as the dualisms Humankind/Nature; object/subject – get diluted, for in the Anthropocene, humanity is able to perform a dual role: it can, at the same time, *observe* and *be part of* the processes that may signal the change from the Holocene to this new geological epoch. Correlatively, as the 'object' becomes blurred and fluid, the Arts are also called on to explore new creative avenues which can not only represent the Anthropocene, but also participate in the production of new knowledge about it.[55]

As a concept with the potential to challenge the grounds of several disciplines, its consequences for international law have also begun to be examined. In particular, a pioneering research project launched by the Fridtjof Nansen Institute in 2011, on "International Law for an Anthropocene Epoch?", introduced a systematic study linking the Anthropocene with several aspects of international law.[56] This stream of research has been further developed by the recently launched FNI research project, on 'Climate Change and Sea-Level Rise in the Anthropocene: Challenges for International Law in the 21st century'.[57] This project seeks to undertake the first comprehensive study of the potential developments in international law in response to the challenges posed by sea-level rise in the overall context of the onset of the Anthropocene. Core paradigms of international law, such as the concept of statehood, are thus expected to be thoroughly re-examined. Yet understanding both the Anthropocene and the challenges posed by it implies trying to understand "a world beyond our reckoning";[58] one that transcends the old understanding of the object world as

---

instance, by Karen N. Scott, 'International Law in the Anthropocene: Responding to the Geoengineering Challenge', *Michigan Journal of International Law*, vol. 34 (winter 2013), 309–358, 315–316. Geopolitical approaches to the Anthropocene have also emerged, such as that of Simon Dalby, 'Anthropocene Geopolitics: Globalisation, Empire, Environment and Critique', Geography Compass', vol. 103 (2007): 1, 103–118.

55 See, for instance, the very recently published pioneer book by Heather Davis and Etienne Turpin (eds.) *Art in the Anthropocene: Encounters Among Aesthetics, Politics, Environments and Epistemologies*, (London: Open Humanities Press, 2015).

56 For the final results of this project, see: Davor Vidas, Ole Kristian Fauchald, Øystein Jensen and Morten Walløe Tvedt, 'International Law for the Anthropocene? Shifting perspectives in Regulation of the Oceans, Environment and Genetic Resources', *Anthropocene*, online 4 July 2015, available at: http://www.sciencedirect.com/science/article/pii/S2213305415300084 (9 August 2015). For further information about this 2011–2014 FNI Anthropocene and International Law research project, see project webpage, available at: http://www.fni.no/projects/anthropocene_law.html (9 August 2015).

57 Website of the Project, available at: http://www.fni.no/projects/climate_change_and_sea_level_rise.html (19 March 2015).

58 Tickell, 369.

constructed by disciplines. It is with this idea in mind that the study of State extinction due to climate change impacts will be approached in this book.

The analysis of the effects of climate change on Pacific Island States is thus carried out through a division of three main dimensions of statehood. Chapter 1 opens this exploration by dealing with the effects of climate change on Pacific Islands' territory – referred to as the 'de-territorialisation challenge' – including how climate change may affect the maritime entitlements of these often called 'Ocean States'. It comprises a genealogical study of the meaning of territory in international law and introduces the question of whether completely de-territorialised nations may still survive as independent sovereign entities.

The continuity of Pacific Islands' statehood, even in extreme cases of total territorial loss, may be justified by the role played by the second dimension of the State, namely, the population. Bearing in mind that de-population may take place before total submergence ensues, chapter 2 addresses the implications of climate change impacts on Pacific Islanders. After a comparative analysis of the climate-induced relocations already accomplished in the region, and a critique of how climate-induced displacement has so far been dealt with in international legal scholarship, this chapter proposes to approach the question of the legal-protection tools for climate-induced displaced people from an alternative approach. The proposal consists on conceptualising the legal protection of Pacific Islanders within a multi-layered legal protection scheme, which axis and main structural variable is the question of the State's continuity.

Although the fundamental and structuring question – whether and when Pacific Islands' statehood may become extinct – remains outstanding, it is clear that in all stages and scenarios covered by the multi-layered legal protection scheme, the action and presence of the political dimension of the State is fundamental. Chapter 3 is finally devoted to a study of the effects of climate change on Pacific Islands' governmental capacities, both nationally and internationally. Considering the literature on 'failed States' and governments in exile, it first deals with the scenario in which governments of Pacific Island States may be obliged to evacuate State territory (referred to as ex situ governments). Ultimately, as the three dimensions of Pacific Islands' statehood may be challenged at any time, this closing chapter explores the role to be played by different normative and political avenues which will determine their continuity as States.

CHAPTER 1

The Challenge of De-Territorialisation

As a ten-year-old, I used to look at the sea with awe, at the seemingly endless supply of fish that I could harvest... now when I look at it, I wonder how far into the new millennium we will be before it overwhelms our coasts.

TAMARI'I TUTANGATA, Former Director of the South Pacific Regional Environment Programme, 2000.

1.0 Introduction

The notions of threat and survival are not unfamiliar to international law. After all, international law is "devoted above all to order and security", as Charles de Visscher recalled when describing how the system operates in practice.[59] Not to mention that, in the classical understanding of the modern European (sovereign and territorial) State, the central function of the State was, precisely, to ensure the security of each State's population (or nation) against external threats. The genealogy of the 'inherent right to self-defence' may for instance illustrate well the ways in which the notion of threat has become anchored in international law. Yet the prime manifestation of the central place but equally polemical nature of such notion lies in the fact that contemporary international law still falls short of regulating a State's behaviour in the most worst-case scenarios or circumstances of extreme vulnerability. Whereas, on the one hand, the determination of the existence of a 'state of necessity' regularly fuels heated debates within the international judiciary and doctrinal circles (as analysed by the International Court of Justice in the *Gabcikovo-Nagymaros* case or, in particular, in the realm of State-investor disputes);[60] on the other hand,

59 Charles De Visscher, *Theory and Reality in Public International Law*, trans. from the French by P.E. Corbett (Princeton, N.J.: Princeton University Press, rev. ed. 1968), 177.

60 The conditions for the state of necessity doctrine to be accepted as grounds precluding responsibility for an international wrongful act were specified in the case *Gabcikovo-Nagymaros Project* (Hungary / Slovakia), Judgment, 1997 ICJ Reports 7, para. 52: "[I]t must have been occasioned by an 'essential interest' of the State which is the author of the act conflicting with one of its international obligations; that interest must have been threatened by a 'grave and imminent peril'; the act being challenged must have been the 'only means' of safeguarding that interest; that act must not have 'seriously impaired an essential interest' of the State towards which the obligation existed; and the State which is the

 | DOI 10.1163/9789004321618_003

when the existence of a national security interest of the State is acknowledged by an international court or tribunal, the so-called *domaine privilégié* of the State often operates as an 'anti-dragon shield', protecting against any possible incursions or limitations that the law may impose on the State's conduct. Indeed, the assurance of State survival has even been considered by the International Court of Justice in its Advisory Opinion on the *Legality of the Use of Nuclear Weapons* as a circumstance blurring the existing rules on the use of force, so much that the very assessment of the legality of the use of nuclear weapons was impeded.[61] These are therefore examples of how the extensive scope of international law is constrained when facing the core protection of its principal and privileged subject – an ambivalence which is arguably connected to the double-edged functionality of the State with regard to the international legal order, as both creator and subject of the rules it has consented to be bound by.

Yet, this notable sensitivity to the heart of the State and the accepted deference towards what John Westlake called "the primitive instinct" of States to preserve their survival seems to have been associated with strong States or even 'big powers'.[62] In fact, it may be that the State, *as a power-based structure*, is what actually animates the recognition of and respect for such an instinct. In contrast, in the cases of this study, States' continuity may be threatened by non-traditional threats, potentially targeting a category of actors – post-colonial small island States – characterised by their inborn physical, political and economic vulnerability and fragility.

It is perhaps the opposite concept of continuity – that is, the notion of *disappearance* – which constitutes the central mind-set of this chapter. As

author of that act must not have 'contributed to the occurrence of the state of necessity'. This conditions reflect customary international law. For a concrete illustration of the tensions induced by the application of the doctrine on the state of necessity in the realm of State-investor disputes, see, for instance, Eric D. Kasenetz, 'Desperate Times call for Desperate Measures: the Aftermath of Argentina's State of Necessity and the Current Fight in the ICSID', *The George Washington International Law Review*; vol. 41 (2010): 3, 709–747".

61 *Legality of the Threat or Use of Nuclear Weapons*, Advisory Opinion, 1996 I.C.J. Report 226, para. 44 [Emphasis added]: "[I]n view of the current state of international law, and of the elements of fact at its disposal, the Court *cannot conclude definitively* whether the threat or use of nuclear weapons would be lawful or unlawful in *an extreme circumstance of self-defence, in which the very survival of a State would be at stake*".

62 See also John Westlake's definition of self-preservation, in Lassa Oppenheim, ed., *The Collected Papers of John Westlake on Public International Law* (Cambridge, UK: Cambridge University Press, 1914), 112.

a complex and multiform concept, the notion of disappearance essentially carries two connotations. On the one hand, it refers to the threat to the tangible, physical, concrete and material dimension of these States' existence (e.g. the physical submergence of their territory or the displacement of their population); on the other hand, it alludes to the threat of their disappearance as political entities which engages with the abstract dimensions of their existence, and necessarily directs the study to the concept of a State as a political construct. Ever since the emergence of the modern State as a new form of social and territorial organisation endowed with sovereign prerogatives, the factual and the abstract dimensions of the State have been inseparable.

Considering that the law of statehood apprehends the world in territorial units,[63] when assessing the effects of climate change impacts on small Pacific Islands' statehood, the spatial unit is the first element to be analysed. The present chapter is therefore devoted to scrutinising the effects of climate change on the spatial dimension of these island States, encompassing both their land territory and the adjoining maritime areas.[64] Section 1.1 of this chapter begins by exploring the meaning of territory in the definition of statehood in classical international law (when it was linked to the modes and titles of acquisition thereof) and contrasts this original meaning with the role it fulfilled when Pacific Islands States came into existence as independent political entities, when the meaning of territory was bound to the end of a history of colonial domination. After considering how the 'territorialisation of nationhood' has been a fundamental factor in the creation of States and extracting

63 Jean d'Aspremont, 'The International Law of Statehood: Craftsmanship for the Elucidation and Regulation of the Birth and Deaths in the International Society', *Connecticut Journal of International Law*, vol. 29 (2014), 201–224, 214. Perhaps the most explicit and lasting trace of influence in international legal scholarship of the constitutive theory is the preference of some scholars to consider the population of a State as the first element of the State, deserving consideration before the territory. Followers of this position include George Scelle, as well as Charles Rousseau, who wrote: "[L]e premier élément de l'État, comme toute société humaine, c'est la population", in *Droit international public* (Paris: Sirey, 1970), 18.

64 For an important distinction between the concept of *territory*, and the concept of *space*, see Paul de La Pradelle, 'Notions de territoire et d'espace dans l'aménagement des rapports internationaux contemporains', *Recueil des Cours de l'Académie de Droit international de La Haye*, vol. 157 (1977), 415–444, 427 [Emphasis in the original text]: "[L]e *territoire* désigne, en droit international, une dépendance de l'ordre géophysique du monde, placée sous le pouvoir de disposition et de contrainte de l'État. L'*espace* désigne, au contact des territoires, les dépendances de l'ordre géophysique du monde qui échappent, par nature, à l'emprise de la souveraineté pour être ouvertes à l'accès et aux libertés d'usage de tous".

its specific meaning in the context of South Pacific State-building, Section 1.2. explores the 'de-territorialisation process' undergone by Pacific Island States. This process is approached by describing first the physical effects of climate change on these States' territory and by focusing then on the jeopardy they constitute for the State as a legal construct – with specific emphasis on their maritime rights. Finally, Section 1.3 addresses the solutions which have been put forward by scholars and policy-makers to limit the negative impacts of climate change on these States' threatened statehood.

1.1 Territorialisation of Nationhood: Fundamental Factor in the Creation of States

1.1.1 *The Meaning of Territory at the Inception of the Modern European State*

A From Feudal Allegiance to the Westphalian Principle of Territorial Sovereignty

Whilst the State operates as a mediator between international law and the concept of territory, the multiple meanings of territory found in international law also result from the influence of premodern legal systems on the Westphalian order.[65] Therefore, to understand the nature of territory and grasp its evolutionary meaning and role in the international legal system, it becomes necessary to adopt the broad approach which, as put by Hannis Taylor, conceives States as a result of the Greek, Roman and Medieval systems, three antecedents whose "individual histories constitute only distinct stages in one unbroken and progressive development".[66]

65 See Julio Barberis, 'Les liens juridiques entre l'État et son territoire: perspectives théoriques et évolution du Droit international', *Annuaire Français de Droit International*, vol. XLV (1999), 132–147, 132: "[L]e concept territoire de l'État est en rapport avec celui de 'souveraineté territoriale' et le fait d'énoncer une définition de 'territoire de l'État' implique, dans une certaine mesure, de donner également une définition de cette dernière".

66 Hannis Taylor, writing immediately after the turn of the 20th century, devotes a whole first chapter of his Treatise on the Ancient and Medieval State-Systems, and considered the ancient State as the city-commonwealth. See Hannis Taylor, *International Public Law: a Treatise* (London: Sweet & Maxwell, Limited, 1902), vi. More recent historical accounts on international law which follow the same in-depth study on the premodern times include Wilhelm Grewe, *The Epochs of International Law*, trans. ed. by Michael Byers (Berlin, New York: Walter de Gruyter, 2000), Chapter 3, 62–81; as well as Arthur Nussbaum, *A Concise History of the Law of Nations* (Michigan: McMillan, rev. ed. 1954), Chapters 1–3, 1–59.

The prominent role of the territory in the configuration of the modern European State is the result of the progressive transformation in the medieval era of the modes by which a leader was entitled to exercise ruling power over the community. The tribal or national concept of sovereignty originally applied by the Teutonic people was not associated with command over a particular portion of the land. Rather, such bonds between the ruler and the members of a people were fundamentally personal and their configuration was conceived around the notion of nation, whose existence was not necessarily associated with a fixed and delimited area.[67] The historical process by which the 'space' gradually became an important factor in the configuration of sovereignty essentially began with the shift from the elective appointment of a head of a people to an hereditary system – according to which the ruler of the nation is the lord of a portion of the earth's surface where the nation lives.

Although feudalism was still a strongly personal system – Max Weber described it as "the ruling power of a militarised nobility with manorial rights"[68] – which did not see the full conversion to a territorial allegiance,[69] the latest stage of the feudal period was characterised by the concern and interest of the seigniorial class in the conservation of their socio-economic position as landlords. Wilhem Grewe thus explains how this evolution impacted the configuration of the legal systems:

> [A]s the strict personalisation of the early medieval tribal codes diminished, areas subject to a single unified law developed, and "the law of the land" replaced the "tribal codes".[70]

Ultimately, the most important outcome of feudalisation was the birth of the principle of territorial sovereignty, defined by Hannis Taylor as the basis of "the modern conception of the State as a nation with fixed geographical boundaries".[71] Beyond the commonly accepted assertion that medieval polities fell short of being built upon the principle of territorial sovereignty as modern European States were, scholarly division arises when it comes to

67 Taylor, vii.

68 Max Weber, *Economy and Society* [*Wirtschaft und Gesellschaft*] (New York: Free Press, 1922), 724.

69 Grewe, 65: "Medieval polities did not know the particular territoriality of the modern State with its sharply defined territory as both a closed legal area and an exclusive sphere of government competence".

70 Grewe, 64.

71 Taylor, 157.

depicting the concept of territory of the late feudal order. Its obscure complexities generally invite consideration of two main different views. On the one hand, the patrimonial concept of the medieval polity views the territory as the object of unrestricted rights of private ownership belonging to the lord or sovereign. This view was predominant among German scholars of the 18th and 19th centuries and is based on the notion of *dominium* originated in Roman law. In contrast, other scholars emphasise the *imperium* or public power – rather than private ownership – which was vested in the lord or sovereign within the territorial limits of the land he ruled. According to Wilhem Grewe though, the reality did not completely correspond to either of these views. As he explains:

> [I]n the Middle Ages dominium referred to a legal institution which belonged to public as well as to private law and which had no precedent in the laws of antiquity, and the power of control the lord exercised over the land was not imperium in the Roman sense either – that is, as "a plenitudo potestatis", a comprehensive unified sovereignty from which all of the singular rights of the sovereign were derived.[72]

Feudalism paved the way for the concept of territory to become central in the configuration of the modern European State. Through the gradual overlap of the personal and the territorial character of a nation's bonds with its ruler, 'spaces' were apprehended to become 'territory', a notion which thereby began to play a role in the legal and political orders of the time.[73] The territory then found in the modern State a privileged habitat, where it became not only an important factor of the organisation of power,[74] but a truly defining feature of the State itself, in its historical and its ontological dimensions.[75] Even authors who consider ancient political entities as States – that is, scholars who recognise a plurality of State forms – agree that the differential characteristic of the modern State compared with its predecessors is the existence of a territorially

72 Grewe, 67.

73 See La Pradelle, 423, who notes that the emergence of the Latin term for space – namely 'spatium' – in the 12th century preceded that of the term territory – namely 'territorium' – which only emerged in the 14th century.

74 Ibid., 425.

75 De Visscher, 205 [Emphasis added]: "[T]he essential place that territory holds in the organisation of the State and its highly symbolic meaning explain *the propensity of authors as well as of State practice to identify the territory with the State*, or at least to regard its spatial delimitation as inseparable from that of sovereignty".

based nation.[76] Most importantly, the integral character of the territory and the modern State became embodied in the principle of territorial sovereignty, recognised in 1648 by the Treaties of Münster and Osnabrück.[77] As the territory became, for the modern State, the "framework of independence and security in the political order",[78] the principle of territorial sovereignty is thus first and foremost meant to protect such a framework from external threats or aggression and, in doing so, to reaffirm the existence of the State independent of any other entities. Max Weber's famous definition of the State as a "human community that (…) claims the monopoly of the legitimate use of physical force within a given territory" is perhaps illustrative of this dimension of the territory.[79] Yet, as the modern State became progressively established between the late 15th century and the early 17th century,[80] other facets of the integral bond between territory and the State also arose; territorial sovereignty was not only involved in the protection of the State's independent existence, but also played a role in asserting both the distinctive identity of the nation the State embodies, and its effective power towards other States.

Though it is difficult to disentangle the historical process by which a community progressively develops an emotional bond to a specific part of the earth's surface, until its identity becomes inseparably associated with that soil, the result of such a process is, in contrast, quite apparent. Within the modern European nation-State, the territory was as much the locus of the exercise of public authority as the body of the nation itself.[81] Resulting from and being fused with the outcome of a process whereby nationhood was 'territorialised', the modern European nation-State embraced the territory as its natural substrate and vested it with a strong and irreplaceable symbolic power. Considering that "the fabric of society is territorial",[82] Thomas Baty eloquently depicted this symbolic dimension when saying that "the absolute

76 See, for instance, Taylor, 26.

77 Thomas Baty, *International law* (New Jersey: The Law Book Exchange Ltd., rev. ed. 2005), 244.

78 De Visscher, 205.

79 Max Weber, 'Politics as Vocation', in Hans H. Gerth and Charles Wright Mills, trans. and ed., *Essays in Sociology* (New York: Oxford University Press, 1946).

80 See, for instance, Mathew Smith Anderson, *The Origins of the Modern European State: 1494–1618* (London, New York: Routledge, 1998), who situates the emergence of the modern State between the French invasion of Italy in 1494 and the start of the Thirty Years War in 1618.

81 See Jens Bartelson, *The Critique of the State* (Cambridge, UK: Cambridge University Press, 2001), 45.

82 Baty, 251.

sacredness of a nation's land is the vital nerve of our present system."[83] The ontological function of the territory in defining both the spatial dimension of the newly emerged independent State entities and their identity as a distinctive political community are already strong reasons explaining the central position of the territory for the conception of the modern State. Such a prominent place continued to grow as States consolidated the cardinal role in the international society they played in modernity and still play today.[84] It is therefore possible to consider that the "obsession with territory", that George Scelle criticised, began to occupy the minds of international scholars particularly at the time of the exponential flowering of modernist thought (from the late 18th century to the late 19th century).[85]

It is generally considered that the main characteristic of modernity was that it brought about a new understanding of history. While Jens Bartelson considered how such new understanding was facilitated by the invention of a whole series of concepts, such as growth, evolution, development and progress, which were used to "define time itself and create sociopolitical temporality",[86] Chenxi Tang recalls and emphasises the subjective historical consciousness and critical self-positioning in time which made possible the creation of such concepts.[87] Most importantly, Tang proves that modern consciousness encompassed as much the position of human existence in time as its position in space; that modernity also had an 'intrinsic spatial dimension' to be found in the 'geographical imagination' of modern society.[88] An important element of such imagination is the association of the territorial features of the State with its power. Although the inextricable link between these two concepts (geography and power) was known since the development of cartography well before the beginning of modernism,[89]

83 Ibid., 245.

84 Hélène Ruiz Fabri, 'Génèse et disparition de l'État à l'époque contemporaine', *Annuaire français de droit international*, vol. 38 (1992), 153–178, 153: "[À] la fois auteurs et sujets de droit international, les États tiennent dans la société internationale, le rôle cardinal. Plus, l'État est devenu une valeur, voire une finalité, au point qu'on a pu parler de statolâtrie".

85 George Scelle, 'L'Obsession du territoire', in Jan H.W. Verzijl and Frederik M. van Asbeck, eds., *Symbolae Verzijl: présentées au Professeur J.H.W. Verzijl à l'occasion de son LXX-ième anniversaire* (The Hague: Nijhoff, 1958), 34.

86 Bartelson, 56.

87 Chenxi Tang, *The Geographical Imagination of Modernity* (Standford: Standford University Press: 2008), 1.

88 Ibid.

89 Ibid., 139.

it was particularly enhanced during modernity as it became the object of the newly born discipline of political geography.

Born in a Prussian cradle and conceptually developed by Ratzel soon after the turn of the 20th century,[90] the German school of political geography (also known as *Geopolitik*) was concerned with the causal relationship between the geography of a State – in particular, its position (*Lage*) and its extent (*Raum*)[91] – and its political power in foreign relations. Social Darwinism and scientific determinism burgeoning at the time significantly influenced the approach of *Geopolitik* to the concept of territory. In the early foundational period of the discipline, the State's land, spiritually bound to the nation it embraced, was represented as a living organism destined to continue growing. Thus, the State's borders were necessarily flexible,[92] and their evolution akin to a mirror of the State's grandeur (if the borders were extended) or of its decline (if the borders remained static or if the territory diminished). Territory thus conceived was both the full expression of the power of the State and the fundamental premise of power as such.

The inherent multidisciplinary nature of the geopolitical discourse attracted scholars from adjacent disciplines, including international law, soon after its inception. Alexander Orakhelashvili recently recalled how the object of study of geopolitics – the interaction between the categories of power, space and right – significantly overlaps the focus of international law.[93] Though geopolitical knowledge remains alive, developed and applied within and outside governmental structures – particularly of hegemonic or emerging countries – the interdisciplinary debate between geopolitics and international legal scholarship seems very limited. Such radical decrease can be explained by the general association of the expansionist policy of the Third Reich with Ratzel's theory of vital space (*Lebensraum*), which was (mis)used by the National-Socialist government to justify the violation of bordering States'

90 Although it is widely known and accepted that the term *Geopolitik* as such was coined by Swedish political scientist Rudolf Kjellén in the late 19th century.

91 Friedrich Ratzel, *Der Lebensraum: Eine biogeographische Studie* (Darmstad: Wissenschaftliche Buchgesellschaft, 1901).

92 Determinism was rejected by the French school of political geography, which shared ideas of flexible boundaries but considered these as the result of human behavior. See Jacques Ancel, 'Les frontières: étude de géographie politique', *Recueil des Cours de l'Académie de Droit international de La Haye*, vol. 55 (1936): 1, 203–298, at 208–210.

93 Alexander Orakhelashvili, 'International Law and Geopolitics: One Object, Conflicting Legitimacies?' *Netherlands Yearbook of International Law*, vol. 39 (2008), 155–204.

territorial integrity.[94] In this context, it does not come as a surprise that scholars of the new post-war international legal order, which was emphatically construed around the general and principled prohibition on the use of force, did not want to be associated with German geopolitical thought. Yet, between the late 19th century and the 20th century inter-war period, the particular impact of political geography on the minds of modernist international lawyers was apparent, animating among them perhaps the most prolific debate on the meaning and legal status of territory in international law, and directly influencing some of their findings.[95]

B Pre-Charter Theories on the Legal Nature of State Territory

Put in historical perspective, the notion of territory has come to serve the modern European State through three main functions: first, it provided the basis for the existence of the State as an independent entity while also representing the identity of its nation (integrative function through unification and representation of a people); secondly, it delimited the boundaries of the exercise of the new form of public authority (organisational function of internal and external competences); and, finally, it played a role as both the premise and the full expression of the realisation of the State's power in its foreign relations (rationalisation function on the use of power). The setting of this three-dimensional functionalist approach to territory helps to explain why, by the turn of the 20th century, the inseparability of State territory (that is, territory in its physical dimension) from statehood was clearly uncontroversial.[96] While widely accepting the existence of such a strong

94 On the (mis)use of geopolitical thought, Mark Bassin, 'Race Contra Space: The Conflict between German "Geopolitik" and National Socialism' *Political Geography Quarterly*, vol. 6 (1987): 2, 115–134; and Oyvind Osterud, 'The Uses and Abuses of Geopolitics', *Journal of Peace Research*, vol. 2 (1988), 191–199.

95 A clear illustration of such impact can be found in the words of Walther Schoenborn, 'Nature juridique du territoire', *Recueil des Cours de l'Académie de Droit international de La Haye*, vol. 30 (1929) : 5, 81–190, 90: "Le droit international peut utiliser, au moins quant aux détails, certaines suggestions précieuses et fécondes résultant de découvertes récemment faites dans le domaine de la géographie politique."

96 Franz Von Listz' consideration – written in 1904 – that supremacy over territory (*Landeshoheit*) is one of the "very indispensable attributes of the State", for instance, illustrates this point. See Franz Von Listz, *Das Völkerrecht: Systematisch Dargestellt* (Berlin: Häring, 1904). Also, some decades later, Sir Robert Jennings would put this issue as follows: "what is intended here is not merely territory in the physical sense but State sovereignty in respect of territory", in *The Acquisition of Territory in International Law* (Manchester, UK: Manchester University Press, 1963), 2.

bond, every attempt to establish a theory on the legal nature of territory in international law has been thwarted by what it implies – namely, that no concept of territory has been able to detach its physical sense from its abstract sense.[97]

Perhaps triggered by the birth of geopolitics as a new school of thought centred on the concepts of space and territory, since the end of the 19th century interest in the concept of territory also came to the forefront of international legal scholarship. Rather than considering the legal nature of the territory as such, international legal theories sought to deal with the nature of State sovereignty in respect of territory.[98] In defining the legal nature of the concept of territory and its place in the general theory of the State, European international scholars of the modernist epoch can be classified by their preference for the patrimonial theory, the constitutive theory, or the competence theory of territory.[99]

The patrimonial theory of territory, which seems to have been predominant until the inter-war period, views the legal bond between territory and the State as a right of property. The Italian school developed the core theoretical foundations of this approach, as illustrated in Donato Donati's landmark work *Stato e territorio*. Donati contended that the right exercised by the State in respect of its territory is "un diritto di dominio" compatible with the existence of a right of property held by the individual persons inhabiting the State.[100] This view seems to be inspired by a loose analogy with the rights that the Lord or King held over his land in the seigniorial epoch of late feudalism which lasted until absolute monarchies were replaced by constitutional systems in early modern times.[101] Both the roots and the lasting effect of this theory – conceiving State sovereignty in respect of territory as a private-law relationship – were undoubtedly facilitated by the significant number of concepts and terminology that international law on the acquisition (and loss) of territory had borrowed from the Roman private law tradition.[102]

97 This point is particularly made by Eric Suy, 'Réflexions sur la distinction entre la souveraineté et la compétence territoriale', in Raymond B. Marcin, ed., *Internationale Festschrift für Aldred Verdross zum 80. Geburststag* (Munich: Flink, 1971), 493.

98 Jennings, 2.

99 This clear and bipartite distinction is adopted by Suy, 493.

100 Donato Donati, *Stato e Territorio* (Rome: Athenaeum, 1924). The patrimonial theory was followed in France by Paul Fauchille, *Traité de droit international public* (Paris: Rousseau & Cnie éditeurs, 8th ed., 1922), vol. 1, 450.

101 Barberis, 135–136.

102 Hersch Lauterpacht, *Private Law Sources and Analogies of International Law* (London: Longman, Greens and Co., 1927), Chapter III; and Jennings, 3.

Sir Robert Jennings, for instance, points out how international law "like the Roman law counterparts", requires the presence of 'a corpus', as much as of an 'animus', for the creation and maintenance of a title of territorial sovereignty. This parallelism explains the great importance given to the possession of the land (through effective control) for the creation and maintenance of a sovereign right over it.[103] Moreover, the effects of Roman private law are not only visible in the meaning of the title to territorial sovereignty itself, but also in the modes by which such title can be acquired (by occupation, prescription, cession, accession or accretion, and conquest) or lost (by cession, operations of nature, subjugation, prescription, dereliction or 'revolt') – all eminent Roman legal institutions. However, by considering the territory as the object of a State's right of property (*dominium*), the patrimonial theory virtually denied the dimension of the State as an entity vested with *imperium*, embodying first and foremost a public authority displayed throughout the territory. Such a viewpoint can be seen as resulting from an anachronistic understanding of statehood, reluctant to dissociate the personality of the State from that of the Head of State. Among the critics of this view was, for instance, John Westlake, who in 1914 upheld that "each State has a sovereignty in and over its territory which presents some points of resemblance to property in land, but more important points of difference",[104] and thus chose to "treat territorial sovereignty as distinct from property and avoid describing it as eminent domain".[105]

Most of the scholarship though took a more ambiguous position and preferred maintaining to a great extent the high regard for Roman private law institutions put forward by the patrimonial theory, whilst also acknowledging the need to consider the mark of the State's *imperium*. The general result of this loose mix was a legally unsatisfactory depiction of the nature of the right

103 The most prominent manifestation of the importance of holding effective control over the land so as to create and defend a right overlies the concept of 'effectivities'. Sir Robert Jennings thus explained that: "[T]he tendency of the law has necessarily been to pay very great regard to the factual possession as creating title and excursions in the realms of an abstract title to sovereignty have been cautious and tentative". To sustain the view that there is widespread rejection of any idea that territorial sovereignty might be abstract, he cites Judge Max Huber in the *Island of Palmas* case, who contended that: "[I]nternational law – the structure of which is not based on any super-state organisation – cannot be presumed to reduce such a right as territorial sovereignty [t]o the category of an abstract right, without concrete manifestations." In Jennings, 5.

104 Lassa Oppenheim, ed., *John Westlake: Collected Papers on Public International Law* (Cambridge, UK: Cambridge University Press, 1914), 131.

105 Ibid.

of a State over its territory – or territorial sovereignty – as being *akin to* a private law relationship.[106] As James Brierly noted, this expression was "in the absence of any better word", "a convenient way of contrasting the fullest rights over territory known to the law with the minor territorial rights to be later mentioned".[107]

The constitutive theory of territory reacted to this far-fetched consequence of the patrimonial theory – namely, the neglect of the State's dimension as a public authority – and, as its name indicates, conceived the territory as a constitutive element of the State rather than as the object of a State's right.[108] This conceptual shift provided an explanation of the nature of territorial sovereignty more consistent with reality. The premises of this theory were, however, strongly criticised. Firstly, it was pointed out that the 'commanding power' of the State cannot logically be exercised over the territory as such, but only over the people inhabiting it.[109] Moreover, opponents argued that, if territory was to be considered as a constitutive element of the State, then any transformation of the territory would necessarily result in the parallel transformation of the nature and identity of the State – a circumstance which, as Eric Suy recalls, only takes place in very rare cases of total succession.[110] Though the constitutive theory proved very useful for the development of the basic tenets of *Geopolitik* and its organic conception of the State, it did not achieve widespread support in international legal scholarship.[111]

106 James L. Brierly, *The Law of Nations: an Introduction to the International Law of Peace* (Oxford: Clarendon Press, 5th ed. 1955), 150.

107 Ibid.

108 This criticism was even acknowledged by Karl Victor Fricker – regarded by Julio Barberis as the founder and most important representative of the constitutive theory – who distinguished the notion of territory (*Gebiet*) from that of land (*Grund, Boden*) and defined the former as the space inhabited by the people over whom the State displays its power of command. See Karl V. Fricker, *Gebiet und Gebietshoheit: Mit Einem Anhang Vom Staatsgebiet* (1901), cited in Barberis, 137.

109 Ibid.

110 See Eric Suy, 494, citing the work of Léon Duguit, *Traité de Droit constitutionnel* (Paris: Fontemoign et Cie., 1921), vol. 1, 48. See also Schoenborn, 116: "[O]n ne peut pas bien se représenter que l'état puisse se séparer en principe de parties de son territoire, c'est-à-dire de parties de son essence! À moins qu'on admet une mutilation du corps de l'État analogue à l'amputation d'un membre du corps humain".

111 One of the rare exceptions was Georg Jellinek, who explicitly embraced it when asserting that "without the existence of human beings, we can't speak of territory, but only of parts of the earth's surface", in *Allgemeine Staatslehre* (Berlin: Haring, 3rd ed. 1914), 174.

At the core of both the patrimonial and the constitutive theories of territory is their common struggle to establish an explanation of the role played by territory, as a physical element, and within the State, as an abstract concept. The theory of competence (*Kompetenztheorie*) emerged as a reaction to this impasse and can be considered as the more ambitious project to detach and free the legal nature of territory from its physical dimension. In seeking to establish a purely juridical concept of territory – and thereby equally "demystify the State by assimilating it with the legal order"[112] – the competence theory operated a change of paradigm.[113] Triggered by the publication in 1906 of Gerard Radnitzky's article *Die rechtliche Natur des Staatsgebiets*,[114] it was developed by Hans Kelsen's specific application of his monistic doctrine of relations between national and international law, and then pursued by Alfred Verdross in the Austrian school.[115] The core contention of the theory of competence is to consider the territory as nothing but the space where a statist legal order is valid[116] and within which the State can exercise its coercive power. Thus, whilst emphasising the idea of competence, this theory downgrades the role of possession which is central to the patrimonial theory; nor does it accept to consider territory as a constitutive element of the State, since it presupposes the latter's existence. The competence theory has proven to be particularly useful in explaining cases in which a State, for instance, exercises its competences in a territory that does not fall under its sovereignty – that is, in identifying the need to differentiate territorial sovereignty from territorial competence.[117] Yet, this differentiation is what permits the patrimonial and constitutive theories, on the one hand, and competence theory, on the other hand, to be regarded as being complementary rather than mutually exclusive. The patrimonial and constitutive theories deal with the creation and maintenance of the right to territorial sovereignty; they

112 Suy, 495.

113 Ibid.

114 Gerard Radniztsky, 'Die rechtliche Natur des Staatsgebiets' [Juridical Nature of the State], *Archiv für öffentliches Recht*, XX (1906), 313.

115 See the Commentary by Benedetto Conforti, 'The Theory of Competence in Verdross', *European Journal of International Law*, vol. 6 (1996): 1, 70–77.

116 Hans Kelsen, *Pure Theory of Law* (Berkeley: University of California Press, 1970).

117 One of the historical illustrations of this situation involves the Austro-Hungarian Empire, which administered Bosnia and Herzegovina between 1878 and 1909, a time when they were provinces of the Ottoman Empire. Although the Austro-Hungarian Empire exercised its competence over Bosnia and Herzegovina during that period, it only acquired the sovereignty of these territories after purchasing them from Turkey, in 1909, for 54 million gold crowns. See Suy, 498.

are concerned with the nature of the right over the territory and how such right is created, maintained or lost. The competence theory operates *after* the title or right to territorial sovereignty over a specific portion of the earth has been established, and instead focuses on explaining what are the consequences of the existence of a right to territorial sovereignty for the validity of the legal order of the State.

To a great extent the legal theories on the bond between the State and its territory seemingly reflect the differing faces of 'territorial functionalism' as revealed in the preceding brief historical account. Yet, since State territory is 'inseparable' from statehood, every attempt to establish a legal theory of the nature of territory in international law also heavily bears the stamp of a particular conception of statehood. The development of these theories of territory only involved European intellectual circles – mostly until the First World War and during the inter-war period. To some extent, these represented the views of different 'national schools' – the Italian School being the house of the patrimonial theory, the German School, that of the constitutive theory, and the Austrian School, that of the competence theory, supported or complemented by French and English legal scholarship. Hence, it does not seem far-fetched to consider that these theories inescapably carried a 'Eurocentric stamp'.

The relevance of this background arises particularly when considering the impact of Article 1 of the 1933 Montevideo Convention on the Rights and Duties of States (hereafter, the Montevideo Convention) on the range of existing understandings of the legal nature of territory in international law at the time.[118] Establishing a definition of the entities with legal personality to which the obligations of the Convention may be applied, Article 1 provided that:

> [T]he state as a person of international law should possess the following qualifications: (a) a permanent population; (b) a defined territory; (c) government; and (d) capacity to enter into relations with the other states.

This provision witnessed ample citation among scholarly circles as well as in international courts and tribunals and, in the absence of a better definition of the State – resulting, for instance, from the International Law Commission – found few obstacles to its being considered as the 'codification of statehood',

118 Convention on the Rights and Duties of States (Montevideo Convention), adopted on 26 December 1933, entered into force on 26 December 1934, 165 L.N.T.S 19. Text reprinted in *American Journal of International Law*, vol. 28 (1974) (Suppl.), 74.

reflecting a general rule of customary international law.[119] This continuing widespread embrace of the content of Article 1 contrasts with the regional origin of the Convention, which was adopted at the Seventh International Conference of American States – by only nineteen States.[120] Commenting on the Montevideo Convention, Thomas Grant has argued that the success story of Article 1 is simply the reflection of a consensus on its content existing at the time when the Convention was adopted. Yet, when interpreting Article 1 as such, as well as its subsequent impact, neither the importance of the wider political context from which the Convention emerged, nor the content of other provisions of the Convention, can be ignored or downgraded.

The Convention reflected the main purposes of pan-Americanism, which marked transatlantic international relations of the late 19th century and early 20th century. Finding its roots in the Monroe Doctrine and explicitly developed in an institutionalised form since 1890, this movement sought to reaffirm the independence of young American States from their former European colonial masters.[121] This principled goal is clearly expressed in Article 3 of the Montevideo Convention, which provides that "the political existence of the State is independent of recognition by the other States". Thereby, along with the criteria for statehood, the signatory States affirmed their acceptance of the declaratory theory of recognition – as opposed to the European-led constitutive theory.[122]

While providing that recognition of a political entity as a State by other States has only a declaratory effect, the Montevideo Convention conceived statehood as a matter of fact, not of law.[123] Article 1 must be read through this prism, as establishing a set of 'criteria' – rather than elements – which can

119 See Thomas D. Grant, 'Defining Statehood: the Montevideo Convention and its Discontents', *Columbia Journal of Transnational Law*, vol. 37 (1999): 2, 403–457, 407. See also James Crawford, 'The Criteria for Statehood in International Law', *British Yearbook of International Law*, vol. 48 (1976–1977), 93–182; and Johan D. Van Der Vyver, 'Statehood in International Law', *Emory International Law Review*, vol. 5 (1991): 1, 9–102.

120 The Montevideo Convention had indeed a regional scope; the small number of signatories contrasts with the weight and importance of its subsequent influence. The original signatories were: Argentina, Bolivia, Brazil, Chile, Colombia, Cuba, Dominican Republic, Ecuador, El Salvador, Guatemala, Haiti, Honduras, Mexico, Nicaragua, Panama, Peru, United States of America, Uruguay, Venezuela.

121 For an historical account of the origins and different stages of development of pan-Americanism, and its effects on the 'codification' of statehood, see Arnulf Becker Lorca, *Mestizo International Law* (Cambridge: Cambridge University Press, 2015), 305–349.

122 Montevideo Convention, Article 6.

123 On the nature and role of State recognition, see chapter 3, section 3.3.1.B below.

be objectively recognised and thus serve to prove the factual existence of a State, rather than to constitute the State itself. The result of this consideration of statehood as a matter of fact and the 'objectivisation' of the criteria enunciated in Article 1, particularly those referring to a 'material existence', such as territory and population, was the correlative decline of the need to establish legal theories of the concept of territory and its functions in international law. The role of the territory for the State and the legal nature of this bond is not as important as its factual presence (for the creation of the State) and possibly also of its permanence (for the continuity of the State).

1.1.2 *The Meaning of Territory in the Creation of Post-Colonial Pacific Island States*

A Decolonisation and the Creation of Pacific Island States

As already pointed out, the legal theories of territory in international law reflect the roles and functions that the territory played in the historical process of formation of the European nation-State. The present section considers the State-building movement resulting from the 20th century decolonisation which brought about a new form of statehood, generally referred to as 'the post-colonial State'. It then correlatively invites to adapt the understanding of the role and place of the concept of territory. When applied to the south Pacific region, this adaptation helps explaining why the limited territorial extent of small Pacific Island States did not impede their creation as independent States.

The consideration of the State as a matter of fact rather than of law that facilitated the independence of American States and was encapsulated in Article 1 of the Montevideo Convention – marked the prelude to the new form of the State which would fully materialise after the World War II. Far from merely constituting a circumstantial political agreement, the new world order was accompanied by a radical shift in the *normative* framework concerning colonies.[124] Decolonisation and its product – the post-colonial State – thus came to embody a new form of statehood; one grounded in a specific paradigm that may find its roots in the fourteen points of Woodrow Wilson's inter-war plea on the right to self-determination.

Although decolonisation ultimately implied the 'global imposition' of the State as the universal form of political organisation, the resulting heterogeneity of the international system cannot be overlooked. The acknowledgment of a typology of States based on a differentiation between the modern and the post-colonial forms can be drawn from Max Weber's ideal types of States. As

124 See Georg Sorensen, *Changes in Statehood: The Transformation of International Relations* (New York: Palgrave McMillan, 2001), 57.

explained by Georg Sorensen, these ideal types constitute analytical constructs or conceptual patterns that accentuate selected aspects of historical reality, rather than expressing concrete empirical findings.[125] Thus, the post-colonial ideal type not only constitutes a new form of statehood, but is also a potential vehicle of a new understanding of statehood, complementing and rendering more complex the previous knowledge traditionally derived from the modern type of State. One manifestation of the contribution of the post-colonial State to the understanding of statehood can be found in, for instance, the rise of independence as a criterion of statehood that was not included in Article 1 of the Montevideo Convention, but eventually came to be considered by prominent authors, such as James Crawford, as the most important criterion.[126]

As the criterion of independence became the cornerstone of post-colonial State-building, other criteria listed in Article 1 of the Montevideo Convention – both material and immaterial – decreased in prominence. Again, Sorensen clearly explains the general consequence of this move when he states that "the normative framework around decolonisation gave the right of independence to ex-colonies no matter what level of actual weakness that they displayed."[127] It is in this context that one can understand how a set of entities with such a limited territorial extension (covering the land as well as the territorial sea) as that of the small Pacific island countries could emerge in the international system in the form of independent States, to which all the attributes of statehood (formal equality, territorial integrity, right to self-defence, etc.) were granted, in spite of their questionable capacity to exercise or protect them.[128] The predominance of the normative framework in which Pacific Island States emerged as independent entities implied that they were subject to different standards to those of the modern European States. In order to grasp the functionality played by the concept of territory in the South Pacific region and explain the emergence of micro-States in this area, one is again compelled to go outside the realm of law and review the historical turning points of the region.

125 Ibid., 73.

126 This position is consistently defended by James Crawford in *The Creation of States in International Law* (Oxford: Oxford University Press, 2nd ed. 2006).

127 Sorensen, 83.

128 European micro-States which preceded the creation of post-colonial Pacific island (micro)-States came into being for different normative purposes and to serve certain interests of European Nation States. For a comprehensive study of these cases, see Jorri K. Duursma, *Self-determination, Statehood and the International Relations of Micro-States: The Cases of Lichteinstein, San Marino, Andorra, Monaco and the Vatican City* (Groningen: Van Denderen, 1994).

What is today referred to as the Pacific Island States corresponds to the division of over 20,000 islands into twelve independent political entities, scattered over an immensely wide oceanic space and surrounded by disproportionally big continental territories – the South American coasts of Peru and Chile to the east, the Oceanic New Zealand and Australian territories to the south and south-west, respectively, and, moving towards the north-west, the Asian coasts of Indonesia, Philippines and Japan. In spite of their recently acquired statehood, the limited extent of the territory of these States– coupled with the considerable distances between the islands that form their respective territories, and with their economic underdevelopment – make them a good example of 'weak players' resulting from decolonisation. Hence, these sometimes condescendingly labelled 'micro-States' are associated with images of remoteness, smallness, vulnerability and, more recently, with a set of actors who, despite being completely devoid of power in the international system, insist on claiming the need for international action to stop the threat that climate change impacts constitute for their survival as States.

And yet, the history of the Pacific Island States suggests exactly the opposite. Far from having been disconnected from the world's history, this region has had close bonds with the colonial powers of each historical epoch ever since the 16th century. The history of this assemblage of scattered islands spread over the biggest and most unpredictable of the world's oceans, the Pacific, is the story of a transformation effected over five different epochs, each of which has had different repercussions that go far beyond Pacific Island States' small geographical size. From being the object of the major powers' geopolitical interests during the first 'ages' of discovery and colonisation, or the theatre of these powers' confrontation in their efforts to establish a balance of power in the 20th century, the newly emerged Pacific Island States have become actors of the international system, themselves owners of their own geopolitical calculations, hence bound to face the specific post-colonial security struggles that make the region highly dependent on different forms of inter-regional and international co-operation.

The 'age of discovery' marked the beginning of the region's connection with European States in the early modern period. As reported by historian Mercedes Maroto Camino, after Juan Sebastián Elcano had completed the circumnavigation initiated by Fernando de Magallanes between 1519 and 1521, three Spanish expeditions, conducted by Álvaro de Mendaña de Neira and Pedro Fernández de Quirós,[129] departed from the west coast of South America to the *Mares*

129 The first expedition, led by Álvaro de Mendaña de Neira in 1567, departed from Lima (Peru) and led to the discovery of the Solomon Islands in 1567. The second expedition,

del Sur (the South Seas) in search of a large Southern Continent referred to as *Terra Australis Incognita* or *Terra Magellanica*.[130] These were followed by Portuguese and Dutch expeditions, which competed with potential Spanish interests, but also helped in the elaboration of the first cartographic images of the south-east Pacific, transforming it into a geo-political space of "geographical, economic, missionary or colonial importance".[131] From that time to the last voyage of James Cook (1776–1779), the European interest in the region would progressively evolve into the mercantilist system in which the Pacific became subject to European colonial domination.

The strength of these colonial bonds increased particularly as the Pacific territories not only constituted a source of trade but also mirrored the changing balance of power occurring in Europe from the 16th century onwards. During the period referred to by Eric Hobsbawm as 'The Age of Empire' (1875–1914), the decline of the Spanish, Portuguese and later Dutch presence in the region was replaced by the increasingly important French, British and then, German naval forces.[132] First, the French incursion into the Pacific began between 1840 and 1850 with the annexation of the Marquesas Islands, New Caledonia (*Nouvelle Calédonie*) and Tahiti, followed by the Loyalty Islands in 1864. The great expansion of the British Empire over the Pacific started in 1874 with the annexation, through the Treaty of Cession with King Catabau, of the roughly 300 islands of Fiji. By the time Archduke Franz Ferdinand of Austria was assassinated in Sarajevo in 1914, Britain had: established its presence in south-eastern New Guinea after a partition agreement with Germany (1884); proclaimed a British Protectorate over Tokelau (1889) and the Gilbert Islands – today Kiribati – (1892); negotiated an Anglo-French Treaty of Condominium of the New Hebrides (today Vanuatu) in 1906; and put an end to the dispute with

---

also led by Mendaña de Neira, who travelled with Pedro Fernández de Quirós as a pilot, took place between 1595 and 1596. The third expedition, this time led by Pedro Fernández de Quirós, left from Lima in 1605, under the command of his wife, Admiral Isabel Barreto, and arrived at an island which they baptised Espirito Santo (today in dispute between Vanuatu and France) and took possession of on 14 May 1606. When leaving the island, they came back to the Philippines via the Torres Strait, between modern Papua New Guinea and Australia. See Mercedes Maroto Camino, *Producing the Pacific: Maps and Narratives of Spanish Exploration (1567–1606)* (Amsterdam, New York: Rodopi, 2005), 15.

130 Ibid., 69.

131 Ibid., 1.

132 Eric J. Hobsbawm, *The Age of Empire: 1875–1914* (London: George Weidenfeld and Nicolson, 1987).

the United States on the recognition of Britain's presence in the Ellice Islands (today Tuvalu), the Solomon Islands and Tonga.[133]

However, by the last decade of the 19th century, this playground of world powers would reflect how British dominance was gradually being challenged by the rise of Prussia.[134] Four years after having taken control of north-east New Guinea (1884), Germany annexed Nauru (1888) and began mining the island's phosphate deposits through a consortium with the British Phosphate Commission. Germany's increasing negotiating power also led to the division of Western Samoa (today Samoa) by agreement with the United States, concluded after the previously failed attempt to reach a tripartite agreement to include Britain. In addition to this imposition of its own space among those of the other already present powers, Prussia took over the very last remnants of the Spanish Empire and purchased the Marianas and Caroline Islands (today Palau) a year after Spain's symbolic loss of Cuba in the Spanish–American War (1898). It also spread its influence on the islands north of the equator by establishing the Protectorate of the Marshall Islands. Meanwhile, the United States maintained a low profile and its timid presence in the region was primarily driven by its concern for the trade route with China and Japan. Thus, right before the turn of the century, the United States acquired Hawaii and the island of Guam, the annexation of which in the same year was also an outcome of the Spanish–American War. A year later, the USA also acquired American Samoa (following the 1890 three-powers Convention, although the USA only formally took power in 1926) and agreed to recognise the British presence in Tonga.

The starting point of the 'age of confrontation' – which essentially matches Eric Hobsbawm's 'short 20th century' (1914–1991) – was the outbreak of the *Grande Guerre*.[135] Though the Pacific Islands were spared direct destruction in World War I, the outcome of the war was nevertheless reflected once more in the region. As the German presence was completely eliminated, the decline of the British Empire in favour of rising regional powers (Australia, New Zealand and Japan) was evident. The presence of Australia, which till the war had been limited to the acquisition, by transfer, of former British New Guinea

133 Several changes in the constitutional status of the islands succeeded, leading especially to a change of their status from protectorates to colonies, as occurred for the Gilbert Islands (Kiribati) in 1915.

134 Eric A. Walker, *The Cambridge History of the British Empire* (Cambridge: Cambridge University Press, 1963).

135 Eric J. Hobsbawm, *The Age of Extremes: The Short Twentieth Century (1914–1991)* (London: Michael Joseph, 1994).

in 1902, grew since its occupation of the phosphate island of Nauru, as well as of German New Guinea, during the war. From 1919 to 1939, Nauru remained under an Australian administration exercised under a League of Nations' mandate, jointly held by Britain, Australia and New Zealand. Likewise, New Zealand's presence in the small Pacific Islands – limited until then to the transfer by Britain of the Cook Islands in 1901 – rose as it became the administering power of the former German Western Samoa in 1918 and received the administration of Tokelau from Great Britain in 1925.

Yet, the deciding shift in the geo-political balance of the region in the aftermath of World War I was the rise of Japan as an important maritime power in the region, which may be said to have started since its victory in the battle of Tsuchima against the Russian Empire in 1905. Japan's growing colonial expansion towards the south Pacific led to its occupation of the islands north of the equator (the Marianas, Caroline and Marshall Islands) during World War I. It then succeeded to these former Prussian colonies which were united and renamed the Pacific Islands Territory and which it administered – supposedly – on behalf of the League of Nations. This would nonetheless prove to be a dead letter, since the islands became key naval and air bases and were used as platforms for attacking the US naval base of Pearl Harbor on 7 December 1941. Hence, from being the playground of world powers, the Pacific Islands became a battlefield of World War II and the place where the turning point of the strife – as well as of world history – took place. For a short period of time during the war, Japan enjoyed a dominant power position in the south Pacific, occupying Nauru, the Solomon and the Gilbert Islands and exercising direct political control of Papua New Guinea. Yet, the act of aggression of the Japanese Imperial Army against the US naval base in Hawaii triggered the direct involvement of the American forces which would be determinant in the victory of the Allies, and was counterbalanced in 1943 by the major American victories against Japan in the islands of Makin and Tarawa (today part of Kiribati). After the war, the Japanese presence in the Pacific Islands was almost completely eliminated. As the settlers were repatriated to Japan and the country was deprived of all its overseas possessions, Japan's new post-war constitution provided that it would renounce the use of force and would only allocate 1% of its GNP to defence spending. The United States retained the exclusive control of all the naval and air bases around Japan (including the Okinawa base) and complemented Japan's minimum self-defence arrangements by the 1951 Mutual Security Agreement. Thus, the succession of the USA of islands formerly under Japanese administration, after the end of World War II, correlatively inaugurated American dominance in the region through the establishment in 1946 of the UN Strategic Trust Territory – covering the Marianas, Caroline and Marshall Islands. Moreover, in parallel with the conclusion of the 1951 San Francisco Peace Treaty with Japan, the USA promoted the creation of ANZUS (Australia, New Zealand, United States Security Treaty), a defence alliance that reaffirmed the position of these regional powers.

The American dominance of the islands through the establishment of the Pacific Trust Territory, which covered an area of over 7.5 million sq. km., became very important during this period, as it was used by the United States in the Korean and Vietnam Wars of 1954 and 1962, respectively. Once again acting as a mirror of the major States' balance of power, the region also witnessed – particularly since the 1980s – the surrounding presence of Soviet naval forces which had begun an expansion towards the South Pacific. Fearing that the Pacific Islands might become 'a new Cuba', the United States unsuccessfully tried to convince Japan to play a greater role in the defence of the Pacific, whilst, in parallel, seeking to reinforce the USA's nuclear presence in the area through the operation of the ANZUS agreement. Yet, the reluctance of Australia and New Zealand to let US nuclear-powered ships and aircrafts use their bases led to the cancellation of the 1985 defence exercise as well as the end of intelligence co-operation with New Zealand. To counterbalance this situation, the USA concluded in 1984 an agreement with Fiji allowing the deployment of nuclear-powered warships from the port of Suva, Fiji's capital.

It is of course within the Cold War context that the Pacific Island States began their 'age of independence'. Although Anthony Christopher indicates that the renaming of French Settlements in the Pacific Ocean as French Polynesia in August 1957 can be considered as "the first official hint of Pacific island ethnic nationalism",[136] the files on this region were not ready to be closed before the early 1970s, with a few exceptions. Considering Pacific Islands as being at "last ripe for decolonisation", David McIntyre emphatically stated that "in all earlier discussions about the future of smaller territories, the Pacific Islands were at the bottom of the list, always in the 'never' category".[137] The reluctance of administering and colonial forces to decolonise Pacific Islands partly illustrated the Anglo-American approach to self-determination, but also stemmed from the fear of dilution of the United Nations by several territories allegedly too small to support their own statehood.[138] Yet, the decolonisation of the region was triggered in 1961 when the United Nations prompted the American government to speed up the constitutional advances

136 Anthony J. Christopher, *Atlas of States: Global Change 1900–2000* (London: John Wiley and Sons, 1999), 180.

137 David W. McIntyre, *British Decolonisation: 1946–1997* (New York: St Martin's, 1998), 68.

138 See Sheila Harden, ed., *Small is Dangerous: Micro-States in a Macro-World* (London: Frances Pinter Publishers, 1985), 179.

in the Pacific Trust Territory. Western Samoa (1962) and Nauru (1968), both under US trusteeship since World War II, were therefore the first to acquire independence in the region. They were followed by the British Protectorates of Tonga (1970) and Fiji (1971).[139] Tonga, Western Samoa and Fiji were thus able to attend the first 'Commonwealth Heads of Government Meeting', in 1971, a new-style gathering that was held for the first time outside London – in Singapore –[140] and where the Commonwealth Declaration of Principles and the Commonwealth Fund for Technical Co-operation were adopted. While, in the mid-1970s, Papua New Guinea acquired independence as a single State (1975),[141] the single British Protectorate over the Gilbert and Ellice Islands was fragmented into two new States following the 1974 referendum. On the one hand, the former Ellice Islands – with a majority Polynesian population – acquired independence as Tuvalu in 1978. It was followed a year later (1979) by Kiribati – with a majority Micronesian population – constituted by the former Gilbert, Phoenix and Line Islands, along with a special status for the phosphate island of Banaba (formerly Ocean Island). By the turn of the 1980s, the former Anglo-French condominium of the New Hebrides was dissolved and became independent as Vanuatu (1980), thus marking the only French retreat from the region.[142] Finally, the UN Pacific Trust Territory composed of the islands north of the equator was fragmented into four parts. In 1986, the Federated States of Micronesia and the Marshall Islands acquired statehood, followed by Palau which, after separating from the Federated States of Micronesia in 1978, also acquired statehood, in 1994.[143] All three chose a Compact of Free Association with the United States as their preferred form of political organisation,[144] while

139 Within Fiji, the island of Rotuma was recognised as having a special status.

140 McIntyre, 68.

141 Later on, between 1988 and 1998, in Papua New Guinea the secessionist Bougainville Revolutionary Army sought independence of Bougainville Island (formerly part of the Solomon Islands) in order to control the copper mine.

142 The French resisted decolonisation in the Pacific just as in the Caribbean, whereas the British assumed that the process was going to take place anyway. The island of Espiritu Santo tried to secede from Vanuatu in the same year. The only major territorial dispute of the region that remains unsettled is between France and Vanuatu, and concerns some islands south of Vanuatu and north of the French territories of Wallis and Futuna.

143 The date of independence was delayed by the approval of the Compact of Free Association, which did not take place until 1993.

144 See Crawford, *The Creation of States*, 719; who explains that: "[T]he classification of common forms of organisation have no legal consequences beyond those arising from the specific agreement or instruments in question, this has no direct bearing on the entity's legal status or lack of it as a state under the criteria referred to [in Article 1 of Montevideo Convention]".

the Marianas Islands, Guam and Hawaii were integrated into the United States. Likewise, the Cook Islands, Niue and Tokelau are today New Zealand territories. Whilst not being UN members, they enjoy a certain level of autonomy which, for instance, enables the Cook Islands and Niue to be Parties to the UN Convention on the Law of the Sea and generate maritime areas of their own.[145]

B Newly Independent Micro-States in the United Nations

This reminder of the decolonisation of Pacific Island States serves the purpose of explaining how statehood could be acquired by political entities with a very small territorial extent, limited population and economically underdeveloped. Simply put, the territorialisation of nationhood in the case of this region can be said to be a function of the exercise of the right to self-determination, an expression of such right. As such, the territory conveys the necessary physical support for these previously colonised 'nations' or 'peoples' to achieve formal independence from colonial control; the territory serves a very specific normative framework, superseding the setbacks that very small newly emerged States may endure in the effective exercise of their competences. If the historical and political context in which the Pacific Island States emerged explains how they could acquire formal independence despite their small geographical size, it is equally important to account for the means by which they were able to become members of the international community. Considering how they could effectively act 'as States' is tantamount to exploring their substantial inequality and to assessing whether, after being freed from colonial rule, these peoples' self-determination attained a full level of realisation.[146]

As in the rest of the world, decolonisation did not put an end to the dependence of now newly emerged Pacific Island States on their former ruling powers. During the new 'age of co-operation', the former colonial dependencies remained functional under the informal control of some world powers, who maintained their presence in the region through economic and political support. The framework for institutional co-operation with former colonial powers in the region was first dominated by the acquisition, by most of the former non-self-governing territories, of membership (either full or associate) of the Commonwealth, which put dominant British presence in the region back on track. This membership included States that had chosen full independence

145 In 2006 and in 2007, a referendum was held in Tokelau on whether to change the status of Tokelau from a territory of New Zealand to an independent State in free association with New Zealand. The results did not reach the threshold needed to approve the change of status.

146 William M. Harris, 'Microstates in the United Nations: A Broader Purpose', *Columbia Journal of Transnational Law*, vol. 9 (1970), 23–53.

as the political form of their organisation (Kiribati, Tuvalu, Solomon Islands, Fiji, Tonga, Vanuatu, Nauru, Papua New Guinea and Western Samoa), States in a Compact of Free Association with another State (Federated States of Micronesia, Palau, Marshall Islands), and the self-governing entities – also in a Compact of Free Association with another State – (Cook Islands, Niue, Tokelau).[147] In the case of these free-association States, irrespective of their Commonwealth membership, the Compacts of Free Association logically created particular and additional bilateral ties. The western – particularly European – presence in the region also materialised through the creation of the South Pacific Commission. Nonetheless, by the time the UK had entered the European Union, in 1973, the South Pacific Commission was replaced by the Pacific Island States Secretariat – which then became the Pacific Islands Forum (PIF), in 2000, to which France, the UK and the USA were not invited to be members. Their leadership in this new Forum was thus replaced by today's main regional powers, namely, Australia and New Zealand. Since its creation, the importance of the PIF as a co-operative framework has increased so as to become a vehicle of financial development through aid flows, technical assistance and defence co-operation (including disaster-relief services) from these two countries to the Pacific Island States. More recently, the European Union's inclination to renew an active collaboration with Pacific Island States seems to be competing not only with Australia and New Zealand, but also with the rising active interest of the Chinese investment sector in creating a zone of influence in Oceania.[148]

The inherently weak governmental and economic structures of Pacific Island States following decolonisation undeniably favoured the maintenance of an informal but critical dependence on established economic and political

147 We exclude from this 'co-operation framework' the territories which were integrated into those of their former colonial rulers, such as: the French DOM-TOM, which were recognised as having a higher degree of autonomy within the centralised French State than the rest of the French *départements* (French Polynesia, New Caledonia, Wallis and Futuna); the American territories like Hawaii, which became the 50th American State on 21 August 1959, Guam, American Samoa; and the New Zealand territories of Cook Islands, Niue and Tokelau. While Tokelau is an integrated territory of New Zealand, Niue and the Cook Islands are self-governing territories and thus enjoy a higher level of administrative and legal autonomy with respect to New Zealand. Yet, they fall short of constituting independent States like the Marshall Islands, the Federated States of Micronesia and Palau, which are in a Compact of Free Association with the United States and yet are also members of the United Nations.

148 Terence Wesley-Smith and Edgar A. Porter, eds., *China in Oceania: Reshaping the Pacific?* (Oxford, New York: Bergham Books, 2010).

relations with a former colonial power.[149] Therefore, qualifying for UN membership unsurprisingly represented, for these States, the assurance that a future release from such informal control was possible and would allow them to finally achieve the right to self-determination. The United Nations was thus seen as a solid structure which could provide an alternative protection;[150] a forum of dialogue between the newly emerged States and their former rulers, where the Pacific Island States could, as W. Harris puts it:

> [P]ublicise grievances and forge public support which may ultimately be the most effective tactic available when a microstate cannot defend itself, or when regional assistance is not available to limit coercion.[151]

Yet, the qualification of micro-States (including Pacific Island States) for UN membership was far from being easily accepted, particularly by the British and American representatives. Indeed, 'being a State' was not enough. The question of UN membership is not a question of recognition of micro-states' independent statehood, but on how to evaluate their participation in the international community within an institutionalised framework.

Besides, as difficult as it is to define a 'State', it is also difficult to define what constitutes a 'micro-State'.[152] The first definition of micro-states was given by UN Secretary-General U-Thant in the 1968 Annual Report of the Secretary-General on the Work of the Organisation, which held that micro-states are entities which are exceptionally small in area, population and human and economic resources, and which were emerging as independent states. Some definitions of 'micro-states' only take into account the criterion of having a population below 1 million. This is, for instance, the view taken by the UNITAR Report directed by Jacques Rapoport, which included States with small

149 Harris, 38.

150 As former UN Secretary-General Dag Hammarskjöld noted, it is not the major powers, but rather "all the others" that need the United Nations for their protection, in Secretary-General, *Annual Report on the Work of the Organisation*, 1960, 15 UN GAOR 332.

151 Harris, 35.

152 As explained by Thomas Grant in 'Micro States', *Max Planck Encyclopedia of Public International Law* (Oxford: Oxford University Press, 2013), online resource, entry last updated February 2013, available at: http://o-opil.ouplaw.com.catalogue.ulrls.lon.ac.uk/view/10.1093/law:epil/9780199231690/law-9780199231690-e1068?rskey=UbAZu5&result=1&prd=EPIL (20 August 2015), para. 1: "[T]he term micro State is sometimes used by legal writers; some very small States even refer to themselves as micro States. But micro State is not generally regarded as a legal term of art. There is as such no authoritative legal definition of micro State".

territories, like Pacific Island States, but also States with larger territories, such as Qatar.[153] Those considering the smallness of the territory do not quantify how small is 'small' but, regardless of the unspecified threshold, all Pacific Island States undoubtedly fall into this category.[154]

While the preliminary question of how to define a 'micro-State' was not completely resolved, the heated debate within the UN on this issue is a vivid reminder of how smallness can be nothing but a *relative* weakness, since, in the context of institutionalised co-operation – particularly in one such as the United Nations, which is based on the principle of formal equality – the leverage of the majority can, at least in principle, supersede the control of the powerful. Thus, aware of the limited capacities of small island States, the view rapidly spread in the corridors of the UN that small could equally be 'dangerous'.[155] By admitting to UN membership a set of small States – including twelve Pacific Island States – large powers would account for only one third of the total UN membership, in spite of providing 90% of the organisation's budget (Oceania would then account for 9% of total membership). In contrast, the 'newcomers' could at least hypothetically use their collective voting strength to influence political decisions and shape the policy in the organs of the United Nations – in particular in the General Assembly, where every States has one vote regardless of its population, territorial extension, level of economic development or political power of leverage. Apart from the political imbalances that could result from the qualification of States for UN membership, their eventual repercussion on the stability of international affairs was also contemplated, fearing that the entrance of these States might provoke an excessive effect of regional issues, so that the East–West bipolar confrontation would, consequently, become multi-polar.

In contrast, other political considerations in favour of the admission of micro-States[156] highlighted the idea that the universalisation of the United

153 See Secretary-General, *Annual Report of the Secretary-General on the Work of the* Organisation, 1968, 22 UN GAOR, Supp. 1 A, document reference: A/6701/Add.1. See Jacques G. Rapoport et al., *Small States and Territories: Status and Problems*, UN Institute for Training and Research (UNITAR), Series nº 3 (1969); and Jacques G. Rapoport, 'The Participation of Ministates in International Affairs', *Proceedings of the American Society of International Law*, vol. 62 (1968), 155–163.

154 Ibid.

155 Harden, ed., 1–13.

156 Daniel Orlow, 'Of Nations Small: The Small State in International Law', *Temple International and Comparative Law Journal*, vol. 9 (1995), 115–140, at 119–123; Orlow distinguishes three types of micro-States: European micro-States (San Marino, Lichtenstein), post-colonial 'enclaves', and international city-states (The Vatican). Pacific Island States likely fall into the second category.

Nations was a goal to be sought for the organisation to be able to transcend its original core role of maintaining international peace and security and fulfil a 'higher' idealist goal, that of becoming an organisation in which nearly all nations "can bring their influence to bear upon problems of mutual concern and in turn are influenced by the policies and needs of other States".[157] This group of supporters relied on the normative argument in favour of the full realisation of the right to self-determination and considered that denying UN membership to these newly independent States could be equated to a 'pro-colonialist' stand. Finally, defenders of micro-State membership of the UN argued that the context of Cold War *détente* was far more appropriate; international stability would be increased by broadening the organisation's membership, and would give "more multidimensional, more pluralistic and perhaps more realistic interpretations of UN developments"[158] whilst, as William Harris explains:

> [S]een from the perspective of the arms race, the faulting of small states for their inability to carry out Charter obligations by rendering military support to the UN becomes less persuasive when the political environment produced by the thermonuclear age is considered.[159]

It is in the context of these competing political calculations that the question of whether micro-States would actually have the capacity to fulfil their membership obligations was raised. Although it may be said that this issue was raised in support of the previously mentioned political considerations opposing their entrance – and thus operates in a highly politicised context – it was a question belonging to the realm of law in so far as it called for the application of Article 4 of the United Nations Charter. This provision contains the conditions necessary to qualify for UN membership and implies that statehood is a necessary but not a sufficient condition; a candidate entity must not only 'be a State', but also "in the judgment of the Organisation", be "able and willing to carry out these obligations [contained in the UN Charter]."[160]

157 Harris, 34.

158 Ibid.

159 Ibid.

160 UN Charter, Article 4. Besides, criteria for UN membership were developed in UN General Assembly Resolution 1514/XV, 'Declaration on the Granting of Independence to Colonial Countries and Peoples', adopted on 14 December 1960, 947th plenary meeting, UN doc.: A/RES/1514 (XV), as well as by the ICJ Advisory Opinions on *Conditions of Admission of a State to Membership in the UN*, (1948) ICJ Report 57; and *Competence of the General Assembly for the Admission of a State*, (1950) ICJ Reports 4.

The introduction of the criterion of capacity hints at the traditional weight of the criterion of effectiveness in the shaping of international rules, which had already been previously revealed in 1928 when Lichtenstein's application for membership to the League of Nations was rejected (before 1960, Luxembourg and Iceland were the only UN micro-state members, all located in Europe). It is also a view consistent with the idea that the UN system constituted a reflection of the prevalence of the so-called 'concept of large powers' in the 1948 San Francisco Conference.[161]

In an attempt to reconcile the core of the opposed political interests – universalisation of the United Nations vis-à-vis preserving a similar allocation of voting power within the General Assembly – proposals for an alternative status to that of full membership were raised and discussed both in academic circles and within UN organs. The possibility of granting micro-States observer status was first raised by former Secretary-General U-Thant in 1965. It is a status with no firm legal basis which rests on UN practice and does not require amendment of the Charter. Being a member of a UN organ or specialised agency and recognised as a State by a majority of UN Member States are the requirements determined by practice. Yet, observers do not have the right to participation in the General Assembly or the Security Council and cannot make proposals in these organs either. The possibility of regional memberships was therefore also raised, according to which a group of States would participate and vote jointly. This proposal had the inconvenience of requiring amendment of the UN Charter and was unlikely to work well in practice, for neighbouring States often have great disparities – especially in the context of intense nationalistic feelings existing in new small States.[162] Yet, the most prominent proposal that was discussed within the UN Security Council was that of creating a form of associate membership. A year after UNITAR conducted its report on the issues of micro-States, the US delegate to the Security Council proposed to grant micro-States an associate member status that would give them the right to attend a meeting of 'special interest' to the micro-State in question, and eventually the right to actively participate – with no right to vote – in meetings in which the micro-State's interests were directly involved.[163] This solution was presented as a "possibility commensurate with its [the micro-

161 On 25 April 1945, delegates of 50 nations gathered in San Francisco (USA) and drew the UN Charter, which would then be adopted on 25 June 1945 in the San Francisco Opera House.

162 Other proposals included the so-called 'Fisher proposal', consisting of a weighted vote and technical and information assistance provided by the permanent secretariat, in Harris, 48.

163 UNSC, Verbatim Record, S/PV.1505 (1969); and UNSC, Verbatim Record, S/PV.L1506 (1969).

states] capabilities" that, in the eyes of the US delegation, had the advantage of not requiring amendment of the Charter, for the General Assembly could create a new category and define the duties, privileges, and benefits of it.[164] As a result of the American proposal, a 'Mini-State Committee' was established to report back to the Security Council, in time for this organ to make recommendations to the General Assembly during its 24th session.[165] The expert Committee on the matter resumed eight times in closed meetings, surrounded by the general feeling that the USA was trying to force the situation and rush the matter, and opposition from other newly independent States which defended the view that imposing new criteria for admission was not acceptable. It was then, within the Committee, that the British delegation proposed an alternative form of associate membership, though this solution did not achieve agreement in the interim report of the Committee that was sent to the Security Council. Finally, both the USA and the UK proposals would be dismissed because, apart from the outstanding issue of the definition of a micro-State pointed out earlier by Rapoport, the UN legal department considered that a Charter amendment would in either of the two cases be required. As a result, in a last attempt to control the entrance of micro-States, the USA sought to reactivate the role of the 'Admissions Committee' as soon as the application of Fiji was filed, in 1970. Although this purpose was achieved, the resurrection of the Admission Committee made no difference in practice to the case of Fiji, nor to that of the other micro-state applications that followed.

The genealogy of the concept of territory presented above has revealed how its evolving meaning reflects the range of functions it has historically been called upon to play, and how such roles have been tackled from three main different legal standpoints. This historical and functionalist approach unveils the double condition of territory as being both a physical reality and a legal construct, two dimensions that are essential to the *creation of States* and which find in State theory a common ground. The three legal theories of territory have also demonstrated that, when assessing 'the role' or 'the meaning' of territory in international rules governing statehood, one ultimately needs to question the nature of the relationship between these two dimensions.

Assessing the effects of transformation or loss of territory of *an existing State entity* – a process referred to as 'de-territorialisation' – correlatively implies the

---

164 These competences derive from Articles 10 and 11 of the UN Charter, together with Article 21 of the General Assembly Rules of Procedure.

165 Michael M. Gunter, 'The Problem of Mini-state Membership in the UN System: Recent Attempts Toward a Solution', *Columbia Journal of Transnational Law*, vol. 12 (1973), 470–472.

existence of the same double-edged meaning of the term 'territory'. Hence, de-territorialisation is to be understood as a bipartite process involving: (1) the set of physical changes in the territory of a State; and (2) the transformation of the territorial rights of the State as a result of such physical changes. It is equally a process that can be divided into two main situations corresponding to different stages of development and increased gravity. The first stage (developed in section 1.2. below) deals with limited or partial de-territorialisation of a State, as currently observed in several Pacific Island States, and involves cases of acute coastal erosion whereby the coastal geography of *some* particularly vulnerable features of an island State – in particular coral reefs and atolls – is transformed. This invites exploring not only the physical manifestations of climate change impacts on the land occupied by a State, but also the legal effects of these changes on the maritime areas adjoining such land – while the territorial sea is also an integral part of a State's territory, beyond the outer limit of the territorial sea only some sovereign rights or exclusive jurisdiction of a State apply.

The second phase (developed in section 1.3. below) contemplates the possibility of total loss of territory, as may be *projected* for at least the three Pacific Island States of a maximum altitude barely a few meters above sea-level: Tuvalu, Kiribati and the Marshall Islands. Outside the Pacific Ocean, the fourth most endangered island State in the world is the Maldives in the Indian Ocean. In the south Pacific, Tokelau could also be included in this study, but since it is a New Zealand territory, its continuity as a State is not really at stake. This situation is to be distinguished from that of the Marshall Islands. In Compact of Free Association with the United States, the Marshall Islands' political organisation does not alter its condition as a State. All other Pacific Island States are a mix of different types of features, and generally count with coral atolls, high coral islands, continental islands and volcanic or high volcanic islands. Therefore, while as a whole the region is undoubtedly vulnerable and directly exposed to various manifestations of the adverse impacts of climate change, the concrete effect of such impacts on the topographical features of an island State is subject to important variations. This – for now – hypothetical worst-case scenario raises the question of whether a completely de-territorialised State can or should survive as a sovereign independent entity (whether taking the form of statehood or not), and calls for an exploration of the devices that would become necessary to ensure the continuity of sovereign independence: from physical re-territorialisation strategies, such as cession by or purchase of land of another State, to a legal fiction creating a *sui generis* sovereign entity through recognition of an 'ex situ nationhood'.

1.2 Partial De-Territorialisation: Maritime Spaces in Jeopardy

1.2.1 *Effects of Coastal Geographical Changes on Maritime Rights*

Since the early stages of climate change regime building, small island developing States have been undeniably confronted with the darkest face of the phenomenon. The vulnerability of these actors was yet never disregarded within mainstream international co-operation on this issue.[166] Integrative approaches to evaluating the 'vulnerability profile' of a region or country take into account a range of factors in which the biophysical attributes of the are studied, its socio-economic and even its cultural characteristics are taken into consideration.[167] Despite difficulties in creating clear climate modelling scenarios, it is possible to assert that the spatial dimension of small island developing States is particularly vulnerable to sea-level rise, ocean acidification, and the increase in ocean temperature.[168] The coasts of these States are indeed the 'locus' where de-territorialisation essentially operates.

The progressively increasing coastal erosion and land loss as a consequence of sea-level rise has long been on the international agenda. It dates back to the campaign launched by the Maldives which sought to raise awareness of this issue at the 1987 Commonwealth Ministerial Meeting which, as recalled in paragraph 13 of the UNFCCC Preamble, was followed by the successful introduction of the item into the work of the General Assembly.[169] As already acknowledged in 1997 by the IPCC in its Report entitled 'Regional Impacts of Climate Change: An Assessment of their Vulnerability', sea-level rise is the

166 IPCC, C.B. Field et al., eds., *Managing the Risks of Extreme Events and Disasters to Advanced Climate Change Adaptation* (Cambridge, New York: Cambridge University Press, 2012), 4.

167 This is the division followed by the Climate Vulnerability Monitor in *The State of the Climate Crisis, 2010 Report of the Climate Vulnerability Initiative* (Madrid: DARA, 2012).

168 For a detailed explanation on the effects of ocean acidification and increase of ocean temperature on coral atolls, see Yamamoto and Esteban, *Atoll Island States and International Law*, 35–103.

169 See Commonwealth Heads of Government Meeting, Speech of President Maumoon Abdul Gayoom, Vancouver, British Colombia, 13–17 October 1987. On that occasion, the statement of Abdul Gayoom, former President of The Maldives, led to the creation of the Commonwealth Expert Group on Climate Change and Sea-Level Rise, which conducted a series of case-studies on adverse effects of climate change in several vulnerable countries, such as Bangladesh, Guyana, Tonga, Tuvalu and Kiribati. It was followed two years later by General Assembly Resolution 44/206, 'Possible Adverse Impacts of Sea-Level Rise on Islands and Coastal Areas, Particularly Low-lying Coastal Areas', UN. Doc.: A/44/206, adopted on 22 December 1989, 85th plenary meeting.

most significant climate-related threat for small islands. At the time of publication of the report, the projected estimate of future global sea-level rise was 5 mm/yr (with a range of 2–9 mm/yr); that is, "2 to 4 times higher than what had been experienced globally over the past 100 year".[170] While the report recognised that the level of vulnerability would vary from island to island, it considered that practically all small island States would be adversely affected by sea-level rise. In the south Pacific region, the twelve independent small island States are indeed of different nature and composition and, while most of them are coral islands or atolls, there are also several continental, old volcanic and high volcanic features in the Pacific region.

Climate change impacts are particularly harmful for the coastal geography of features based on coral skeletons. First of all, low-lying coral atolls with an altitude of less than 5 metres are undoubtedly the most directly exposed to sea-level rise – inundations and sea flooding having already been projected in 1997 for the Marshall Islands and Kiribati.[171] Secondly, coral atolls below 5 metres in altitude and raised coral islands with an average altitude of 60 metres, along with the coral reefs surrounding them, are extremely vulnerable to progressive bleaching primarily due to increased ocean temperature and ocean acidification by carbon dioxide.[172] Recent studies have categorised South-East Asia as the region with both the highest level of biodiversity and the greatest degree of threat to reefs (followed by the Caribbean),[173] since in many parts

170 IPCC, Robert T. Watson et al., eds., *Regional Impacts of Climate Change: an Assessment of their Vulnerability* (Cambridge: Cambridge University Press, 1997), chapter 9.

171 Ibid.

172 According to the Global Island Database, islands host 65% of the world's coral reefs and include 14 of the 20 largest coral-reef countries. The livelihoods of islanders are inextricably linked to coral reefs, as they provide a source of food, protection from natural hazards, and a significant attraction for tourists. The increase in the atmospheric carbon concentration and in the sea-surface temperature fosters coral bleaching. A compilation of the global distribution of reefs can be found in UNEP-World Conservation Monitoring Centre (WCMC), *Global Distribution of Coral Reefs – Extracted from Version 7.0 of the Global 1km Raster Dataset Compiled by the UNEP-World Conservation Monitoring Centre* (Cambridge: ReefBase, 2002). Detailed information on the location and degree of bleaching can be found in the database Information on Global Coral Bleaching, ReefBase Project, WorldFish Center. This dataset is built upon an existing bleaching database developed at UNEP-WCMC, and has been maintained and regularly updated by ReefBase since 2002 from a variety of sources. The data indicate the severity of bleaching, categorised as low (1–10% corals bleached), medium (10–30% of corals bleached), or high (>30% of corals bleached).

173 Lauretta Burke et al., *Reefs at Risk in Southeast Asia* (*RRSEA*) (Washington D.C.: World Resources Institute, 2002). Lauretta Burke and Jon Maidens, *Reefs at Risk in the Caribbean* (Washington D.C.: World Resources Institute, 2004).

of the tropics species of corals live near their temperature tolerance limits (25–29°C).[174]

The socio-economic and ecological impacts of the *physical loss of land* due to coastal erosion of these features are undisputable. Yet, when considering the legal consequences of coastal geographical transformation, the focus shifts towards Pacific Island States' *rights over maritime spaces.* In a partial de-territorialisation scenario, an assessment of how the spatial dimension of Pacific Island States is affected by the adverse impacts of climate change therefore implies studying how the transformation of the coastal geography of several coral features may jeopardise these States' present rights over valuable ocean space. The following map shows the current state of Pacific Island States' maritime boundaries.[175]

A Sea-Level Rise and Undelimited Maritime Boundaries

In contrast to the extreme smallness of their land, Pacific Island States soon benefited from the 'expansionist' evolution in the distribution of sovereignty and jurisdiction over maritime spaces that the development of the Law of the Sea initiated after World War II, and which was eventually codified in the 1982 United Nations Convention on the Law of the Sea (UNCLOS).[176] The consolidation, at the Third UN Conference on the Law of the Sea (1973–1982), of the width of the territorial sea up to 12 nautical miles from the baselines and, most importantly, the possibility for coastal States to extend their sovereign rights and exclusive jurisdiction to the resources of the newly created Exclusive Economic Zone (EEZ) and extended continental shelves, compensated the land limitations of Pacific Island States with access to disproportionately vast maritime spaces. All Pacific Island States are Parties to the Convention.[177] Other than signifying an objectively added value to their maritime dimension,

174 IPCC, *Regional Impacts Report*. See also, UNEP–WCMC. *Annual Sea-Surface Temperature 2003–2007 – Derived from the NASA 'OceanColor' Database* (Cambridge: Monitoring Centre, 2008).

175 *Maritime Claims and Boundaries in the South Pacific*, map prepared for the author by Dr. I Made Andi Arsana, Universitas Gadjah Mada Gedung Pusat, Indonesia.

176 United Nations Convention on the Law of the Sea (UNCLOS), adopted at Montego Bay (Jamaica), on 10 December 1982; entered into force on 16 November 1994, 1833 U.N.T.S. 3.

177 In chronological order of ratification (R) or accession (A) to UNCLOS: Fiji (R, 1982); Federated States of Micronesia (A, 1991); Marshall Islands (A, 1991); Cook Islands (R, 1995); Tonga (R, 1995); Samoa (A, 1995); Nauru (R, 1996); Palau (A, 1996); Papua New Guinea (R, 1997); Solomon Islands (R, 1997); Vanuatu (R, 1999); Tuvalu (R, 2002); Kiribati (A, 2003); and Niue (R, 2006). For the status of UNCLOS, see webpage of the UN Secretariat Division for Ocean Affairs and the Law of the Sea, available at: www.un.org/depts/los/ (13 March 2015).

and despite the fact that their capacity to effectively exploit and control these vast spaces is sometimes limited,[178] this maritime extension of certain sovereign rights and jurisdiction provided for a source of income and livelihoods that would prove significantly important for the economic development and continued habitability of these States. Therefore, any adverse effects of climate change on the maritime zones of Pacific Island States should be considered both from the standpoint of their territorial rights – regarding the territorial sea- as well as from the perspective of the economic viability of the State regarding their EEZs and continental shelves.

UNCLOS provides that the normal baseline is the low-water line along the coast,[179] though in the case of very irregular coasts, States can establish straight baselines.[180] Special rules also govern the determination of baselines for reefs[181] as well as for island States eligible to claim archipelagic status, of particular interest and importance for Pacific Island States (see table 1 below).[182] After determining the baseline in accordance with the rule corresponding to the configuration of the coast, coastal States may claim from that line the territorial sea up to 12 nautical miles. They can also exercise control in the contiguous zone (up to 24 nautical miles from the baseline),[183] and have sovereign rights and jurisdiction in the EEZ up to 200 nautical miles. Moreover, the continental shelf in some cases may be extended far beyond 200 nautical miles from the baselines up to 350 nautical miles, since the decisive criteria for the 'outer' continental shelf is the outer edge of the continental margin.[184]

178 Apart from some Pacific Island States which have coastguards, most of them do not have military forces of their own and depend on the services provided primarily by Australia and New Zealand.

179 UNCLOS, Article 5 – on normal baselines.

180 UNCLOS, Article 7 – on straight baselines.

181 UNCLOS, Article 6 – on reefs.

182 UNCLOS, Articles 46–47 – on archipelagic States. The role played by the Group of Archipelagic States in pushing forward the recognition of a special method of drawing archipelagic straight baselines during the Third Conference of UNCLOS, of which Fiji was a member, cannot be disregarded. Likewise, Fiji, Papua New Guinea, Samoa, Tonga, along with Federated States of Micronesia, Palau and Marshall Islands (at the time constituting the Trust Territories of the South Pacific), composed the Oceania Group, which sought to ensure that islands were not precluded from establishing EEZs and continental shelves. See also Myron H. Nordquist, ed., *United Nations Convention on the Law of the Sea 1982: A Commentary* (Dordrech, Boston, Lancaster: Martinus Nijoff, 1985), vol. 1, 77–79.

183 UNCLOS, Article 33 – on the contiguous zone.

184 UNCLOS, Article 76 for the relevant criteria, and especially paras (5) and (6) for the maximum extent of the 'outer' continental shelf.

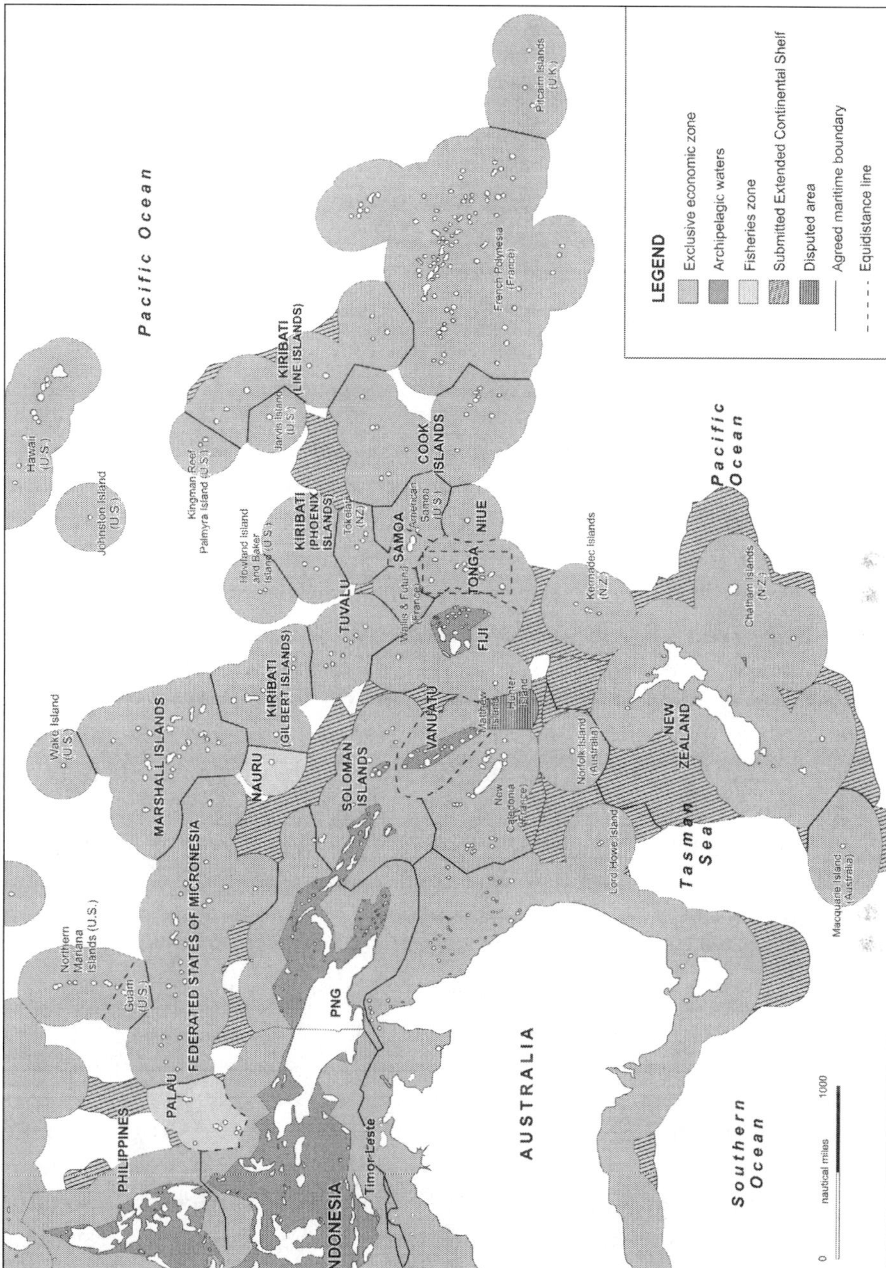

FIGURE 1 *Maritime claims and boundaries in the South Pacific.*

The outer limits of these zones are to be found, first of all, in the unilateral claims of the coastal States, generally embodied in national legislation. Yet, given the extraordinary projection of Pacific Island States' rights oceanwards (partly resulting from the benefits of claiming the archipelagic status), and the correlative reduction of high-seas, the limits of up to 48 EEZs of neighbouring States in this region overlap. Until 2012, most of these boundaries had not been subject to delimitation agreements. The assistance of SOPAC, coupled with a clear governmental will to change this situation, has since yet prompted the conclusion of nine treaties in the region between 2012 and 2014, the details of which are presented in section 3.1.2. below.

Table 1 includes all the political entities composing the Pacific region. This includes twelve independent island States (Federated States of Micronesia, Fiji, Kiribati, Marshall Islands, Nauru, Palau, Papua New Guinea, Samoa, Solomon Islands, Tonga, Tuvalu and Vanuatu) as well as the territories of New Zealand which enjoy a certain degree of administrative autonomy – namely, Cook Islands, Tokelau and Niue. The table shows that nine out of twelve independent Pacific Island States have claimed archipelagic status in their national legislations, though this position has not always been followed by the drawing of archipelagic baselines. Thus, while Fiji, Papua New Guinea, the Solomon Islands and Vanuatu incorporated the coordinates of their archipelagic lines as soon as their maritime-zones legislation was enacted, and Palau included in its Constitution the direction of its archipelagic lines by specifying the names of the features involved, the most underdeveloped States of the region – namely, the Federated States of Micronesia, the Marshall Islands, Kiribati and Tuvalu – until very recently lacked the economic resources necessary to conduct scientific surveys and draw official charts. These are also, precisely, the States directly affected by sea-level rise, for they are mostly (as the Federated States of Micronesia) or entirely (as Tuvalu, Kiribati, Marshall Islands) built on coral islands and atolls.

As the effects of the lack of specification of the position of the archipelagic baselines on the right to claim archipelagic status remain unclear, these three most vulnerable States operated for long in a context of legal uncertainty. Indeed, whilst failing to produce archipelagic baselines and to ensure their due publicity by depositing the list of coordinates and charts with the UN Secretary-General may arguably not affect the existence of the right in itself, the omission is not only in breach of Article 47 (9) of UNCLOS, but can also preclude the State from being able to prove any offence or violation of its sovereign archipelagic waters by foreign vessels. Against this background, the development brought about by the enactment of Tuvalu's 2012 Maritimes Zones Act, accompanied by the Declaration of Archipelagic Baselines in which the coordinates of all archipelagic basepoints and corresponding charts are attached, is to be welcomed. Although Kiribati's 2011 Marine Zones Declaration Act still falls short of drawing its archipelagic baselines, between 2012 and 2014 the country has focused on negotiating maritime delimitation agreements (not yet in force) with its bordering States (see table 2 below). Both developments constitute a clear indication of the current political commitment of low-lying Pacific Island States to put into operation as many legal safeguards as the existing law offers to preserve the extent of their maritime spaces.[185]

The determination of baselines in the region is also problematic because of the presence of important coral reefs in the region. This situation has, for instance, been explicitly covered by maritime-zone legislation of two New Zealand territories: Tokelau's legislation specifies that its baseline is the low-water line along the edge of the reef and the straight line joining the points where the reef is broken, whereas that of Niue provides that British Admiralty Charts should be used to draw the low-water lines and coral-reef lines. All in all, the general rule of Article 5 of UNCLOS, establishing that the normal baseline is the low-water line drawn along the coast, is not often applied by Pacific Island States and territories. Yet, notwithstanding the geographical and socio-economic complexities of drawing baselines in this region, all Pacific Island States have claimed in their national legislation the maximum breadth permitted by customary international law and by treaty law, since the entry into force of UNCLOS, in 1994, of the territorial sea and of the exclusive economic zone. Only six Pacific Island States explicitly claim fishery rights within the 24 nautical miles of the contiguous zone (Kiribati, Marshall Islands, Nauru, Samoa, Tuvalu and Vanuatu), presumably because the rights granted in the contiguous zone were subsumed in the breadth of their respective EEZs.

Besides, it is also worth noting that Pacific Island States have become particularly active with regard to their claims over their respective continental shelves.[186] On 16 April 2009, the Cook Islands inaugurated the series of submissions lodged by Pacific Island States to the Commission on the Limits of

185 On the legal regime of archipelagic States, see Mohamed Munavvar, *Ocean States: Archipelagic Regimes in the Law of the Sea* (Dordrecht: Martinus Nijhoff, 1995); UNDOALOS, *The Law of the Sea: Practice of Archipelagic States* (New York: United Nations, 1992); and Daniel P. O'Connell, 'Mid-Ocean Archipelagos in International Law, *British Yearbook of International Law*, vol. 45 (1971), 1–77.

186 More details on submissions and submissions of preliminary information by Pacific Island States to the CLCS can be found in Clive Schofield, Andi Arsana and Robert van de Poll, 'The Outer Continental Shelf in the Asia-Pacific Region: Progress and Prospects', in Davor Vidas, ed., *Law, Technology and Science for Oceans in Globalisation* (Leiden and Boston: Martinus Nijhoff Publishers, 2010), 539–576, 555–565.

the Continental Shelf (CLCS). In its submission, the Cook Islands argued for an extension of its continental shelf of 400,000 sq. km. – almost reaching the Manihiki Plateau.[187] Four days later, Vanuatu, the Solomon Islands and Fiji lodged a joint submission of preliminary information regarding the North Fiji Basin, stating that it was without prejudice to future maritime delimitation among the three States.[188] It was followed by the partial submission of Fiji regarding the Lau Ridge – which includes areas that had already been part of a submission by New Zealand and received recommendations of the CLCS in 2008.[189] Then, a second individual submission of preliminary information was submitted by Vanuatu regarding the extension to an area still in dispute with France, close to the Leka and Umaenupae islands (Matthew and Hunter islands for France, respectively).[190] On 5 May 2009, the Solomon Islands submitted preliminary information regarding a high-seas enclave surrounded by the EEZs of the Solomon Islands, Papua New Guinea and Australia.[191] Likewise, Papua New Guinea lodged preliminary information on the Mussau Ridge and the Eauripik Rise, which are located between Papua New Guinea and the Federated States of Micronesia.[192] Unsurprisingly, the Federated States of

187 Commission on the Limits of the Continental Shelf (CLCS), 'Outer limits of the continental shelf beyond 200 nautical miles from the baselines: Submissions to the Commission: Submission by the Cook Islands', 16 April 2009, available at: http://www.un.org/depts/los/clcs_new/submissions_files/submission_cok_23_2009.htm (7 October 2014).

188 CLCS, 'Preliminary information submitted by the Republic of Vanuatu, Solomon Islands and the Republic of Fiji Islands on the North Fiji Basin', 19 April 2009, available at: http://www.un.org/depts/los/clcs_new/submissions_files/preliminary/fji_slb_vut_2009_preliminaryinfo.pdf (7 October 2014).

189 CLCS, 'A Partial Submission by the Republic of the Fiji Islands for the Establishment of the Outer Limits of the Continental Shelf of Fiji Pursuant to Article 76, paragraph 8 of the United Nations Convention on the Law of the Sea', 20 April 2009, available at: http://www.un.org/depts/los/clcs_new/submissions_files/fji24_09/fji_2009exsummary.pdf (19 April 2015).

190 CLCS, 'Preliminary information submitted by the Republic of Vanuatu to the Commission on the Limits of the Continental Shelf', 10 August 2009, available at: http://www.un.org/depts/los/clcs_new/submissions_files/preliminary/vut_2009_preliminaryinfo.pdf (7 October 2014).

191 Solomon Islands Mission to the United Nations, 'Note 13/09', 4 May 2009, available at: http://www.un.org/depts/los/clcs_new/submissions_files/preliminary/slb_preliminaryinfo.pdf (19 April 2015).

192 CLCS, 'Preliminary information Indicative of the Outer Limits of the Continental Shelf Beyond 200 Nautical Miles for the Mussau Ridge and Eauripik Rise Areas submitted by Papua New Guinea', available at: http://www.un.org/depts/los/clcs_new/submissions_files/preliminary/fsm_preliminaryinfo.pdf (19 April 2015).

Micronesia also made a submission of preliminary information concerning the same areas.[193] In addition to these three individual submissions of preliminary information, the three States – Federated States of Micronesia, Papua New Guinea and the Solomon Islands – made on the same day a joint submission for extended continental shelf related to the Ontong Java Plateau and covering approximately 600,000 sq. km.[194] Likewise, Tonga also made a submission related to an extended continental shelf in the eastern part of the Kermadec Ridge.[195] Finally, the year 2009 closed with the submission of Palau for three different areas of outer continental shelf, the largest area being located to the north and covering around 243,000 sq. km. It appears that the delimitation of a maritime boundary with Japan regarding the Southern-Kyusyu-Palau Ridge will be necessary.[196]

Moreover, in spite of the geophysical composition of Kiribati and Tuvalu, exclusively composed of coral atolls – which tend to result in narrow continental margins – recent discoveries of manganese beds in their respective EEZs have prompted these countries to pay attention to the breadth of their continental shelves. Thus, in December 2012 – following the enactment of the 2012 Maritime Zones Act and the Declaration on Archipelagic Baselines – Tuvalu filed a submission with Tokelau (New Zealand) and Wallis and Futuna (France) requesting a joint area of additional sea-bed territory of 18,000 sq. km.[197]

---

193 CLCS, 'Preliminary information Indicative of the Outer Limits of the Continental Shelf Beyond 200 Nautical Miles for the Eauripik and Mussau Ridge Areas, submitted by the Federated States of Micronesia', available at: http://www.un.org/depts/los/clcs_new/commission_preliminary.htm (15 September 2015). See also Schofield, Arsana and van der Poll, 565.

194 CLCS, 'Joint Submission by Federated States of Micronesia, Papua New Guinea and the Solomon Islands', 5 May 2009, available at: http://www.un.org/depts/los/clcs_new/submissions_files/submission_fmpgsb_32_2009.htm (7 October 2014).

195 CLCS, 'A Partial Submission of Data and Information of the Outer Limits of the Continental Shelf of the Kingdom of Tonga Pursuant to Part VI and Annex II to the United Nations Convention on the Law of the Sea', May 2009. See also Schofield, Arsana and van del Poll, 561.

196 CLCS, 'Republic of Palau, Submission on the Limits of the Continental Shelf Pursuant to Article 76 of the United Nations Convention on the Law of the Sea', available at: http://www.un.org/depts/los/clcs_new/submissions_files/submission_plw_41_2009.htm (9 September 2015).

197 CLCS, 'Joint Submission by Tuvalu, France and Tokelau (New Zealand)', 18 December 2012, available at: http://www.un.org/depts/los/clcs_new/submissions_files/submission_tft_62_2012.htm (7 October 2014).

TABLE 1 *Unilateral claims of maritime zones and continental shelf by Pacific Island States and territories*

Name of State	National Legislation	Territorial Sea	Contiguous Zone	Exclusive Economic Zone	Continental Shelf	Archipelagic Status
Cook Islands	Territorial Sea and Exclusive Economic Zone Act nº 16/1977	12M	Not specified	200M	200M Regulated by Continental Shelf Act 1964	No
Federated States of Micronesia	Territory, Economic Zones and Port Entry, FSM Consolidated Legislation Title 18/Chap. 1	12M	Not specified	200M	Not specified	Yes Claimed but not drawn Article 1 Federal Constitution refers to the 'Districts of the Micronesian Archipelago'
Fiji	Marine Spaces Act [CAP 158A], Act nº 18/1977, Amended by nº 15/1978, and Legal Notice nº 117 of 1981	12M	Not specified	200M	200M Regulated in Continental Shelf Act [CAP 149], nº 9/1970	Yes Claimed and drawn in schedule (Coordinates) 2 Archipelagos: Fiji and Rotuma
Kiribati	Marine Zones (Declaration) Act 2011	12M	24M	200M	200M	Yes Claimed but not drawn

Marshall Islands	Marine Zones (Declaration) Act 1984 [33MIRC Chap. 1]	12M	24M	200M	Not specified	Yes Claimed but not drawn
Nauru	Sea Boundaries Act 1997	12M	24M	200M	200M	No
Niue (New Zealand)	Territorial Sea and Exclusive Economic Zone Act nº 220/1996 (as amended by the Maritime Zones Act 2013)	12M	Not specified	200M	Not specified	No Charts for low-water lines and coral reefs from British Admiralty Chart
Tokelau (New Zealand)	Territorial Sea and Exclusive Economic Zone Act 1977	12M	Not specified	200M	200M	No Baseline is low-water line along the edge of the reef and straight line joining the points where the reef is broken
Palau	Constitution of the Republic of Palau 1979, Article 1, as amended on 15 July 2005	12M	Not specified	200M	Over 200M 2005 Constitutional Amendment 2007 and ICCS Claim of Extended CS	Yes Claimed and drawn in 2005 Amendment (Name of Features)
Papua New Guinea	National Seas Act 1977 Offshore Seas Proclamation 1978 (declaration of baselines location of archipelagic lines)	12M	Not specified	200M	200M Regulated in Continental Shelf (Living Resources) Act, Chap 210 of Consolidated Legislation; and by the Oil and Gas Act 1998	Yes Claimed and drawn in Act (Coordinates) 3 Archipelagos: Principal; Tauu; and Nukumanu

TABLE 1 *Unilateral claims of maritime zones and continental shelf by Pacific Island States and territories* (cont.)

Name of State	National Legislation	Territorial Sea	Contiguous Zone	Exclusive Economic Zone	Continental Shelf	Archipelagic Status
Samoa	Maritime Zones Act nº 18/1999; Amendment Act nº 13/2004	12M	24M	200M	Not specified	No
Solomon Islands	Delimitation of Marine Waters Act, Chap 95 of Consolidated Legislation; Declaration of Archipelagic Lines 1979	12M	Not specified	200M	200M Regulated in the Continental Shelf Act, Chap 94 of the Consolidated Legislation	Yes Claimed and drawn in Declaration (Coordinates) 4 Archipelagos: Rennell, Bellona and Indispensable Reef Atoll; Ontong Java Group; Santa Cruz Islands; and Duff Islands
Tonga	Territorial Sea and Exclusive Economic Zone Act nº 30 1978	12M	Not specified	200M	200M Mentioned in Amended 1979 Schedule	No

Tuvalu	Tuvalu Maritime Zones Act 2012 Declaration of Archipelagic Baselines 2012	12M	24M	200M	200M (from archipelagic baselines)	Yes Claimed and drawn in Schedule (Coordinates and Chart) 1 Archipelago
Vanuatu	Maritime Zones Act (nº 6/2010)	12M	24M	200M	200M or outer edge of continental margin	Yes Claimed and drawn in Schedule (Coordinates) 1 Archipelago

SOURCE: DATA FROM THE PACIFIC ISLANDS STATES LEGAL INSTITUTE.

In the same month – and soon after having concluded maritime delimitation agreements with six of its neighbouring States – Kiribati mirrored Tuvalu's action and filed a submission for an area of extended continental shelf in the Line Islands of 100,000 sq. km.[198] These submissions have been followed by the latest second submissions of both the Federated States of Micronesia and Tonga. Whilst, in 2013, the Federated States of Micronesia requested an extension of 144,000 sq. km. on the Eauripik Rise;[199] in April 2014, Tonga's second submission concerned the western part of the Lau-Corville Ridge.[200]

On the basis of the available information it is possible to list the thirty-two maritime boundaries overlapping in the Pacific region and which are not the object of a delimitation agreement *yet in force*.[201]

From the twelve independent Pacific Island States, and due to the immense area of their maritime zones, Kiribati leads the list as the State with the highest number of neighbouring States with overlapping outer EEZ boundaries. Its maritime spaces extend over almost 3.5 million sq. km. and it is the only State in the world which is part of the four hemispheres at the same time. The ten borders of Kiribati not subject to an agreement *in force* are (North-West to South-West): Kiribati/Marshall Islands; Kiribati/United Kingdom (Baker and Howland Islands); Kiribati/United States (Jarvis Island); Kiribati/United States (Palmyra Atoll and Kingman Reef); Kiribati/France (French Polynesia); Kiribati/New Zealand (Tokelau); Kiribati/Cook Islands; Kiribati/New Zealand (Niue); Kiribati/Tuvalu; and Kiribati/Nauru. Yet it is important to note that up to nine new maritime delimitation agreements have been signed between 2012 and 2014 in the region, of which seven correspond to Kiribati's maritime boundaries with its neighbouring States (see table 3 below). Their entry into

198 CLCS, 'Submission by Kiribati, 24 December 2012', available at: http://www.un.org/depts/los/clcs_new/submissions_files/submission_kir_64_2012.htm (7 October 2014).

199 CLSC, 'Submission by the Federated States of Micronesia', 30 August 2013, available at: http://www.un.org/depts/los/clcs_new/submissions_files/submission_fsm_67_2013.htm (7 October 2014).

200 CLCS, 'Submission by Tonga, 25 April 2014', available at: http://www.un.org/depts/los/clcs_new/submissions_files/submission_ton_73_2014.htm (7 October 2014).

201 Victor Prescott and Clive Schofield, *The Maritime Boundaries of the World* (Leiden and Boston: Martinus Nijhoff Publishers, 2nd ed., 2005), chapter 17, 397–418. This work serves as a complement to the present analysis, particularly for the identification of the names of features involved in the tracing of the equidistant or median lines of the undelimited maritime boundaries of the region. For an analysis on how the law of the sea operates in the region, beyond issues of maritime delimitation, see James Crawford and Donald R. Rothwell, eds., *The Law of the Sea in the Asian Pacific Region: Developments and Prospects*, Publications on Ocean Development, vol. nº 21 (Dordrecht, Boston and London: Martinus Nijhoff Publishers, 1995).

force in the near future will certainly prompt substantial changes in the legal landscape regarding the maritime spaces of the region.

Kiribati's case is followed by that of the Federated States of Micronesia (FSM), another Pacific Island State with vast maritime areas, whose overlapping EEZ boundaries are: FSM/Marshall Islands; FSM/United States (Guam); and FSM/Papua New Guinea. Fiji and the Marshall Islands also count with four overlapping and undelimited EEZ boundaries. On the one hand, there are Fiji/Tonga; Fiji/France (New Caledonia); Fiji/Tuvalu; and Fiji/France (Wallis and Futuna); on the other hand, there are the Marshall Islands/Federated States of Micronesia; Marshall Islands/Kiribati; Marshall Islands/Nauru; and Marshall Islands/United States (Wake Island). Likewise, the Solomon Islands have four EEZ boundaries: Solomon Islands/Vanuatu; Solomon Islands/France (New Caledonia); Solomon Islands/Papua New Guinea; and Solomon Islands/Australia. These are followed by Tuvalu, which has three EEZ boundaries: Tuvalu/Kiribati; Tuvalu/Fiji; and Tuvalu/France (Wallis and Futuna). Similarly, the three EEZ boundaries generated by The Cook Islands are: Cook Islands/Kiribati; Cook Islands/New Zealand (Niue); and Cook Islands/New Zealand (Tokelau). Vanuatu's EEZ boundaries are Vanuatu/France (French Caledonia); and Vanuatu/Solomon Islands. Tonga's two boundaries are: Tonga/Fiji and Tonga/American Samoa. Samoa has two boundaries: Samoa/American Samoa; and Samoa/New Zealand (Tokelau); Nauru also has two boundaries: Nauru/Marshall Islands and Nauru/Kiribati. Finally, the richest State of the region, Papua New Guinea, has only two undelimited boundaries: Papua New Guinea/Federated States of Micronesia and Papua New Guinea/Solomon Islands, (the rest having been made subject to treaty delimitation currently in force). However, the regional powers also have some undelimited maritime boundaries: Australia/New Zealand; New Zealand (Niue)/United States (American Samoa).

Articles 74(1) and 83(1) of UNCLOS establish the same rule for the delimitation of the EEZ and the continental shelf, respectively, between States with opposite and adjacent coasts. They provide that the delimitation in these cases shall be effected by agreement 'in order to achieve an equitable solution'. This rule was born out of the need to reach a compromise, during the Third UN Conference on the Law of the Sea, between supporters of 'equidistance' – who were in favour of establishing a compulsory, third-party system of settlement of international disputes – and supporters of the 'equitable principles' generally against compulsory judicial procedures. It has the disadvantage of not formulating any method of delimitation.[202] Besides, in cases where no

202 See, for instance, Joshifumi Tanaka, *The International Law of the Sea* (Cambridge: Cambridge University Press, 2012), 191; and Robin R. Churchill and Alan V. Lowe, *The Law of the Sea* (Manchester: Manchester University Press, 3rd ed. 1999), 181–198.

agreement has been reached on the delimitation of an EEZ or continental shelf between neighbouring States –and in contrast to the delimitation rule for the territorial sea – Articles 74(1) and 83(1) of UNCLOS do not provide a subsidiary rule.[203] Thus, in the absence of agreement, maritime delimitation disputes over the EEZ or the continental shelf shall be settled by international courts and tribunals (resort to the procedures of Part XV of UNCLOS on the settlement of disputes). Hence, given the absence of a delimitation method explicitly established in UNCLOS, the interpretation of Articles 74(1) and 83(1) shall follow the evolutions of the law on maritime delimitation effected by the international judiciary.[204] In the progressive evolution towards greater predictability, the International Court of Justice has through a series of judgments consolidated the use of neutral, objective and legal criteria, as opposed to subjective views. The Court's case law has gone through two main distinct phases.[205] Under the initial 'result-oriented equity approach' of the earlier jurisprudence of the ICJ – applied, for instance, in the 1969 *North Sea Continental Shelf*, the 1982 *Tunisia v. Libya* or the 1984 *Gulf of Maine* judgments,[206] the Court gave preference to having full flexibility to choose the method of delimitation it considered most suitable to achieve, in the concrete case, an equitable solution. The shift by the ICJ to a 'corrective equity approach' since the 1993 *Greenland and Jan Mayen* judgment introduced more predictability into the field of maritime delimitation.[207] Under this

203 Mirroring Article 12(1) of the 1958 Convention on the Territorial Sea and the Contiguous Zone, Article 15 UNCLOS provides that the delimitation of the territorial sea shall be effected by agreement or, by default, by the median line (equidistance), unless special circumstances make it necessary to apply a variant of such rule.

204 An illustration of the analysis made on this point can be found in Tullio Scovazzi, 'Maritime Delimitation Cases before International Courts and Tribunals', in Rudiger Wolfrum, ed., *Max Planck Encyclopedia of Public International Law* (Oxford: Oxford University Press, 2006); and Vladimir-Duro Degan, 'Consolidation of Legal Principles on Maritime Delimitation: Implications for the Dispute between Slovenia and Croatia in the North Adriatic', *Chinese Journal of International Law*, vol. 6 (2007), 601–634.

205 Tanaka, 192–197.

206 *North Sea Continental Shelf* (Federal Republic of Germany v. Denmark; Federal Republic of Germany v. Netherlands), Judgment, 1969 ICJ Reports 3, para. 55–56; *Continental Shelf* (Tunisia v. Libyan Arab Jamahiriya), Judgment, 1982 ICJ Reports 18, para. 76; *Delimitation of the Maritime Boundary in the Gulf of Maine Area* (Canada/USA), Judgment, 1984 ICJ Reports 246, para. 122.

207 *Maritime Delimitation in the Area between Greenland and Jan Mayen* (Denmark v. Norway), Judgment, 1993 ICJ Reports 38, para. 46. This initial effort to introduce more objectivity into judicial decisions on maritime delimitation was then pursued by the Court in *Maritime Delimitation and Territorial Questions between Qatar and Bahrain* (Qatar v. Bahrain), Merits, Judgment, 2001 ICJ Reports 40, para. 176–177; and *Land and Maritime*

approach, the Court first establishes a provisional equidistance line, and then examines whether the circumstances of the case call for an adjustment of such line with a view to achieving an equitable result. The 2006 Award in the Barbados/Trinidad and Tobago arbitration contributed to this trend, as the Arbitral Tribunal stated that "the quest for neutral criteria of geographical character prevailed in the end over area-specific criteria such as geomorphological aspects of resource-specific criteria".[208] Although in the 2007 *Nicaragua v. Honduras* case, the ICJ established a maritime boundary by applying the bisector method and therefore shed doubts as to whether the Court was – at least partly- departing from the general rule of equidistance,[209] the 2009 *Romania v. Ukraine* judgment more recently confirmed that the corrective-equity approach is still en vogue.[210]

In order to evaluate how the adverse effects of climate change (particularly sea-level rise and coastal erosion) are likely to affect these thirty-two maritime boundaries in the Pacific region not subject to a delimitation agreement yet in force, and considering the current status of the equidistance method in international courts and tribunals, it is essential to identify the features involved in the tracing of the equidistant or median delimitation lines.[211] These correspond to the basepoints used for drawing the relevant baselines of each State. While acknowledging the difficulty of the task (due primarily to the lack of official charts indicating the direction of the archipelagic lines and the presence of coral reefs and indented coasts), table 2 seeks to provide an *illustration* of the effects of climate change on the maritime spatial dimension and extent of Pacific Island States, before embarking on a legal analysis of the data. This table has been assembled on the basis of Prescott and Schofield's indications, who enumerated the features that would likely be involved in the delimitation

Boundary between Cameroon and Nigeria (Cameroon v. Nigeria: Equatorial Guinea intervening), Judgment, 2002 ICJ Reports 303, para. 288–290.

208 *Award of the Arbitral Tribunal in the Matter of an Arbitration between Barbados and the Republic of the Trinidad and Tobago*, The Hague, 11 April 2006, para. 228.

209 *Territorial and Maritime Dispute between Nicaragua and Honduras in the Caribbean Sea* (Nicaragua v. Honduras), Judgment, 2007 ICJ Reports 659, para. 281; despite the controversial application of the bisector method, the Court nonetheless indicated that "remains the general rule". Rejecting all the 'compelling reasons' upheld in this case by the Court to apply the bisector method instead of a provisional equidistance line adjusted with straight lines when required, see *Dissenting Opinion of Judge ad-hoc Torres Bernárdez*, para. 119–132.

210 *Delimitation in the Black Sea* (Romania v. Ukraine), 2009 ICJ Reports 61, para. 115–122.

211 Section 1.2.1.B. below dwells into how regional State practice often has recourse to the corrective-equity approach (equidistance/adjustment) in maritime delimitation agreements, as currently followed in international jurisprudence.

of these boundaries.[212] As developed in section 1.2.B below, in the past few years a historical move towards the conclusion of maritime delimitation agreements has radically changed the regional landscape regarding maritime boundaries of Pacific Island States. As the majority of these agreements are not yet in force, the information provided by Prescott and Schofield will still be used, notwithstanding that in the agreements recently concluded but not yet in force, some of the features referred to by these authors may have not in practice been used by the relevant Parties. A search on the geophysical characteristics of these features (including their maximum altitude) has then been incorporated by using the Global Island Database of the United Nations Environment Programme, so as to distinguish five different categories of features. This geophysical information serves two purposes. On the one hand, each category has been associated with a vulnerability factor marked from 1 to 5 (1 being the level of lowest vulnerability, and 5, the highest). On the other hand, the nature and geological composition of these features will be important in the legal analysis developed below in order to determine their corresponding legal regimes. Finally, table 2 indicates the maritime boundary that would likely be affected by sea-level rise and coastal erosion.

TABLE 2 *Sea-level rise and key geographical features in maritime boundaries of Pacific Island States not subject to a delimitation agreement yet in force*

Name of Feature	Country	Geomorphology (max. altitude)	Degree of vulnerability (1–5)	Potentially Affected Boundary
Abaiang	Kiribati	Coral Atoll (<5m)	5	Kiribati/Marshall Is.
Ahe	French Polynesia	Coral Atoll (<5m)	5	French Polynesia/Kiribati
Aneityum	New Caledonia (France)	Volcanic Island (852m)	1	New Caledonia/Vanuatu
Arorae	Kiribati	Coral Atoll (<5m)	5	Kiribati/Tuvalu

212 Prescott and Schofield, *Maritime Political Boundaries of the World*, chapter 17, 397–418.

Name of Feature	Country	Geomorphology (max. altitude)	Degree of vulnerability (1–5)	Potentially Affected Boundary
Astrolabe	New Caledonia (France)/ Vanuatu	Reef	5	New Cal/Vanuatu
Ata	Tonga	Volcanic Island	1	Tonga/Fiji
Atafu	Tokelau (NZ)	Coral Atoll (4m)	5	Tokelau (NZ)/Kiribati; Tokelau (NZ)/Samoa; Tokelau (NZ)/(France)
Baker	USA territory	Coral Atoll, Guano (8m)	5	Baker and Howland Islands (USA)/Kiribati
Banaba	Kiribati	Raised Coral (Phosphate) Island (81m)	3	Kiribati/Marshall Islands.; Kiribati/Nauru
Beautemps-Beauprès	Vanuatu/ New Caledonia (France)	Oceanic Bank/ Atoll(1–3m)	5	New Caledonia/Vanuatu
Bellinghausen	French Polynesia (France)	Coral Atoll (5m)	5	French Polynesia/Kiribati
Bougainville	Papua New Guinea	High Volcanic Island (2792m)	1	Papua New Guinea/ Solomon Islands
Bukatatanoa	Fiji	Barrier Reef	5	Fiji/Tonga
Butaritari	Kiribati	Coral Atoll (1m)	5	Kiribati/Marshall Islands
Cakau-Levu	Fiji	Coral Reef	5	Fiji/Tonga
Canton	USA territory/ Kiribati	Coral Atoll (5m)	5	Kiribati/Baker and Howland Islands (USA)
Caroline	Kiribati	Coral Atoll (5m)	5	Kiribati/French Polynesia
Ceva-i-ra	Fiji	Coral Atoll (<5m)	5	Fiji/New Caledonia
Ebon	Marshall Islands	Coral Atoll (<5m)	5	Marshall Is./Kiribati; Marshall Is./Nauru; Marshall Is./FSM
Efate (Vate)	New Caledonia (France)/ Vanuatu	Volcanic/Raised Coral (647m)	2	New Caledonia/Vanuatu
Eromanga (Erromango)	New Caledonia (France)/ Vanuatu	Volcanic/Raised Coral Island (886m)	2	New Caledonia/Vanuatu

TABLE 2 *Sea-level rise and key geographical features in maritime boundaries of Pacific Island States not subject to a delimitation agreement yet in force* (cont.)

Name of Feature	Country	Geomorphology (max. altitude)	Degree of vulnerability (1–5)	Potentially Affected Boundary
Espiritu Santo	New Caledonia (France)/ Vanuatu	Volcanic/Raised Coral Island (1879m)	2	New Caledonia/Vanuatu; Vanuatu/Solomon Islands
Fakaofo	Tokelau (New Zealand)	Coral Atoll (4m)	5	Tokelau (NZ)/Cook Islands; Tokelau (NZ)/ Kiribati
Fanning (Tabuaeran)	Kiribati	Coral Atoll (4m)	5	Kiribati/Jarvis Island (USA)
Fataka (Mitre)	Solomon Islands	Volcanic Island (122m)	2	Solomon Islands/Vanuatu
Flint	Kiribati	Coral Island (8m)	4	Kiribati/French Polynesia
Fonulalei	Tonga	Volcanic Island	1	Tonga/Fiji
Futuna	French territory	High Volcanic Island (765m)	1	Wallis and Futuna (France)/Tuvalu
Gardner (Nikumaroro)	Kiribati	Coral Atoll (<5m)	5	Kiribati/Baker and Howland Islands; Kiribati/ Tokelau (NZ)
Guam	USA territory	Coral Atoll (<5m)	5	Guam (USA)/FSM
Hiu	Vanuatu	Raised Coral Island (366m)	3	Vanuatu/Solomon Islands
Hull (Orona)	Kiribati	Coral Atoll (9m)	5	Kiribati/Tokelau (NZ)
Hunga Ha'apai	Tonga	Volcanic Island	1	Tonga/Fiji
Hunter (Fearn; Umaenaeg)	French territory/ Vanuatu	Volcanic Island (260m)	1	New Caledonia/Fiji
Huon	Vanuatu	Continental Island (332m)	1	New Caledonia/Vanuatu
Indispensable	Solomon Islands	Reef – Low-tide elevation	5	Solomon Islands/Australia
Jaluit	Marshall Islands	Coral Atoll (<5m)	5	Marshall Islands/Kiribati
Jarvis	USA territory	Coral Island (8m)	4	Jarvis Island (USA)/ Kiribati
Kapingamarangi	Federated States of Micronesia (FSM)	Coral Atoll (<5m)	5	FSM/Papua New Guinea

Name of Feature	Country	Geomorphology (max. altitude)	Degree of vulnerability (1–5)	Potentially Affected Boundary
Kings (Three Kings)	New Zealand territory	(280m)	2	New Zealand/Australia
Kingman	USA territory	Reef	5	Palmyra Atoll and Kingman Reef (USA)/ Kiribati
Kiritimati-E	Kiribati	Coral Atoll (5m)	5	Kiribati/Jarvis Island (USA)
Knox	Marshall Islands	Coral Atoll (<5m)	5	Marshall Islands/Kiribati
Kosrae	FSM	High Volcanic Island (628m)	1	FSM/Marshall Islands
Late	Tonga	High Volcanic Island (519m)	1	Tonga/Fiji
Lifou	Vanuatu/New Caledonia (France)	Raised Coral Island (85m)	4	New Caledonia/Vanuatu
Maewo	Vanuatu	Volcanic and Raised Coral Island (811m)	2	Vanuatu/Solomon Islands
Mahur	Papua New Guinea (PNG)	Continental and Raised Coral Island (67m)	2	Papua New Guinea/FSM
Makin	Kiribati	Coral Atoll (<5m)	5	Kiribati/Marshall Islands
Malden	Kiribati	Coral Island (8m)	4	Kiribati/Jarvis Island (US)
Malekula	New Caledonia (France)/ Vanuatu	Volcanic and Raised Coral Island (879m)	1	New Caledonia/Vanuatu
Mare	Vanuatu/ New Caledonia (France)	Raised Coral and Volcanic Island (129m)	2	New Caledonia/Vanuatu
Matahiva	French Polynesia (France)	Coral Atoll (<5m)	5	French Polynesia/Kiribati
McKean	USA territory/ Kiribati	Coral Island (5m)	4	Baker and Howland Islands/Kiribati
Mellish	Australia	Reef	5	Australia/Solomon Islands

TABLE 2 *Sea-level rise and key geographical features in maritime boundaries of Pacific Island States not subject to a delimitation agreement yet in force* (cont.)

Name of Feature	Country	Geomorphology (max. altitude)	Degree of vulnerability (1–5)	Potentially Affected Boundary
Mera Lava	Vanuatu	Volcanic Cone (1017m)	1	Vanuatu/Solomon Islands
Mili	Marshall Islands	Coral Atoll (<5m)	5	Marshall Islands/ Kiribati
Mokil	FSM	Coral Atoll (<5m)	5	FSM/Marshall Islands
Mota Lava	Vanuatu	Volcanic Island/ Steep Cliffs (411m)	1	Vanuatu/Solomon Islands
Namorik	Marshall Islands	Coral Atoll (<5m)	5	Marshall Islands/FSM
Namumea	Tuvalu	Coral Atoll (<5m)	5	Tuvalu/Kiribati
Nauru	Nauru	Raised Coral Island (70m)	3	Nauru/Marshall Islands; Nauru/Kiribati
Nendo	Solomon Islands	Volcanic Island (549m)	1	Solomon Islands/Vanuatu
Niguria	Papua New Guinea	Coral Atoll (<5m)	5	Papua New Guinea/FSM
Niutao	Tuvalu	Coral Island (<5m)	4	Tuvalu/Kiribati
Niuafo'ou	Tonga	Volcanic Crater (205m)	1	Tonga/Fiji
Niuatoputapu	Tonga	Volcanic Island (146m)	1	Tonga/American Samoa
Niue	New Zealand territory	Raised Coral Island (73m)	3	Niue/Cook Islands
Niulakita (Nurakita)	Tuvalu	Coral Island (<5m)	5	Tuvalu/Wallis and Futuna (France); Tuvalu/Fiji
Norfolk	Australia	Old Volcanic and Limestone (318m)	1	Australia/New Zealand
North Minerva	Tonga/Fiji	Reef	5	Tonga/Fiji
Nugarba	Papua New Guinea	Rock	2	Papua New Guinea/FSM
Nui	Tuvalu	Coral Atoll (<5m)	5	Tuvalu/Fiji
Nuku Cikobia	Fiji			Fiji/Tonga

Name of Feature	Country	Geomorphology (max. altitude)	Degree of vulnerability (1–5)	Potentially Affected Boundary
Nukufetau	Tuvalu	Coral Atoll (<5m)	5	Tuvalu/Fiji
Nukulaelae	Tuvalu	Coral Atoll (<5m)	5	Tuvalu/Wallis and Futuna (France)
Nukunono	Tokelau (New Zealand)	Coral Atoll (4m)	5	Tokelau (NZ)/Samoa
Ofu	American Samoa (USA)	Volcanic Island (494m)	1	American Samoa/Samoa
Ono-i-lau	Fiji	Volcanic/Low Coral Island (113m)	2	Fiji/Tonga
Ouvea	Vanuatu	Raised Coral Island (39m)	3	New Caledonia/Vanuatu
Pakin	FSM	Coral Atoll (<5m)	5	FSM/Marshall Islands
Palmerston	Cook Islands	Coral Atoll (<5m)	5	Cook Islands/Niue
Palmyra	USA territory	Coral Atoll (6m)	5	Palmyra Atoll and Kingman Reef (USA)/ Kiribati
Penrhyn	Cook Islands	Coral Atoll (<5m)	5	Cook Islands/Kiribati
Petrie	New Caledonia (France)	Reef	5	New Caledonia/Vanuatu
Phillip	Australia	Volcanic Island (280m)	1	Australia/New Zealand
Pingelap	FSM	Coral Atoll (<5m)	5	FSM/Marshall Islands
Pocklington	Papua New Guinea	Reef	5	Papua New Guinea/ Solomon Islands
Pohnpei	FSM	Volcanic Island (791m)	1	FSM/Marshall Islands
Rangiroa	French Polynesia	Coral Atoll (<5m)	5	French Polynesia/Kiribati
Reid	Fiji	Reef	1	Fiji/Tonga
Rennell (Mu Nggava)	Solomon Islands	Raised Coral Island (154m)	3	Solomon Islands/Vanuatu
Rotuma	Fiji	Volcanic Island (256m)	1	Fiji/Tuvalu
Santa Catalina (Owakiri)	Solomon Islands	Coral Atoll (<5m)	5	Solomon Islands/Vanuatu

TABLE 2 *Sea-level rise and key geographical features in maritime boundaries of Pacific Island States not subject to a delimitation agreement yet in force* (cont.)

Name of Feature	Country	Geomorphology (max. altitude)	Degree of vulnerability (1–5)	Potentially Affected Boundary
Savaii	Samoa	High Volcanic Island (1858m)	1	Samoa/American Samoa; Samoa/Tokelau (NZ)
Shortland	Solomon Islands	Volcanic and Raised Coral Island	5	Solomon Islands/Papua New Guinea
Simberi	Papua New Guinea	Volcanic Island (…)	1	Papua New Guinea/FSM
South Indispensable	Solomon Islands	Reef	5	Solomon Islands/Australia; Solomon Islands/Vanuatu
South Minerva	Tonga/Fiji	Reef	5	Tonga/Fiji
Starbuck	Kiribati	Coral Island (5m)	5	Kiribati/Cook Islands
Swains	American Samoa (USA)	Coral Atoll (6m)	5	American Samoa/Samoa
Sydney	Kiribati	Coral Island (6m)	5	Kiribati/Tokelau (NZ)
Tabiteuea	Kiribati	Coral Islet (<5m)	5	Kiribati/Tuvalu
Tafahi	Tonga	Volcanic Island (519m)	1	Tonga/American Samoa
Tamana	Kiribati	Coral Island (<5m)	5	Kiribati/Tuvalu
Tanna	New Caledonia (France)/ Vanuatu	Volcanic/Raised Coral (1084m)	1	New Caledonia/Vanuatu
Taongi	Marshall Islands	Coral Atoll (<5m)	5	Marshall Islands/USA territory
Tench	Papua New Guinea	Coral Atoll (<5m)	5	Papua New Guinea/FSM
Tiga (Toga corr)	Vanuatu	Raised Coral (240m)	3	Vanuatu/New Caledonia
Tikehau	French Polynesia	Coral Atoll (<5m)	5	French Polynesia/Kiribati
Tikopia	Solomon Islands	Volcanic Islet (397m)	1	Solomon Islands/Vanuatu
Tofua	Tonga	High Volcanic Island (505m)	1	Tonga/Fiji

Name of Feature	Country	Geomorphology (max. altitude)	Degree of vulnerability (1–5)	Potentially Affected Boundary
Tongatapu	Tonga	Raised Coral Island (82m)	3	Tonga/Fiji
Tupai	French Polynesia	Coral Atoll (<5m)	5	French Polynesia/Kiribati
Tutuila	American Samoa (USA)	High Volcanic Island (653m)	1	American Samoa/Tonga; American Samoa/Samoa
Tuvana-i-colo	Fiji	Coral Atoll (<5m)	5	Fiji/Tonga
Tuvana-i-ra	Fiji	Coral Atoll (<5m)	5	Fiji/Tonga
Ujae	Marshall Islands	Coral Atoll (<5m)	5	Marshall Islands/FSM
Ujelang	Marshall Islands	Coral Atoll (<5m)	5	Marshall Islands/FSM
Upolu	Samoa	High Volcanic Island (1143m)	1	Samoa/American Samoa
Utupua	Solomon Islands	Volcanic/Raised Coral Island (380m)	2	Solomon Islands/Vanuatu
Vanikoro (Vanikolo)	Solomon Islands	Recent Volcano (924m)	1	Solomon Islands/Vanuatu
Vanuabalavu	Fiji	Raised Coral/ Volcanic Island (283m)	2	Fiji/Tonga
Vat Ganai	Vanuatu	Rock (64m)	1	Vanuatu/Solomon Islands
Vostok	Kiribati	Coral Island (5m)	5	Kiribati/Cook Islands
Vuata Vatoa	Fiji	Reef	5	Fiji/Tonga
Wake	USA territory	Coral Atoll (6m)	5	USA territory/Marshall Islands
Wallis	French territory	High Volcanic Island (765m)	1	Wallis (France)/Tokelau (NZ); Tuvalu/Wallis and Futuna (France)
Walpole	New Caledonia (France)	Raised Coral Island (100m)	3	New Caledonia/Vanuatu
Washington (Teraina)	Kiribati	Coral Island (5m)	5	Kiribati/Jarvis Island (US)

SOURCES: DATA FROM THE UNITED NATIONS ENVIRONMENT PROGRAMME GLOBAL ISLAND DATABASE; PRESCOTT AND SCHOFIELD, 2005.

As one of the pioneer scholars pointing out to the effects of sea-level rise on maritime boundaries, David Caron noted:

> [T]he crucial point to recognise is that [t]he law of baselines authorises, quite intentionally, the most insubstantial, sometimes ephemeral and transient, geographic features to serve as anchors for baselines, thus maximising for each coastal state the reach of their oceanic zones into the ocean.[213]

Table 2 perfectly illustrates the extent of this fundamental characteristic of the law of baselines,[214] for around 53% of the features involved in the equidistant lines of the maritime boundaries in the Pacific region not subject to a delimitation agreement yet in force could be significantly affected by a sea-level rise of 1 metre. These have been marked in table 2 as level 5 of vulnerability and correspond to 60 coral atolls or islets (maximum altitude of 5 metres) and coral islands (maximum altitude of 8 metres) as well as 10 coral reefs – that is, 70 extremely vulnerable features out of the 130 features listed in the table. When non-volcanic raised coral islands are added to this list (6 features marked as level 4 vulnerability), the proportion of affected features is 58%.

All in all, thirty-two maritime boundaries not subject to an agreement *yet in force* in the region are potentially affected by the transformation (or even total disappearance) of these geographical features.[215] The effects of sea-level rise and coastal erosion on the boundaries are a reflection of their effects on the baselines in which these features are involved. Therefore, the

213 David D. Caron, 'Climate Change, Sea-Level Rise and the Coming Uncertainty in Oceanic Boundaries: A Proposal to Avoid Conflict', in Seoung-Yong Hong and Jon M. Van Dyke, eds., *Maritime Boundary Disputes, Settlement Processes, and the Law of the Sea*, Publications on Ocean Development, vol. 65 (Leiden and Boston: Martinus Nijhoff, 2009), 5.

214 See Jon M. Van Dyke and Robert A. Brooks, 'Uninhabited Islands: Their Impact on the Ownership of the Oceans' Resources', *Ocean Development and International Law Journal*, vol. 12 (1983): 3–4, 265–300.

215 Unsurprisingly, and from the data in table 2, the boundaries most affected are those of Kiribati: Kiribati/Marshall Islands; Kiribati/French Polynesia; Kiribati/Tuvalu; Kiribati/Tokelau (New Zealand); Kiribati/Baker and Howland Islands (USA); Kiribati/Jarvis Island (USA); Kiribati/Palmyra Atoll and Kingman Reef (USA); Kiribati/Nauru; Kiribati/Cook Islands; followed by that of Wallis and Futuna (France)/Tokelau (New Zealand); New Caledonia (France)/Vanuatu; Marshall Islands/Nauru; Marshall Islands/FSM; Marshall Islands/ USA territory; FSM/Guam (USA); FSM/Palau; FSM/PNG; Cook Islands/Tokelau; Cook Islands/Niue (New Zealand); Fiji/Tonga; Fiji/New Caledonia (France); Tuvalu/Wallis and Futuna (France); Tuvalu/Fiji; Samoa/Tokelau (New Zealand); Samoa/American Samoa.

de-territorialisation of Pacific Island States' maritime spaces raises two legal questions: (1) how does the physical transformation of the various features change their legal characterisation and modify the corresponding regime (e.g. their capacity to generate maritime spaces)?; and (2) do these changes in the legal nature and regime of these features *necessarily* generate the recession of the baselines? In other words, may the geographical features' capacity to generate maritime spaces be limited by a fundamental characteristic of the law of baselines?

Effect nº 1: Change of Features' Legal Category and Corresponding Legal Regime

During the Third Conference on the Law of the Sea, two types of competing interests clearly crystallised. The will of many States (including Pacific Island States) of using the presence of islands in their territories to secure an advantage and generate extensive maritime claims collided with reactions seeking to limit claims based on remote topographical features that would significantly diminish the extent of the high seas and the international seabed area, respectively, available for navigation and exploitation by all. As recalled by Jon Van Dyke and Robert Brooks, discussions among States over the legal definition of the term 'island' and the extension of the maritime area attached to it can be traced back to the 1930 League of Nations Codification Conference held at The Hague, when the diplomatic debate centred on how to find criteria that could limit the effects of islands in cases where delimitations obstructed by their presence impeded reaching an equitable result.[216]

Since the consolidation of the EEZ implies that one single island could generate around 430,000 sq.km. of associated maritime space, the tension between these two standpoints resulted in active and long negotiations on this matter. Yet, Van Dyke also points out that the drafting of Article 121 of UNCLOS was achieved soon after the opening of the Third Conference and was never amended afterwards.[217] This provision, comprising three paragraphs, establishes a complex and controversial classification of two types of features, each able to generate maritime rights of a different nature and scope.[218]

216 Van Dyke and Rooks, 271.

217 For a detailed account of the negotiating history of Article 121 UNCLOS, see Nordquist, *a Commentary*, vol. 3, Articles 86 to 132 and documentary annexes, pp. 59–455.

218 Thorough studies on this matter – though published before sea-level rise became noticeable as a potential source of instability – include Derek W. Bowett, *The Legal Regime of Islands in International Law* (New York: Oceana, 1979); and Hiran W. Jayewardene, *The Regime of Islands in International Law* (Dordrech: Martinus Nijhoff, 1990).

Paragraph 1 of Article 121 provides a legal definition of island as "a naturally formed area of land, surrounded by water, which is above water at high tide"; paragraph 2 specifies the regime applicable to legal islands and lists the maritime rights generated by them; and paragraph 3 establishes an exception to this general rule for features falling into a second category, namely "rocks which cannot sustain human habitation or economic life of their own". The differences arising out of the classification of a feature in one of the two categories are significant: while features under Article 121(1) give full effect to the territorial sea, the contiguous zone, the exclusive economic zone and the continental shelf, the exceptions considered under Article 121(3) can only generate territorial sea. Furthermore, bearing in mind the nature of the adverse effects of climate change on the region, the intent of Article 121 should be complemented by Article 13, defining low-tide elevations as "a naturally formed area of land which is surrounded by and above water at low tide but submerged at high tide," and which do not generate any sort of maritime rights. A wide range of scholarly and diplomatic interpretations has arisen, since the adoption of the Convention, on various aspects of this triple classification of features and is to be kept in mind when considering how the physical effects of sea-level rise could transform a feature originally falling under the definition of 'island' into either a rock or a low-lying elevation.

The definition of 'island' provided in Article 121(1) is perhaps the least contentious of the three types of features. It is generally accepted that features must fulfil two conditions to be considered an island in the legal sense. In conformity with Article 60(8), which provides that artificial islands do not generate territorial sea of their own,[219] legal islands must first of all be *naturally formed.* Nonetheless, State practice seems to indicate that artificial means may be used to *maintain* the legal status of a feature by protecting it from severe erosion and sea-level rise, as has been done, for instance, by Japan with Okinotorishima Island.[220]

219 Artificial islands generally take the form of platforms, some of which have been built up to claim the creation of a new State, the most notorious examples probably being that of the Principality of Sealand (located off the coast of the United Kingdom in the North Sea, and based on a former World War II sea fort), and the Republic of Rose Island (located off the coast of Italy in the Adriatic Sea, and based on an artificial man-made platform). See Nikos Papadakis, *The International Legal Regime of Artificial Islands* (Leiden: Nijhoff, 1977).

220 See, for instance, Andrew L. Silverstein, 'Okinotorishima: Artificial Preservation of a Speck of Sovereignty', *Brooklyn Journal of International Law*, vol. 16 (1990): 2, 409–432; and Guifang Xue, 'How Much Can a Rock Get? A Reflection on the Okinotorishima Rock', in Myron H. Nordquist, ed., *Law of the Sea Convention: US Accession and Globalisation*, (Charlottesville, Va.: Center for Oceans Law and Policy, 2012), 341–388.

Secondly, features must be above water at high tide, though it has been considered that exceptional episodes of sea flooding do not preclude a feature from being classified as an island.[221] This criterion is what distinguishes islands from low-tide elevations – features submerged at high tide – and becomes all the more crucial in the Pacific region now that episodes of sea flooding by the so-called 'king tide' are gradually becoming more frequent and harmful – to the extent that, in some cases, forced population displacement is fostered. Islands which have become low-lying elevations would thus lose all their maritime zones (though those located within the 12 nautical miles of the main or remaining islands may still be used as basepoints).[222]

The observed displacement movements resulting from the increased frequency of king tides and other phenomena has also prompted several authors to consider that legal islands may turn into the category of "rocks which cannot sustain human habitation or economic life of their own" described in Article 121(3), a highly controversial provision which has generated many divergent interpretations. A closer look into the meaning and scope of this Article reveals that the applicability of this provision for severely eroded legal islands could be disregarded on several procedural and substantive grounds. First of all, bearing in mind that the purpose of this exception to the general regime of islands was to limit potentially inequitable delimitations resulting from claims over remote features,[223] Article 121(3) is arguably only applicable to features relevant for unilateral claims or bilateral delimitation purposes which are not close to a main or indisputable island.

Secondly, given that Article 121 does not constitute *lex specialis* to be applied with preference over the law of baselines, these features can be protected from the applicability of Article 121(3) in cases where the State has drawn archipelagic or straight lines in accordance with Article 47 and Article 7, respectively. Some authors, such as Jonathan Charney, even consider that the

221 Preparatory documents for the deliberations of the 1958 Convention on the Territorial Sea and Contiguous Zone contained a definition which included the phrase "which in normal circumstances is permanently above high water". This conflicting wording led the USA to call for the inclusion of a reference to seasonal tidal action as well as tidal action that was subnormal or abnormal. Article 5 of UNCLOS finally and consistently states that the "normal baseline is the low-water line along the coast as marked on large scale charts officially recognised by the coastal state."

222 UNCLOS, Article 13.

223 Alex O. Elferink, 'Clarifying Article 121(3) of the Law of the Sea Convention: the Limits set by the Nature of the International Legal Process', *Boundary and Security Bulletin*, vol. 6 (1998): 2, 58–68, who indicates that islands which might qualify as a rock can be included in a system of baselines and as such be used to establish the outer limits of maritime zones.

'freezing' effect on the legal classification of these features could be defended by States which have claimed archipelagic status but have not yet drawn their archipelagic baselines.[224] Such interpretation would thus protect features that are part of Fiji, Palau, Papua New Guinea, Solomon Islands, Tuvalu and Vanuatu – all States which have specified in their national legislation either the names or the coordinates of their archipelagic basepoints).[225] It would also benefit the mostly affected low-lying coral island States of Kiribati, the Marshall Islands and the Federated States of Micronesia (which have claimed archipelagic status but have not yet produced charts with archipelagic baselines).

Moreover, if these two procedural arguments were contested, it would still be possible to defend on substantive grounds – related to the interpretation of the meaning of the term "rock that cannot sustain human habitation" – that the applicability of Article 121(3) on 'transformed legal islands' would in any case be reserved for very exceptional cases. The grounds for this view are not to be found in case law, but rather in different doctrinal positions as well as in concurrent State practice. The first interpretative strategic position to be held regards the definition of the term 'rock'. Much controversy has arisen over whether it refers to the geo-morphological characteristics of the feature concerned and should therefore be defined in scientific terms as a hard mass, solid part of the earth's crust; or whether the term results from a test of the socio-economic viability of the feature and should, in contrast, be defined in cultural–geographical terms as a "feature capable of supporting the stable community who use the ocean space surrounding it."[226] The position on this preliminary point is essential, for it determines the order of the methodological steps to be followed subsequently. When the geo-morphological interpretation is adopted, two questions arise: (1) is it a rock or an island? ; and (2) if it is a rock, can it sustain human habitation or an economic life of its own? Adopting the geographical–cultural interpretation implies raising one single question: can the feature sustain human habitation or economic life of its own (for, if it cannot, it is a 'rock' in the sense of Article 121(3))? Authors such as Robert Kolb have argued in favour of a wide interpretation and consider that the use of the term 'rock' simply resulted from the will to emphasise the distinction between 'rocks' and the 'islands' – the latter being a type of features falling under paragraph 1 of Article 121 – but did not foresee any need to restrict its application on

224 Jonathan I. Charney, 'Rocks that Cannot Sustain Human Habitation', *American Journal of International Law*, vol. 93 (1999): 4, 863–878.

225 See table 1 above.

226 Barbara Kwiatkowska and Alfred H.A. Soons, 'Entitlement to Maritime Areas of Rocks Which Cannot Sustain Human Habitation or Economic Life of their Own', *Netherlands Yearbook of International Law*, vol. 21 (1990), 139–181, 153.

the basis of the geological composition of the features.[227] Barbara Kwiatkowska and Alfred Soons have followed a similar line and consider that the essential element of the definition is not the term 'rock' – which can be replaced by island or islet – but the fact that the features concerned *cannot sustain human habitation or economic life of their own.*[228] In contrast, other scholars, such as Jonathan Charney, upholding the narrower interpretation of Article 121(3), and have argued that the purpose of UNCLOS was to distinguish different types of features according to their sise. Thus, Charney explains:

> [O]riginally islets and small islands used to define the features that would fall within the provision that ultimately became Article 121(3) and yet, the ultimate redaction of Art. 121(3) seems to apply to an even narrower range of small features than these: only 'rocks'.

All in all, no uniform rule seems to be available for these features wherever they are encountered. Thus, in the absence of a clear-cut understanding of the meaning and scope of Article 121(3), States affected by it could defend the preservation of their maritime claims by upholding the narrower interpretation of the term 'rock' – either considering that it refers to the geological composition of the feature or, as proposed by Charney, to the most insignificant features.

It is only when this interpretation of the meaning and scope of the term 'rock' fails to allow suspension of the application of Article 121(3) that an important island of the affected State – particularly in the Pacific region – may risk being re-classified as a rock which cannot sustain human habitation or an economic life of its own. The test of the socio-economic viability of the feature concerned is problematic because of its two ambivalent characteristics. Despite the fact that its purpose is to measure the *capacity* of the feature to sustain human population or economic life autonomously, rather than its actual condition (namely, whether or not it is inhabitable rather than being inhabited), its inter-temporal scope[229] implies that even when the feature's

---

227 Robert Kolb, 'L'interprétation de l'article 121, paragraphe 3, de la Convention de Montego Bay sur le droit de la mer: les "rochers" qui ne se prêtent pas à l'habitation humaine ou à une vie économique propre', *Annuaire Français de Droit International*, vol. XL (1994), 876–909. Kolb upholds this interpretation after a close examination of the *travaux préparatoires* of UNCLOS, and considers that the wording of Article 121 originated from the Turkish and Romanian proposals.

228 Kwiatkowska and Soons, 153.

229 On the inter-temporal application of Article 121(3), see Charney, 'Rocks that Cannot Sustain Human Habitation', 865: "[C]onsequently, the application of Art. 121(3) to a feature may vary over time, just as an ambulatory baseline might move in response to geographical changes."

past capacity can be evidenced, present claims must prove that such capacity continues to exist.[230] This may be an obstacle difficult to overcome in most features of Pacific Island States, which, for the vast majority, have always been uninhabited and, amongst those which have been inhabited, current displacement of human populations does not provide particularly favourable evidence of the present continuity of the features' capacity to sustain human habitation or economic life of their own. To avoid the inter-temporal application of Article 121(3), one could again have recourse to the second procedural argument mentioned above and consider, as pointed out by Charney, that the legal classification of a feature was 'frozen' at a certain date, corresponding to the drawing of the archipelagic baselines with indication of the coordinates of the basepoints and the respective charts. Yet, it is uncertain and quite doubtful that this claim may serve to preclude the inter-temporal application of the capacity test in cases where archipelagic status has been claimed but the archipelagic baselines have not been defined.[231] In these cases, as well as in situations of features pertaining to States which have not even claimed archipelagic status at all, the inter-temporal application of Article 121(3) would most likely generate a reclassification of the feature concerned and consequent redrawing of the baselines to which such a feature was anchored.

Effect nº 2: Recession of Baselines and Retreat of the Affected Boundaries

Once the anchor of the baseline is either submerged or susceptible of being reclassified as a rock or as a low-tide elevation beyond 12 nautical miles, the baseline is to be redrawn on the basis of the still valid basepoints, which in turn implies that the outer maritime boundary generated by the previous baseline is also accordingly modified.[232] Alfred Soons indicates that the distance over which the baseline shifts landwards "depends on the gradient of the land surface in the area involved: the lesser the gradient, the greater the distance."[233] Thereby, in some areas, a rise of the sea level by half a metre can provoke a shift in the baseline of ten kilometres. It is generally explained that this chain of changes is due to the fact that "both the baseline and the

230 Kwiatkowska and Soons, 162.

231 See section 1.2.2.B below, on the discussions about the difference between applying the 'actual baseline', as opposed to the 'charted baseline'.

232 Caron, 'Climate Change, Sea-Level Rise, and the Coming Uncertainty in Oceanic Boundaries', 9.

233 Alfred H.A. Soons, 'The Effects of a Rising Sea Level on Maritime Limits and Boundaries', *Netherlands International Law Review*, vol. 37 (1990), 207–232, 216.

boundary generated by that baseline are ambulatory,"[234] a position which was recently expressed by the ILA Committee on Baselines under the International Law of the Sea.[235]

This characteristic seems, *prima facie*, consistent with, and resulting from, the subsidiary and derived nature of maritime rights which, as Prosper Weil recalls, "have no independent existence, but are an extension of the pre-existing territorial sea."[236] Although the principle of equity introduced into the delimitation process operates as a "bridge between nature and the law",[237] the fundamental premise that the 'land dominates the sea' as has been stressed on many occasions by the International Court of Justice, implies that maritime delimitations are in most cases primarily driven by the configuration of the coast.[238] It is equally applicable to maritime limits which, as explained by Alfred Soons on the basis of Articles 7 and 47:

> [T]he drawing of straight or archipelagic baselines must not depart to any appreciable extent from the general direction of the coast, and the sea areas lying within the lines must be closely linked to the land domain.[239]

---

234 Caron, 'Climate Change, Sea-Level Rise, and the Coming Uncertainty of Oceanic Boundaries', 9.

235 Committee on Baselines under International Law of the Sea, *Final Report* (Sofia Conference: ILA, 2012), 31.

236 Prosper Weil, 'Geographic Considerations in Maritime Delimitation', in Jonathan J. Charney and Leslie M. Alexander, eds., *International Maritime Boundaries* (Dordrecht: Nijhoff, 1993), 115–130, 115. This principle has been consistently recognised by the International Court of Justice, as for instance in the *North Sea* Judgment, para. 96: "[T]he land is the legal source of the power which a State may exercise over territorial extensions to seaward."

237 Words of Judge Lachs, President of the Guinea v. Guinea Bissau Arbitration Tribunal, cited in Prosper Weil, *The Law of Maritime Delimitation: Reflections* (Cambridge, UK: Grotius Publications, 1989), 88.

238 See for instance, *North Sea* Judgment, para. 91: "[T]here can never be a question of completely refashioning nature"; *Continental Shelf* (Lybia Arab Yamahiriya v. Malta), Judgment, 1985 ICJ Reports 13, para. 46: "[E]quity does not necessarily imply equality [...] nor does it seek to make equal what nature has made unequal". See Weil, *Law of Maritime Delimitation*, 85–90. Since its formulation in the *North Sea* case, the principle of never refashioning nature has not been questioned, not even after the shift operated by the ICJ from a result-oriented equity approach to the adoption of the equidistance/adjustment method or 'corrective equity approach' currently in use (see comments on the evolution of the jurisprudence regarding maritime delimitation methods in section 1.2.1.B. below).

239 Soons, 'Effects of a Rising Sea-Level on Maritime Limits and Boundaries', 211. See Article 7 of UNCLOS on straight baselines, completed by Article 47 on the special ratios of archipelagic baselines.

Nonetheless, while the close correlation between the shape of the coastline and the direction of the baseline is undisputed, David Caron recalls that when the law of baselines was first formulated in the 1930s and then revised in the 1970s, sea-level rise or the transformation of features due to the adverse impacts of climate change were not in the minds of the authors. Indeed, as he holds, "despite [the fact] that coastline erosion is common, they apparently viewed such changes as rare and isolated."[240]

Although UNCLOS does not explicitly specify the ambulatory nature of baselines (nor the outer maritime boundaries), the majority of scholars seem to agree that the correlation between the shape of the coastline and the baseline is inter-temporal and implies its necessary adaptation to the new coastal geography of the land. For instance, the ILA Committee on Baselines under the International Law of the Sea recently concluded that if the legal baseline "changes with human-induced expansions of the actual low-water line to seaward (e.g. cases of artificial conservation or even extension of the baseline through permanent coastal works), then it must also change with contractions of the actual low-water line to landward".[241] This conclusion can be solidly grounded in three arguments. First of all, as a corollary of the fundamental premise that, the land dominates the sea', baselines depend on whether the State's title to the land survives, is maintained, or sustained.[242] Moreover, it has been argued that Article 76(9) of UNCLOS, which establishes that (under certain conditions) the outer limits of the continental shelf are permanently described ('permanence' being understood in relation to Art 76(8), and the process which involves the CLCS),[243] allows the *a contrario* conclusion that the outer limits of all other maritime zones might not be final.[244] The reason for fixing the outer boundaries of the continental shelf was that seabed exploitation implies high levels of investment. This provision was intended to permanently fix the boundary between the continental shelf and the

240 Caron, 'Climate Change, Sea-level Rise, and the Coming Uncertainty of Oceanic Boundaries', 5.

241 ILA Committee on Baselines under the International Law of the Sea, *Final Report* (ILA: Sofia Conference, 2012), 1–32, 28.

242 This is the position of Soons in 'The Effects of a Rising Sea Level on Maritime Limits and Boundaries'.

243 UNCLOS, Article 76(9) – on the continental shelf.

244 Jenny Grote Stoutenburg, 'Implementing a New Regime of Stable Maritime Zones to Ensure the (Economic) Survival of Small Island States Threatened by Sea-Level Rise', *International Journal of Maritime and Coastal Law*, vol. 26 (2011), 263–311, at 269.

international seabed area in view of the legal security question which is of great importance for the holders of mining concessions.[245]

All in all, while the ambulatory or changing nature of baselines and outer maritime boundaries seems uncontroversial, the eventual level of protection against such modifications offered by claiming archipelagic status without drawn archipelagic lines remains uncertain. In contrast, those States which have published their archipelagic lines may be protected from changes as long as the information of the charts provided to the UN Secretary-General for publicity in accordance with Article 47(9) of UNCLOS is not updated. However, whether charting a line and formally depositing a chart may protect from future *substantial* change is yet highly debatable.

B Sea-Level Rise and Maritime Delimitation Agreements

As opposed to the establishment of maritime limits, which are established by a unilateral act and consists on drawing lines that define the maritime spaces of one single State; the delimitation of maritime boundaries is effected between two or more States and seeks to separate overlapping areas where legal titles of coastal States compete.[246]

As illustrated in the previous section, most Pacific Island States have overlapping EEZs and continental shelves. Yet, as much as Pacific Island States have unilaterally declared the extent of their maritime spaces and limits to it; many maritime boundaries in the South Pacific remain undelimited; or, at best, have been the object of maritime delimitation agreements which are not yet fully operative as the respective Parties have not yet perfected their consent. Indeed, of the twenty-four bilateral maritime delimitation agreements that have been signed in the region (including one preliminary exchange of notes between France/ Wallis and Futuna, and Tuvalu; and the agreed minutes of the 2014 meeting by Fiji, Wallis and Futuna (France) and Tuvalu), only nine have been ratified and are currently in force.[247] Until recently, these treaties necessarily involved at least one former colonial or administrating power willing to clarify, after decolonisation, the maritime boundaries generated by them just prior to the end of their colonial presence in the region. In cases where the other Party is not France, the United Kingdom, the United States of America, New Zealand or Australia, the agreements involve Pacific Island States likely to generate important economic or political interests, such as those of Papua

245 Soons, 'The Effects of a Rising Sea-Level on Maritime Limits and Boundaries', 217. See also, Nordquist, ed., *Commentary*, vol. 2, 882–883.

246 Tanaka, 187.

247 See table 3 below.

New Guinea (mineral producer and the richest country of the region), and Fiji (second most populated Pacific Island State); or those remaining politically linked to their former colonial or administrating power under a Compact of Free Association, such as the Cook Islands, or the territories of Niue or Tokelau.

Considering the economic burden that scientific surveys of the coastal zones represent, the newly decolonised and underdeveloped Pacific Island States did not initiate negotiations on this matter until very recently. With support from the South Pacific Applied Geoscience Commission (SOPAC) through the Regional Maritime Boundaries Project Initiative, sub-regional co-operation among the three States formerly integrated into the UN Strategic Pacific Trust Territory led to the conclusion, in 2006, of two maritime delimitation agreements between the Federated States of Micronesia and its respective western and eastern neighbours, Palau and the Marshall Islands. The former entered into force in 2008. Similarly, direct legal and technical assistance from the Special Advisory Services Commission (SASC) of the Commonwealth Secretariat and from the Australian Attorney-General's Office and the Pacific Islands Forum very recently fostered the historical signing, on 30 August 2012, of seven maritime delimitation agreements among Pacific Island States. Six of them were bilateral agreements between Kiribati – the Pacific island State with the widest maritime area of the region – and each of its neighbours: (1) Cook Islands (New Zealand); (2) Niue (New Zealand); (3) Tokelau (New Zealand); and (4) the Marshall Islands; (5) Nauru; as well as a trilateral agreement (6) among Kiribati, Nauru and the Marshall Islands, concerning the determination of a tri-junction point in the zone covered by the intersection of the EEZs of these three States. An additional bilateral maritime delimitation agreement (7) between Nauru and the Marshall Islands was also concluded. This unique signing took place during the 43rd Pacific Islands Forum on 'Large Ocean Island States: the Pacific Challenge', held in Rarotonga (Cook Islands) on 30 August 2012.[248] It was followed by the conclusion of a delimitation agreement between (8) Kiribati and the Palmyra Atoll, Kingman Reef, Jarvis and Baker Islands (USA) in 2013; as well as by an agreement between (9) the Federated States of Micronesia and Guam (USA) in 2014. On-going negotiations between (10) Fiji, Wallis and Futuna and Tuvalu on the geodesic coordinates of a tri-junction point have led the Parties to agree on the Minutes of their 2014 meeting. Although, of the nine maritime delimitation agreements involving exclusively underdeveloped Pacific Island States rather than former colonial powers, only one (Federated States of Micronesia/Palau) has come into force,

248 Official Press release from the Commonwealth Secretariat available at: <http://www.thecommonwealth.org/news/249480/290812maritimeboundaries.htm> (1 October 2014).

this move is already indicates how serious steps are being taken by Kiribati to defend its territorial boundaries.

Other than the 1978 Torres Strait agreement between Papua New Guinea and Australia, which achieved a comprehensive settlement on a wide range of complex issues,[249] the primary objective of these agreements is to delimit their overlapping EEZs and continental shelf, in view of the economic merits of the zone for exploitation of the living and seabed mineral resources.[250] On some occasions, political, historical and strategic reasons have also played an important role, as in the 1980 Treaty between the Cook Islands and American Samoa in which the United States recognised the sovereignty of the Cook Islands over four islands (Pukapuka, Manahiki, Rakatanga and Penrhyn) formerly administered by New Zealand and claimed by the USA since the 19th century. Another prominent example can be found in the 1983 Agreement between Fiji and Wallis and Futuna (France), which has not yet come into force because of its association with the only outstanding territorial dispute of the region, between Vanuatu and France, on sovereignty over the Matthew and Hunter Islands.[251] Finally, some agreements embrace wider purposes, establishing a framework for sub-regional co-operation in environmental matters or the protection of indigenous peoples' traditional areas of fishing. These can be found in the two most recent agreements, concluded by the Federated States of Micronesia with Palau and with the Marshall Islands, which interestingly introduced clauses responding to the eventual adverse effects of climate change on their delimitation line.

A compilation of the twenty-four existing maritime delimitation agreements of Pacific Island States (including the 1985 interim arrangement between Wallis and Futuna (France) and Tuvalu; and the 2014 Agreed Minutes of Fiji, Wallis and Futuna (France) and Tuvalu), of which only nine are in force, is provided in table 3, below. Based on the texts of the agreements and the explanations provided in the reports of Jonathan Charney and Lewis Alexander,

249 Treaty between Australia and the Independent State of Papua New Guinea Concerning Sovereignty and Maritime Boundaries in the Area between the Two Countries, Including the Area Known as Torres Strait, and Related Matters, signed 18 December 1978; entered into force 15 February 1985.

250 Including transboundary, highly migratory species such as tuna (Federated States of Micronesia/Marshall Islands and Federated States of Micronesia/Palau). For instance, an outstanding dispute is that of Tonga and Fiji over the North and South Minerva Reefs: while Tonga claimed sovereignty in 1972, renaming them Teleki Tokelau and Teleki Tonga, respectively, Fiji has placed them on the edge of its own EEZ.

251 While the conclusion seems to strengthen the French claim over the Hunter and Matthew Islands, France accepted in return the use by Fiji of the 2-metre-high cay of Theva-i-ra (formerly Conway Reef) as one of its base points.

TABLE 3 *Maritime delimitation agreements of the South Pacific region*

Agreement (Signature – Entry into Force)	Length of Line	Zones Delimited	Nº of Segments	Method(s) of delimitation	BASEPOINTS Name	Type	Risk (1–5)	Location of Delimitation Points
1978–1985 Australia/Papua New Guinea	…	Territorial Sea Seabed Fisheries Protected Zone	20	Adjusted Equidistance + Straight lines (Torres Strait)	Warrior Reef (AUS)	Reef	5	Geodetic Coordinates
					Pearcy Cay (AUS)	Cay	5	
					Kawa (PNG)	Coral Atoll (<5m)	5	
					Mata Kawa (PNG)	Coral Atoll (<5m)	5	
					Kussa (PNG)	Coral Island (19m)	5	
1980–1980 France (Wallis and Futuna)/Tonga	220M	EEZ	2	Equidistance	Futuna (FR)	Volcanic Island (765m)	1	No Exact Location of Delimitation Points Article 3.c): 'Technical corrections' would be made 'at some future date', as necessary to update these data (triggered by exchange of letters)
					Niua Fo'ou (TG)	Volcanic Crater (205m)	1	
					Minerva Reefs (TG)	Reef	5	

1980–1983 Cook Islands/USA (American Samoa)	566M	All-Purpose Boundary EEZ (Since 1983 US claim)	24	Equidistance + Full Effect of all Features	Pukapuka (CI) Nassau (CI) Suwarrow (CI) Palmerston (CI) Rose (USA) Manua (USA) Swains (USA)	Coral Atoll (<5m) Coral Atoll (<5m) Coral Atoll (<5m) Coral Atoll (<5m) Coral Atoll (<5m) Continental Island (964m) Coral Atoll (<5m)	5 5 5 5 5 1 5	Geodesic Coordinates
1982–1983 Australia/France (New Caledonia)	1200M	Fisheries Zone (Australia 1979 claim) EEZ (France 1978 claim)	21	Partially Adjusted Equidistance Straight + Lines + Full Effect of Middleton Reef	Middleton Reef (AUS) Norfolk (AUS) Walpole (FR) Matthew (FR) Hunter (FR)	Low-Tide Elevation Old Volcanic Island (318m) Raised Coral Island (100m) Active Volcanic Island (117m) Active Volcanic Island (266m)	5 1 4 1 1	Geodesic Coordinates
1983–** Fiji/France (New Caledonia, Wallis and Futuna)	SW Line: 320M NE Line: 240M	EEZ	SW Line: 1NE Line: 4	Slighted Adjusted Equidistance	Theva-I-Ra (FJ) Hunter (FR) Futuna (FR)	Island Active Volcanic Island (266m) Volcanic Island	5 1 1	Geodesic Coordinates

TABLE 3 *Maritime delimitation agreements of the South Pacific region* (cont.)

Agreement (Signature – Entry into Force)	Length of Line	Zones Delimited	Nº of Segments	Method(s) of delimitation	BASEPOINTS Name	Type	Risk (1–5)	Location of Delimitation Points
1983–1984 France (French Polynesia)/ UK (Pitcairn, Henderson, Ducie and Oeno Islands)	350M	EEZ	5	Strict Equidistance	Temoe (FR) Marutea (FR) Pitcairn (UK) Oeno (UK)	Coral Atoll (<5m) Coral Atoll (<5m) Volcanic Island (347m) Coral Atoll (<5m)	5 5 1 5	Geodesic Coordinates + Loxodrome Lines Doubts about precise location of certain islands
1985–1985 (Exchange of notes, Interim Arrangement) France (Wallis and Futuna)/Tuvalu	300M	EEZ Extended CS Possible (Tuvalu)	To be fixed	Equidistance	Not Specified	Not Specified	…	No Exact Location of Delimination Points Absence of Appropriate Charts Relevant Baselines to be Defines by SOPAC

1988–** Australia/Solomon Islands	150M	Fisheries Zone EEZ	2	Equidistance	Mellish Reef (AUS) Indispensable Reef (SI)	Reef Reef/Low-Tide Elevation	5 5	Geodesic Coordinates
1989–** Papua New Guinea/Solomon Islands	1000M	Territorial Sea EEZ CS Protected Zones	12	Adjusted Equidistance	Pocklington (PNG) Bougainville (PNG) Shortland (SI)	Reef High Volcanic Island (2792m) Volcanic and Raised Coral Island (234m)	5 1 5	Geodetic Coordinates
1990–1990 France (New Caledonia)/ Solomon Islands	260M	EEZ	4	Equidistance + Straight Lines	Ile Huon (FR) Recif Bampton (FR) South Reef (SI)	Coral Atoll (<5m) Reef Reef	5 5 5	Loxodromes and Geodetic Coordinates
1997–** Niue (New Zealand)/USA (American Samoa)	279M	All-Purpose Boundary	18	Equidistance	Niue Not specified	Raised Coral Island (76m) …	4 …	Geodetic Coordinates New coastal geodetic positioning required to update information on location of base points

TABLE 3 *Maritime delimitation agreements of the South Pacific region* (cont.)

Agreement (Signature – Entry into Force)	Length of Line	Zones Delimited	Nº of Segments	Method(s) of delimitation	BASEPOINTS Name	Type	Risk (1–5)	Location of Delimitation Points
2003–2003 France (Wallis and Futuna)/New Zealand (Tokelau)	101M	All-Purpose Boundary	2	Equidistance	Wallis (FR)	High Volcanic Island (765m)	1	Geodetic Coordinates
					Futuna (FR)	High Volcanic Island (765m)	1	
					Nukunonu (NZ)	Coral Atoll (<5m)	5	
					Atafu (NZ)	Coral Atoll (<5m)	5	
2006–** Federated States of Micronesia/ Marshall Islands	755M	EEZ CS	10	Adjusted Equidistance	Kosrae (FSM)	High Volcanic Island (628m)	1	Geodetic Coordinates (Article 1.5) Significant shifts in the geographic location of islands used as base points may imply revision of coordinates of the agreed line
					Pingelap (FSM)	Coral Atoll (<5m)	5	
					Mwokil (FSM)	Coral Atoll (<5m)	5	
					Pohnpei (MI)	Volcanic Island (721 m)	1	
					Ebon (MI)	Coral Atoll (<5m)	5	
					Namidrik (MI)	Coral Atoll (<5m)	5	
					Ujae (MI)	Coral Atoll (<5m)	5	
					Ujelang (MI)	Coral Atoll (<5m)	5	

2006–2008 Federated States of Micronesia/Palau	…	EEZ CS	50	Equidistance	Ngulu (FMS)	Coral Atoll (<5m)	5	Geodesic Coordinates Use of reefs by both Parties although Palau 2003 legislation imposes that baselines must be drawn from specific features
					Yap (FSM)	Continental/Old Volcanic Island (173m)	1	
					Northern Islands (PL)	Diversity of Features, Including Unspecified Reefs	5	
2012–** Cook Islands/ Kiribati	…	EEZ CS	13	Adjusted Equidistance	Starbuck (KR)	Low coral island (5m)	5	Geodesic coordinates
					Vostok (KR)	Low coral island (5m)	5	
					Flint Island (KR)	Low coral island (7m)	5	
					Penrhyn Islands (CI)	Atoll	5	
2012–** Kiribati/Tokelau (NZ)	…	EEZ CS	4	Equidistance	Nikumaroro (KR)	Atoll	5	Geodesic coordinates
					Orona (KR)	Atoll (9m)	5	
					Manra (KR)	Low coral island (6m)	5	
					Atafu (NZ)	Atoll (<5m)	5	
					Fakaofo (NF)	Atoll	5	

TABLE 3 *Maritime delimitation agreements of the South Pacific region* (cont.)

Agreement (Signature – Entry into Force)	Length of Line	Zones Delimited	Nº of Segments	Method(s) of delimitation	BASEPOINTS			Location of Delimitation Points
					Name	Type	Risk (1–5)	
2012–** Kiribati/Niue (NZ)	Information not available							
2012–** Kiribati/Nauru	…	EEZ CS	15	Equidistance	Banaba (KR)	Raised coral island (81 m)	3	Geodesic coordinates Point 1 corresponds to the agreed provisional location of the tri-jonction point between Kiribati, Nauru and the Marshall Islands
					Nauru	Raised coral island (70 m)	3	
2012–** Kiribati/ Marshall Islands	…	EEZ CS	23	Equidistance	Banaba (KR)	Raised coral island (81m)	3	Geodesic coordinates
					Makin (KR)	Low coral island	5	
					Ebon (MS)	Atoll	5	
					Narik (MS)	Atoll (<5m)	5	

2012–** Kiribati/Nauru/ Marshall Islands	...	...	...	Equidistance	Banaba (KR) Makin (KR) Butaritari (KR) Ebon (MS) Narik (MS) Nauru	Raised coral island Low coral island Atoll Atoll Atoll Raised coral island (70m)	3 5 5 5 5 3	Geodesic coordinates of the tri-jonction point between the maritime boundaries delimiting the EEZ and CS of the three countries.
2012–** Marshall Islands/ Nauru	...	EEZ CS	10	Equidistance	Ebon (MI) Nauru	Atoll Raised coral island (70m)	5 3	Geodesic coordinates Point 1 corresponds to the agreed provisional tri-junction point between Marshall Islands, Kiribati and Nauru

TABLE 3 *Maritime delimitation agreements of the South Pacific region* (cont.)

Agreement (Signature – Entry into Force)	Length of Line	Zones Delimited	Nº of Segments	Method(s) of delimitation	BASEPOINTS Name	Type	Risk (1–5)	Location of Delimitation Points
2013–** Palmyra Atoll, Kingman Reef, Jarvis Island and Baker Island (USA)/Kiribati	Segment No. 1: 332M Segment No. 2: 548M Segment No. 3: 383M	All-purpose boundary	Segment No.1: 5 Segment No. 2: 9 Segment No. 3: 4	Strict equidistance	Kanton (KR)	Atoll (5m)	5	Geodesic coordinates
					McKean (KR)	Low coral island (5m)	5	
					Nikumaroro (KR)	Atoll	5	
					Teraina (KR)	Low coral island (5m)	5	
					Tabuaeran (KR)	Atoll	5	
					Kiritimati (KR)	Atoll (<5m)	5	
					Malden (KR)	Low coral island (8m)	5	
					Starbuck (KR)	Low coral island	5	
					Jarvis (USA)	Low coral island (8m)	5	
					Baker (USA)	Low coral island	5	
					Palmyra (USA)	Atoll (<5m)	5	
					Kingman (USA)	Reef	5	

2014–** Guam (USA)/ Federated States of Micronesia	Information not available							
2014–** Agreed Minutes Fiji/Wallis and Futuna (France)/ Tuvalu	…	…	…	Equidistance	…	…	…	Geodesic coordinates of Tri-junction Point

* *Data from the Global Island Database (UNEP Island Directory): UN-UNEP System-wide Earth Watch and International Maritime Boundaries.*

table 3 summarises the main characteristics of each of them, including the method of delimitation, the name and geological composition and degree of vulnerability (according to the same vulnerability test, given in table 2, of the features which served as base points for the delimitation line), and whether the Parties to the agreements have specified the exact location of the boundary by listing the geodetic coordinates of the points composing such line.[252]

The following analysis is based on the information available, which concerns only twenty-two of the twenty-four signed agreements listed above. Two agreements signed between 2012 and 2014 (Kiribati/Niue; and Guam/Federated States of Micronesia, respectively) have not yet been publicly available. Of the seventy-three features listed in table 3 which served as basepoints in the twenty-two delimitation agreements of which the information is available, fifty-eight can be considered as highly vulnerable to sea-level rise and coastal erosion. These changes are, in principle, likely to provoke important redrawing of the baselines used for the boundary agreements – although the extent of their impact on the baseline may be alleviated in cases where the Parties have agreed not to give full effect to the features at stake. Yet, in contrast to situations involving undelimited maritime boundaries, it can be argued that the modification of the baselines may not have a mirroring effect on agreed delimitation boundaries. As developed below, this would require accepting that maritime delimitation agreements have a 'freezing' effect on the geographical location of both the baselines and, correlatively, of the outer maritime boundaries (see section 1.2.2.B below). Such an effect would, nonetheless, be relative, as it would have to remain in accordance with the axiom of the law of the sea, that 'the land dominates the sea'.

As recalled in the previous section, neighbouring coastal States with overlapping maritime boundaries are prone to reach an agreement on the delimitation of their respective maritime zones. Articles 74(1) and 83(1) of UNCLOS, which respectively establish the same rule for the delimitation of the EEZ and the continental shelf between States with opposite and adjacent coasts, provide that the delimitation in these cases shall be effected "by agreement on the basis of international law, as referred to in Article 38 of the Statute of the ICJ, in order to achieve an equitable solution".[253] Table 3 above shows

252 Jonathan J. Charney and Lewis M. Alexander/ David A. Colson and Robert W. Smith, eds., *International Maritime Boundaries* (Dordrecht: Martinus Nijhoff, 1991–2005), vols. 1–6.

253 Note that the rule formulated in UNCLOS for the delimitation of the territorial sea differs from that concerning the EEZ and the continental shelf. Mirroring Article 12(1) of the 1958 Convention on the Territorial Sea and the Contiguous Zone, Article 15 of UNCLOS provides that for the delimitation of the territorial sea [Emphasis added]: "[N]either of

how State practice on maritime delimitation agreements in the South Pacific follows the 'equidistance/adjustment' or 'corrective-equity approach' (see section 1.2.1.A). Of the twenty-two agreements in the Pacific region the content of which is available, only seven have departed from the strict equidistant line (of these seven, only three are in force). Secondly, Parties generally specify the geodetic coordinates of a list of points composing the delimitation line and include such data in the main body of the Treaty or in an annex, irrespective of whether the delimitation line agreed upon is strictly equidistant or has undergone adjustments. In doing so, the delimitation line is virtually fixed or 'frozen', as in the majority of the agreements listed (twenty out of the twenty-two agreements for which information is available, and of which seven are in force). In contrast, the Parties may decide to merely make a reference in the agreement to the equidistant or median line as constituting the agreed boundary line, without providing the geodetic data of the delimitations points. In these cases, the boundary is not fixed and can be presumed to fluctuate in accordance with the changes undergone by the relevant baselines.[254] This is the case in the agreements between France (Wallis and Futuna)/Tonga, and Federated States of Micronesia/Palau (both in force). In the case of Federated States of Micronesia/Marshall Islands (not yet in force), the Parties provided for the coordinates of the boundary, yet they also agreed that the coordinates of the agreed line may be revised if significant shifts in the geographic location of islands used as base points take place (see Article 1.5 of the Federated States of Micronesia/Marshall Islands Agreement).

Therefore, the consequences of sea-level rise and coastal erosion for the maritime entitlements of Pacific Island States may radically differ, from treaties with fixed coordinates to agreements not specifying coordinates. On the one hand, the change in the location of the delimitation lines of the France/Tonga, FSM/Marshall Islands and FSM/Palau agreements may be simply triggered by a request of one of the Parties to update the scientific information of both Parties' coastline. On the other hand, all other Treaties with

the two [opposing of adjacent] States is entitled, failing *agreement* between them to the contrary, to extend its territorial sea beyond *the median line.* [T]he above provision does not apply, however, where it is necessary by reason of historic title or other *special circumstances* to delimit the territorial seas of the two States in a way which is at variance therewith".

254 See Soons, 227: "[I]n cases where the delimitation agreement explicitly refers to the median line the boundary may shift as a result of sea level rise: asymmetrical changes of the baselines of both states will lead to changes in the location of the median line. The states concerned have opted for a potentially fluctuating boundary line".

fixed delimitation points establish permanent and acquired territorial rights which cannot be modified unless the Treaty is terminated by lawful means – as well as to the extent that the principle that 'the land dominates the sea' is not overturned.

Some authors have raised the question of whether a Party can claim a fundamental change of circumstances to terminate a maritime delimitation agreement, as recognised in general international law and codified in Article 62(1) of the Vienna Convention on the Law of Treaties (VCLT). In principle, this could be raised by a Party that considers that its counterpart's coastline has undergone substantial geographical modifications which decompensate the original balance of the delimitation line, and prevent the claimant from potentially increasing the scope of its maritime entitlements. It would therefore seem to promote 'aggressive' or 'expansionist' claims rather than to serve as a protective or defensive device against the loss of already acquired maritime entitlements, a trend which could possibly generate interstate conflicts.[255] It is therefore understandable why treaties establishing a boundary explicitly exclude the possibility of claiming a fundamental change of circumstances under Article 62(2) of the VCLT. Moreover, the applicability of this exception is, in any case, subject to very stringent conditions which do not seem to be met in these cases. First of all, the circumstances motivating the claim must not have been foreseen by the Parties at the time of the conclusion of the Treaty. Considering that the adverse effects of sea-level rise on small island States were without doubt known by Pacific Island States by 1987, when the Maldives clearly introduced the issue into the Commonwealth agenda and then into that of the UN General Assembly, State Parties to delimitation agreements concluded that, after 1984, they could have taken it into consideration during the negotiations. Secondly, the changes must affect original circumstances "the existence of which constituted an essential basis of the consent of parties to be bound by the treaty" (Article 62(2) VCLT). Such a condition introduces de facto a *probatio diabolica*, particularly in cases where the agreed delimitation line is simply the strict median or equidistant line. Finally, the changes involved must be 'fundamental', so as to transform the extent of the obligations still to be performed under the Treaty. Perhaps this last condition is the only one that

255 Jonathan Lusthaus, 'Shifting Sands: Sea-Level Rise, Maritime Boundaries and Inter-State Conflict' *Politics*, vol. 30 (2010): 2, 113–118; and David D. Caron, 'Climate Change, Sea-Level Rise and the Coming Uncertainty in Oceanic Boundaries: A Proposal to Avoid Conflict', in Seoung Yong Hong, and Jon M. Van Dyke, *Maritime Boundary Disputes, Settlement Processes and the Law of the Sea* (Leiden and Boston: Martinus Nijhoff, 2009), chapter 1, 1–17.

could be met in all cases, since the disappearance of an island or severe coastal recession can substantially affect the baseline.

Having discarded this possibility, it can be concluded that delimited maritime boundaries shield the maritime entitlements of the Parties involved – in cases where the Parties decide to establish the coordinates of the delimitation points – or at least enable Pacific Island States to decide and control what effects they recognise as being due to foreseen adverse impacts of climate change – in cases where the agreement does not provide the coordinates of the delimitation line. The freezing effect of the delimitation agreement is all the more important given that, as Jenny Grote Stoutenburg recalls, it falls within the 'objective' or the 'status' of contracts which are an exception to the *pacta tertii rule* of Article 34 of the VCLT and which produces effects beyond the treaty Parties involved.[256] Of the twenty-two agreements signed in the region, the contents of which are available, only three may really see the location of the boundary modified: France (Wallis and Futuna)/Tonga, Federated States of Micronesia/Palau and Federated States of Micronesia/Marshall Islands. Given that only the first two of these three agreements are in force, it can be concluded that only the maritime entitlements of four States (France, Tonga, FSM and Palau) may be more imminently at stake, and that 'partial de-territorialisation' would arise for the latter three Pacific Island States. Yet, despite the fact that the remaining agreements have arguably 'fixed' maritime delimitation lines, present awareness of the impacts of sea-level rise and coastal erosion may delay even more or perhaps even prevent the six agreements not yet in force and which were signed before August 2012, from being ratified.[257] Eventual movements of these delimitation lines can, in some cases, become dangerous points of international dispute. For instance, the boundary that was agreed by the Solomon Islands and Papua New Guinea, signed but not yet ratified, involves the Bougainville Island, the centre of an important secessionist movement which drove Papua New Guinea into four years of civil strife. Likewise, the still pending ratification of the agreement between France (New Caledonia) and Fiji – which apparently favours French claims against Vanuatu over the Mathew and Hunter Islands, may have an impact on this still outstanding territorial dispute in the region.

256 Grote Stoutenburg, 280.

257 Agreements signed *in and after* August 2012 were presumably concluded not only with the acknowledgement of the effects of climate change, but actually as a means to react to its adverse effects on Pacific Island States' territorial dimension.

1.2.2 *Reactions to the Loss of Maritime Spaces*

A Protecting the Coasts

Physical and/or legal actions seeking to prevent de-territorialisation are a reflection of the tension between the need for minimum physical and economic viability of a political entity to continue to be considered a State, on the one hand; and the ethical, political and legal will to find avenues by which to stretch those minimum standards as far as possible, on the other hand. Once the maximum extension of such minimum standards is achieved, finding alternative approaches to statehood – that is, not only looking for ways of preserving statehood, but also exploring other forms of international subjectivity – may be required. When dealing with the partial de-territorialisation scenario described above, measures seeking either to prevent or to react against the effects of sea-level rise and coastal erosion on Pacific Island States' maritime entitlements are mixed in nature: whilst physical actions respond to the evolution of the geographical transformation of the coastline, legal proposals seek to preserve the maritime rights of Pacific Island States in spite of the geographical transformation of their coastlines.

There is a wide range of technological tools and socio-economic policies which can be implemented for coastal protection or shoreline preservation without necessarily affecting the consideration of the features as islands under Article 121(1) of UNCLOS. As recalled by the ILA Committee on Baselines under the International Law of the Sea, in practice, permanent structures constructed on the coast with a view to protecting the coast from erosion, have been assimilated to 'harbour works' (Article 11 of UNCLOS) and thus considered to be part of the coast.[258] Similarly, permanent structures that seek land reclamation (that is, an artificial extension of the baseline), as erected by The Netherlands, United Arab Emirates, Singapore and Japan, "seem to receive a similar treatment".[259] Conceptually, coastal protection or extension devices can be considered as adaptive measures to minimise the adverse impacts of climate change on Pacific Island States' maritime entitlements. They can be divided into two main types.

The category of technological tools includes soft engineering measures, such as beach replenishment and artificial enhancement or elevation of beaches, proposed since 1990 by the IPCC Coastal Management Subgroup as Strategies

---

258 ILA Committee on Baselines, *Final Report*, 26–27.

259 Ibid., 28.

for Adaptation to Sea-Level Rise.[260] They are followed by more aggressive engineering measures, such as the construction of sea walls, as effectively put in place in the Maldives around its capital atoll Malé, or even artificial platforms within the State's EEZ, as developed by Japan to prevent Okinotorishima from becoming submerged.[261] These hard engineering measures have nonetheless been criticised for various reasons. As pointed out by Grote Stoutenburg, the construction of sea walls is a very limited solution, given that it cannot be extended to all the features which serve as base points and constitutes an "economically inefficient behaviour".[262] She also highlights how this solution may be counterproductive both from a socio-economic and an environmental perspective. On the one hand, sea walls can lead to the erosion of beaches and other sediments. Thus, while stabilising the high-water mark, they may foster the landward displacement of the low-water mark – determining the baselines – to the foot of the sea wall. On the other hand, the construction of sea walls on larger islands may affect their appeal as tourist destinations, an economic sector which Tuvalu, for instance, has been particularly keen to develop in the last decade. Moreover, other than the negative side-effect on the economic development of an island subject to the construction of a sea wall, an important decrease in the economic viability of the feature could arguably be used against the interested State to call into question the continued inclusion of the feature in the island category (as defined by Article 121(1) of UNCLOS), and to argue that it should be reclassified as either a "rock which cannot sustain human habitation or economic life of its own" (Article 121(3) of UNCLOS) or as an artificial island (Article 60 of UNCLOS).[263]

Considering the high costs generated by such technological tools, as well as the potential side-effects they may generate from an environmental, economic and legal perspective, the development of wider socio-economic frameworks appears all the more fundamental. As indicated by Charles di Leva and Sachito

260 IPCC Coastal Zone Management Subgroup, Job Dronkers et al., *Strategies for Adaptation to Sea Level Rise* (Geneva: Intergovernmental Panel on Climate Change, 1990), 1–148.

261 Article 60 of UNCLOS indicates that these do not generate any maritime zones. Artificial platforms can be more controversial. See Charles Di Leva and Sachito Morita, 'Maritime Rights of Coastal States and Climate Change: Should States Adapt to Submerged Boundaries?', *World Bank, Law and Development Working Paper Series*, nº 5 (2008), 1–52.

262 Grote Stoutenburg, 277.

263 This is the position of, for instance, Van Dyke on Okinotorishima, who defends it for an artificial island when total disappearance of the island is possible. This however can be fought from a legal perspective by endorsing the narrower interpretations of Article 121(1) of UNCLOS.

Morita, adaptation policies may thus include early warning mechanisms and the introduction of coastal zoning and risk-management plans into national development plans and into appropriate national climate change adaptation plans (NAPAs). It may even be advocated that more radical changes in the economic development models of these island States should be pursued, as recently implemented by the Maldives which has attracted green technology investment and aspires to become the first carbon-neutral economy of the planet.

B 'Freezing' the Baselines

Considering the economic burden that the physical and policy devices depicted above constitute for underdeveloped and mostly agricultural countries, such as the Pacific Island States, scholars concerned with the potential loss of maritime spaces have started to advocate different legal solutions, which could be raised by the concerned States, as a possible complement to the artificial means of shoreline protection and irrespective of the cost and difficulty of their implementation. It is generally admitted that the *existing* law on baselines – which, as already mentioned, conceives baselines as being ambulatory – does not offer an adequate solution to prevent or halt the potential loss of maritime spaces due to sea-level rise.[264] Therefore, arguments seeking to offer legal solutions that may protect the maritime spaces of the affected States mainly focus on ways to develop *new law* or interpret grey areas of the law currently in force – an endeavour which comes into play to seemingly serve the interests of small island States with regard to vast areas and resources.

The legal protection of Pacific Islands' maritime spaces could first be fostered, Alfred Soons suggests, through the creation of a new rule of customary international law allowing low-lying coastal States to maintain the original outer maritime limits of their zones.[265] This strategy has nevertheless the inconvenience of requiring the approval of the relevant international fora. Even assuming that no general objection to this practice would be raised – for instance, by States interested in an enlargement of the scope of the high seas in the South Pacific region, the lengthy process of creation of customary rules of international law radically contrasts with the urgency and the gravity of the situation in which low-lying Pacific Island States *already* find themselves. Solutions that permit argument more imminently in favour of low-lying island States are therefore necessary.

---

264 ILA Committee on Baselines, *Final Report*, 30.

265 Soons, 'Effects of the Rising Sea-Level on Maritime Limits and Boundaries', 231.

A second option thus involves the conclusion of maritime delimitation agreements, a strategy that enables the States concerned to take quick action, lead the negotiations, and even maintain a higher degree of control over the relevant law-making process than what a customary rule would ever allow. As previously seen, in section 1.2.1.B. above, in recent years this form of State practice has been highly developed in the region. Indeed, table 3 above shows that, whilst the first twelve maritime agreements were concluded over a period of 25 years – between 1978 and 2003–, in the last eight years the number has almost doubled, as twelve new maritime delimitation agreements (including the 2014 Agreed Minutes between Fiji, Wallis and Futuna (France) and Tuvalu) were adopted between 2006 and 2014. In cases where a maritime delimitation agreement has been signed, the most immediate legal device to ensure the final preservation of the maritime spaces of the Parties involved is the ratification of a treaty by all Parties so that it can enter into force.

This is particularly important in cases where the Parties involved are Kiribati, Tuvalu or the Marshall Islands – three countries composed exclusively of coral atolls, since the base points used to trace the boundary are most likely to be highly vulnerable to climate change impacts and susceptible to disappearance or to severe coastal recession. The signature and adoption of the seven maritime agreements between Kiribati and (1) Nauru, (2) Cook Islands (New Zealand), (3) Niue (New Zealand), (4) Tokelau (New Zealand), and (5) Palmyra Atoll, Kingman Reef, Jarvis Island and Baker Island (USA); (6) Marshall Islands; (7) one trilateral Treaty involving Kiribati, Nauru and the Marshall Islands; (8) one bilateral agreement between Nauru and the Marshall Islands; and two bilateral agreements between (9) the Federated States of Micronesia and the Marshall Islands, as well as (10) the Federated States of Micronesia and Guam (USA), already generate the obligation of the Parties involved to refrain from acts which may defeat the object and purpose of the treaties, in accordance with Article 18 of the VCLT. Their final consent to the obligations set out in the treaties involved would make a very important difference, as it would arguably 'freeze' or fix the baselines and, consequently, the maritime boundaries of the respective States.

The third means of protection, in the absence of a maritime delimitation agreement setting the boundary between two facing or adjacent States, is the publication, as soon as possible, of the location of the baselines in official charts, to then refrain from updating the data in new charts in which coastal geographical changes due to sea-level rise would be incorporated. This strategy, referred to by Grote Stoutenburg as 'masterly inactivity', would arguably operate as a *de facto* stabilisation of the original location of the baselines, and the scope of the maritime zones of the State concerned.[266]

The weakness of this strategy therefore lies in the fact that it does not preclude other States from challenging the baselines depicted in the charts for not reflecting the real coastal geographical situation. Indeed, in two recent inter-State disputes, the question of whether the chartered baselines should prevail over the actual or geographical baseline was raised,[267] prompting the creation of the ILA Committee on Baselines under the International Law of the Law of the Sea.[268] After extensive deliberations, the Committee considered, first of all, that the 'legal baseline' is the actual or geographical baseline, *as indicated on the official charts recognised by the coastal State*. It then recognised that, while charts may be challenged, "there is a presumption in favour of the chartered low-water line", and that further clarification is needed regarding the strength of the presumption in favour of the charted line and the evidence that may be necessary to effectively challenge such presumption.[269] Truly enough, giving relative preference to the 'charted baseline' over the 'actual baseline' implies that, what begins with the legal depiction of the geographical characteristics of a coastal State, released in an official format and important for the safety of navigation, may eventually become an artificial legal construct which sustains a factual reality that may no longer match the real state of the coastal area concerned. Such a legal fiction is, in addition, called to operate to favour 'weak' States in a context where vast areas and resources are in question. Yet, if such legal construct had a chance to become operative, it could at least give the affected States *temporary protection*, effective as long as their official charts are not appropriately challenged, and that the axiom that the 'land dominates the sea' is not overruled.

Finally, another politico-legal solution, proposed for instance by Moritaka Hayashi and Jenny Grote Stoutenburg, involves a higher level of international co-operation.[270] These authors suggest the development of collective

266 Grote Stoutenburg, 279.

267 *Nicaragua v. Honduras* (ICJ, 2007); and *Guyana v. Suriname*, UN Law of the Sea Annex VII Arbitration Tribunal, 17 September 2007, 47 ILM 166 (2008).

268 ILA, *Proposal for the Establishment of a New Committee on Baselines* (ILA: 2008). See also, ILA Committee on Baselines under the International Law of the Sea, *Draft Report* (Internal Document), 2010, pp. 1–28.

269 Ibid., Draft Report, 28.

270 Moritaka Hayashi, 'Sea-Level Rise and the Law of the Sea: Legal and Policy Options', Proceedings of the International Symposium on Islands and Oceans, *Ocean Policy and Research Foundation*, 2009, 78–90, and from the same author, who developed further this analysis, see 'Sea-level Rise and the Law of the Sea: Future Opinions', in Davor Vidas and Peter J. Schei, ed., *The World Ocean in Globalisation: Climate Change, Sustainable Fisheries,*

implementation mechanisms of a new regime of stable maritime zones, through either a UN General Assembly resolution on stable maritime zones, or the adoption of an Implementation Agreement on Sea-Level Rise which would provide for stable maritime zones and could complement the 1994 Implementation Agreement of UNCLOS.[271] While this proposal inconveniently – but necessarily – relies on gathering sufficient political support, which may take a long time, it would certainly be beneficial to Pacific Island States that do not have official charts showing the outer boundaries of their archipelagic lines, like the Marshall Islands, granting them a *de iure* protection of their baselines – while ratification of the set of treaties remains pending.

1.3 Total De-Territorialisation: From State Continuity to New Forms of Subjectivity

While the effects of sea-level rise and coastal erosion on Pacific Island States' maritime dimension constitute the most immediate territorial issue they have to face, what the region genuinely fears is the prospect of a total territorial loss. If partial de-territorialisation – when climate change effects provoke the retreat of the shorelines and may even submerge entire topographical features – affects the terrestrial and maritime spatial dimension of the twelve independent States of the region and can potentially occur in the Caribbean and the Indian Ocean as well, the prospect of total de-territorialisation affects (at least for the time being) a more restricted number of States. Of the approximately 195 States that currently exist, only four are undeniably under the threat of total loss of their territory, for they consist mostly of low-lying features with an altitude of around 5 metres. Three of these States are in the Pacific – the Marshall Islands, Kiribati and Tuvalu; the fourth is the Maldives in the Indian Ocean. The prospect of these four cases of total territorial disappearance brings back to the fore the central question of the role of the territory in international law (dealt with in section 1.1. above).

The opening section of this chapter explored the role of the concept of territory in the *creation of States* and stressed its contingent character by differentiating the historical context of the modern European State from that of the post-colonial Pacific Island State. Three main legal theories on the territory, all stemming from the concept of the modern European State,

---

Biodiversity, Shipping, Regional Issues (Leiden: Nijhoff, 2011), 187–206. See also, Grote Stoutenburg, 293–295.

271 By extension or *lex lata* of Article 76(9) of UNCLOS on the permanent outer limits of the continental shelf.

were analysed to explain the specific role of territory in the Pacific following decolonisation. In contrast, the present section deals with the role of territory in the *continuity of the State* by questioning whether the total disappearance of the spatial dimension of a State necessarily implies the loss of statehood. Bearing in mind that the natural bias of the international legal order towards stability and order generally favours the maintenance of existing structures, the functions of, and legal theories on, territory attached to the creation of States are not automatically mirrored in the consideration of the place of territory in the continuity of States.

Thereby, what may appear as farfetched proposals seeking to defend the continuity of Pacific Island States merits thorough consideration. These solutions can be divided into two main types. First of all, scholars have pointed to strategies seeking to maintain and even produce some sort of territorial dimension of the State, which have referred to as 're-territorialisation strategies', and generally involve the use of territory of other States under different legal forms. Secondly, considering the impossible maintenance of a State's spatial dimension, *ultima ratio* proposals advocate the continuity of de-territorialised 'non-State sovereign entities', and generally refer to the Holy See and the Sovereign Military Hospitalier Order of St. John of Jerusalem of Rhodes and of Malta (hereafter referred to as the Order of Malta) as 'precedents' of these exceptional cases. Other than considering and balancing the merits of these two types of solutions from the standpoint of their feasibility, the present chapter argues that they must be approached by the functionalist approach as applied to the role of territory in the creation of States.

The fundamental question is not *how* Kiribati, Tuvalu, the Marshall Islands may preserve – in theory and in practice – their international legal personality despite having been deprived of their spatial dimension, but *why* should their continuity be assured in these extreme circumstances. Which specific purpose(s) would be fulfilled, should the survival of these political entities either as States, as non-State sovereign entities, or as any other innovative form of international subjectivity, be acknowledged by the international community? At the heart of this issue lies one of the most long-standing riddles of modern political theory, namely, the question: What is the function of the State?

1.3.1 *Applying New Strategies: 'Re-Territorialisation'*

A Cession or Purchase of Land from another State

Alongside and complementary to measures already mentioned for shoreline protection, or construction of artificial platforms, the first type of scholarly proposals for preserving and producing the spatial dimension of the threatened States includes either acquisition of new land from, or a merger with, another State. Consistent with the influence of the Roman law tradition, the cession or the transfer of a part of the territory between States (by a treaty provision) is among the modes of acquisition and transfer of territory traditionally recognised as lawful under international law.[272] The Pacific region is not unfamiliar with either of these two modalities.

Wishing to ensure its presence in the area, Prussia, in 1898, purchased the Marianas and Caroline Islands from Spain, thereby lowering the boom on a massive colonial empire that had lasted for over three centuries.[273] A second interesting precedent in the region involved the Fijian island of Rabi, purchased in 1941 by the British Phosphate Commission (BPC) for the purpose of relocating the inhabitants of the island of Banaba (formerly known as Ocean Island, today part of Kiribati) after the discovery of phosphate resources there. During World War II, the Japanese forces in the Pacific occupied the phosphate islands of Banaba and Nauru and deported their inhabitants to other islands of Micronesia. When the Japanese surrendered to Australian troops, the 'accident of war' which had already made the Banabans a displaced people was soon seen by the BPC as a golden occasion to settle them in a different location. Hence, the Banabans were told that the island had become uninhabitable because of war damage and were collected from different prisoner-of-war camps in Micronesia, reassembled in Tarawa, and quickly transferred to Rabi (Fiji). They were forcibly relocated under conditions which then proved to be flawed, including the condition that the relocation was for a period of only two years, at the end of which anyone wishing to return to Banaba would have their transport arranged at the expense of the Gilbert and Ellice Islands Colony Government. Most importantly, they were told that the resettlement would not impair their land rights in Banaba, nor their right to access to any of the Banaban Funds. The inadequate settlement of roughly 4,000 Banabans in Rabi made them suffer from, inter alia, extreme lack of adequate shelter and severe undernourishment. Besides, following the respective independence of Fiji (1970) and Kiribati (1978), the legal status of the Banabans in both countries was unclear, leading to a situation of de facto statelessness. In Fiji, Banabans of Rabi were equated to Fijian Indians, until, in 2005, a 3-month period was opened for Banabans willing to gain citizenship by naturalisation, to benefit from a waiver of the usual conditions and fees. In Kiribati, a seat is specifically

272 Jennings, 6.

273 Today, the Marianas have been integrated into the USA, while the Caroline Islands are part of the Federated States of Micronesia.

reserved in the Parliament for a Banaban representative. Curiously enough, the original inhabitants of Rabi had previously been relocated to another Fijian island (Taveuni) in the mid-19th century. Their descendants claim the right to return to their island, threatening the Banabans again with the prospect of resettlement.[274]

While the first case involved the transmission of a colonial territory from one State (Spain) to another (Prussia), the second case was more complex in that it involved the three partner governments of BPC (Australia, New Zealand and Britain) purchasing an island part of Fiji at a time when Fiji was not yet an independent entity, but was still part of the British Empire. The precedents of purchase of land in the region took place within an overall context of colonial exploitation of natural resources and preceded the independence of the relevant political entities.

Nowadays, the idea is back on track within a different political context. The possibility of purchasing land from another State has been suggested as a means to compensate for the territorial loss of low-lying island nations threatened by sea-level rise, and thus indirectly help ensure the continuity of the territorial substratum of their statehood. First formulated in 2008 by the former president of the Maldives, Mohammed Nasheed, the idea was announced immediately after his election. The prospective creation of a Sovereign Wealth Fund in the Indian Ocean Island State, a sort of ultimate insurance policy sourced from tourist revenues, would finance the purchase of land abroad. Countries with similar cultural, religious traditions and meteorological conditions, such as India and Sri Lanka, were first suggested and appeared in the media as the possible future lands on the most paradigmatic face of the global environmental disaster. The term 'climate refugees' was coined, updating the early interest that UNEP had shown in the issue of environmental displacements,[275] and informal talks were also said to have taken place between the Maldives and the governments of Australia and New Zealand.[276] Such a fascinating and

274 See Stacey King and Raobeia K. Sigrah, 'Te rii ni Banaba', 2001, at: http://www.banaba.com (15 September 2015).

275 In 1988, the United Nations Environment Programme released a first report on environmental refugees, at the time when the precedent on international environmental security was being dealt with in the United Nations. Essam El-Hinnawi, *Environmental Refugees* (Nairobi: UNEP, 1988).

276 See for instance Ben Doherty, 'Climate Change Castaways Consider Moving to Australia', *Sydney Morning Herald Tribune*, 7 January 2012; and, from the same newspaper, Brian Merchant, 'To Escape Rising Seas, Maldives President May Move his Entire Island Nation to Australia', 6 January 2012.

appealing policy was one of the elements of a wider strategy put to international organisations, aiming at the introduction of climate change into the agendas of human rights institutions and, vice versa, of promoting a human rights approach into climate change negotiations. Yet, as the Australian government denied the existence of any sort of informal agreement, and the Sovereign Wealth Fund was yet to be settled two years after the idea was first formulated, doubts about the real intentions of ex-president Nasheed on this matter started to flourish.[277] It was later confirmed that, instead of pursuing this road, the Maldives had given preference to a new strategy, consisting of becoming the first totally carbon-neutral country on the planet.[278]

Notwithstanding this change of policy in this Indian Ocean country, the idea has been mirrored in the Pacific Ocean. Very recently, Kiribati concluded the purchase in Fiji of 2000 hectares of land from the Church of England for 8.77 million USD, with funds from Kiribati's Revenue Equalisation Reserve Fund, set up in 1956 from phosphate revenues. The land in question, covered by dense forest, is located in Viti Levu, Fiji's main island, about 2000 km away from Kiribati. Although so far Kiribati's President, Antote Tong, noted that the purpose of the purchase would be to develop coastal agriculture and thus enhance food security for Kiribati's people, the possibility that it may become a relocation site for the 103,000 inhabitants of Kiribati has been openly discussed.[279] In fact, a few months prior to the purchase, the President of Fiji, Ratu Epeli Nailatikau, publicly expressed the support of his country to the people of Kiribati, who "would be able to migrate to Fiji with dignity if need

277 See for instance, Ashby Monk, 'Did the Maldives Pretend to Set Up a Sovereign Wealth Fund?' 20 February 2010, available at: <http://oxfordswfproject.com/2010/01/20/did-the-maldives-pretend-to-set-up-a-swf/> (1 October 2014), citing an interview with President Nasheed in which he vaguely stated: "[T]he fund is now formulated. We will have to save for a rainy day. And during the worst-case scenarios, as responsible politicians, we should be able to tap funds and money set aside for a rainy day. So the fund is going on, and hopefully we will have something when the going gets very bad".

278 Costing the Earth – TheMovieChannel.com., 22 March 2009, available at: <http://www.emailwire.com/release/20736-Costing-the-Earth-TheMoveChannelcom.html> (1 October 2014): "[A]s the Maldivian economy relies so heavily on tourism. [...] The Government is also hoping that the carbon-neutral promise will attract more environmentally conscious tourists. The Government is working with climate energy experts to replace fossil fuels with renewable energy sources".

279 Laurence Caramel, 'Besieged by the Rising Tides of Climate Change, Kiribati Buys Land in Fiji, The Guardian', 1 July 2014, available at: http://www.theguardian.com/environment/2014/jul/01/kiribati-climate change-fiji-vanua-levu (1 October 2014).

arose".[280] Referring to the presence of I-Kiribati population in Banaba, the President also stated that "these people now live in Fiji but have their own seat in the parliament of Kiribati and, if necessary, we will do it again".[281]

The two parties involved in this commercial transaction are the State of Kiribati on the one hand, acting in its public capacity (i.e. as a sovereign State), and the Church of England, which has the status of a private personality. Thus, in principle, the purchase of the land would not involve the States of Kiribati and Fiji as such. Since the land remains under Fijian sovereignty, the laws applicable in that area would remain the laws of Fiji, ruling both the public and private spheres of life (criminal law, property law, family law, etc.). This situation could eventually be transformed, should there be a special agreement between Fiji and Kiribati, in which Fiji would delegate the exercise of some competences to the State of Kiribati (e.g. matters of family or criminal law, police, commercial matters resolved by customary rules, etc.). Micro-States around the world provide particularly illustrative examples of delegation of competences. Saving the differences with Kiribati's present situation, The Vatican State has, for instance, delegated the exercise of police security to Italy, by virtue of the Lateran Agreement, just as the majority of Pacific Island States have done with Australia, New Zealand, the United States and the United Kingdom. Such delegation of competences implies that the authorities of a foreign country may act and intervene lawfully within the territorial boundaries of a State to restore or maintain order, though in no case is it tantamount to a delegation of sovereignty as such. Moreover, the agreements not involving a delegation of competences may simply recognise immunity from execution or even of jurisdiction over the goods and properties of the land in question.

In either of these scenarios, 're-territorialisation' strategies through purchase of land seems to be a limited solution to defend the continuity of a State in a total de-territorialisation scenario, subject in any case to the will of the States involved in possible transactions. Although Fiji could theoretically renounce its territorial sovereignty over the 2,000 hectares of land in Viti Levu purchased by Kiribati, it is difficult to see that this will be crystallised in practice, particularly given that Viti Levu is the main island of the Fijian archipelago. The same may be said about cases of cession of territorial sovereignty by one State to another, which are theoretically possible but remain rather unlikely in practice (particularly in post-colonial States which have very recently acquired

280 'Fiji Supports Kiribati on Sea-Level Rise', Press Release SUVA (Fiji), 11 February 2014, available at: http://www.climate.gov.ki/tag/government-of-kiribati/ (1 October 2014).

281 Ibid.

independence and already have a very limited territory). It may, nonetheless, constitute a temporary solution for a partial de-territorialisation scenario – as is currently developing in Kiribati, Tuvalu and the Marshall Islands. Limited population flows from geographical features most at risk could be relocated in the purchased Fijian lands (under Fijian sovereignty).[282] Rosemary Rayfuse has suggested that in their struggle for survival, the country of origin of the displaced people puts in place the necessary mechanisms to 'freeze' their original maritime entitlements (by entry into force of the maritime delimitation agreements signed, or by fixing their baselines through the drawing of archipelagic baselines), so that the relocated populations can at least benefit from the exploitation of the marine resources (see section 1.3.4.B below on the recognition of 'water States').[283]

B Merger with Another State

Given the practical and legal difficulties associated with the idea of re-territorialising disappearing States through the purchase or cession of land, some authors have also raised the possibility of merging two or more States. Putting aside the practical difficulties of making such a proposal a reality, the main remaining (theoretical) difficulty arising out of such a possibility depends on the form of political organisation to be given to the new resultant political entity. If, on the one hand, the chosen form of merger is a federation or free association, the continuity of the State undergoing severe de-territorialisation ought to be admissible (like the Cook Islands, having the status of a self-governing territory of New Zealand). This may not constitute a problem as long as the low-lying State remains partly de-territorialised, but it may be a difficult position to sustain once the complete submersion of the land occurs.

Only two legal fictions could sustain these forms of merger in a total de-territorialisation scenario: (1) acknowledgement that one of the merged States comprises exclusively maritime territory or is a 'water State' (an option that would remain subject to the prior defence of the permanence of their original maritime entitlements – which would turn the axiom of the present law of the sea – 'the land dominates the sea' – meaningless; or (2) acknowledgement that the statehood of the submerged State continues in spite of the inexistence of

282 It is important to consider here only displacements of a limited number of people affected, for, as will be seen in chapter 2 below, total de-population scenarios can arguably have the same effects as total de-territorialisation scenarios.

283 Rosemary Rayfuse, 'International Law and Disappearing States: Utilising Maritime Entitlements to Overcome the Statehood Dilemma', 1–13.

any sort of spatial dimension. If neither of these legal forms is acknowledged, the merger of the States would necessarily take the form of an integration, whereby the international legal personality of one of them (presumably the State most at risk of disappearing) is subsumed in that of the other State and therefore disappears in its former independent version.

1.3.2 *Coining a New Category: De-Territorialised Pacific Island States*

A The Holy See and the Order of Malta: Existing Forms of 'Non-State Sovereign Entities'

As explained by Sir Robert Jennings and recalled by Judge Huber, while it is widely admitted that territorial sovereignty subsists even when divorced from possession, "excursions into the realm of an abstract title to sovereignty have been cautious and tentative."[284] There are, nonetheless, two prominent – but very particular – exceptions to this principled rule, whereby sovereign entities with international legal personality have existed (and subsisted) in spite of having virtually lost a territory specifically ascribed to the exercise of their sovereign powers. These are good examples of Brierly's remark:

> [T]here are other elements of international law that mean these specific criteria [set up in Article 1 of the Montevideo Convention] need not be slavishly followed. It was these "other elements" that determined the inclusion of both the Holy See and the Order of Malta among the sovereign subjects of international law.[285]

The Holy See, supreme organ of the Catholic Church, headed by the Pope according to the Code of Canon Law, and composed of the College of Cardinals governing the Church,[286] is a non-territorial religious entity. In spite of this characteristic, inherent in its temporal nature and functions, the Holy See has known different moments in its history since the fall of the Roman Empire

---

284 Jennings (quoting Judge Huber), 5.

285 Brierly, 150.

286 Holy See is a designation under Canon Law for the highest offices of the Church. The concept finds its roots in apostolic succession according to which an unbroken chain is formed of representatives of the Christ from the Apostle St Peter to the present day. See Josef L. Kunz, 'The Status of the Holy See in International Law', *American Journal of International Law*, vol. 46 (1952), 308–314; and Bo J. Theutenberg, 'The Holy See, the Order of Malta and International Law', 2003, 1–18, 12: http://www.theutenberg.se/pdf/the_holy_see_the_order_of_malta_and_international_law.pdf (1 October 2014).

in which it had a territorial basis. This situation has proven to be a fruitful source of confusion about its State or non-State nature since the inception of the modern European State, and of uncertainty, in the 20th century, about the eligibility of the Holy See for membership of the United Nations. The Holy See was first associated with the territorial extent of the Papal States, which had been created in the 9th century by Pépin le Bref and Charlemagne (in the context of the Holy Roman Empire). Such territorial connection vanished in 1870 with the annexation by Italian troops of the Papal States, whereby the Apostolic Palaces were granted freedom. In spite of the annexation, the Holy See pursued its international political influence and activity, maintaining its diplomatic presence in many States, and concluding treaties. This ambiguous situation lasted throughout the 19th century and even through World War I, until, in 1929, the Lateran Agreement concluded between Italy and the Holy See established that the State of the Vatican City would fall under the sovereignty of the Supreme Pontiff, while Rome would remain under the jurisdiction of Italy.[287] Although the Holy See reacquired territorial visibility through the institution of the Vatican City, neither of the two entities has been considered a State. The Vatican City does not exist to support, maximise the welfare or ensure the security of the Holy See, nor does it respond to a legitimate right to self-determination of its 800 residents and 400 citizens (the majority of which are church officials on a non-permanent basis). Italy provides a police force to patrol the Vatican City, administers punishment of crimes within the City, maintains the water and railways systems, and ensures freedom of communications and transportation. Therefore, the Vatican City virtually acts as a 'vassal State', serving as the base for the administration of the Roman Catholic Church. To some extent, it also operates as an artificial construct that provides the Holy See with some sort of 'claim to territorial integrity', which was relevant later on to the determination of the form of the Holy See's participation in the United Nations, for which it was eligible for. After the initial tentative in 1944 to present the application of the State of the Vatican City for full membership of the United Nations, which failed essentially because of the principled policy of neutrality provided in Article 24 of the Lateran Treaty, the Holy See – and not the Vatican City – under Pope Paul VI, established, in 1964, a 'permanent observer' mission to the United Nations with offices in the UN headquarters.

287 Roberto Ago, 'Occupazione Bellica dell' Italia e Trattato Lateranense,' *Comunicazioni e Studi*, (1946), 129ff.

The case of the Order of Malta differs from that of the Holy See, for it has not until now regained any territorial basis, such as that which the Vatican City to some extent provides for the Holy See.[288] Currently with 11,000 members, the Order of Malta has its headquarters in Via Condotti, in Rome. Therefore, in this sense, it is perhaps the best example of a contemporary de-territorialised sovereign entity. The origins of the Order date back to the 11th century when Blessed Gerard founded St John's Hospital along with other fellow Dominican monks. The Hospital was annexed to the Monastery of St John the Baptist of Jerusalem and was built up to look after the dead and wounded of the Crusades that were taking place between Christians and Muslims. The reputation of St. John's Hospital soon attracted the attention of the Pope, to the extent that, in 1113, through the *Bull Pie postulatis voluntatis*, the group headed by Blessed Gerard received official recognition as Christianity's oldest religious Order, a status which granted the Order of Malta a set of privileges, including the right to appoint its own Grand Master. After Jerusalem surrendered to the forces of Sultan Saladin in 1187, the Order moved to Cyprus (1291–1309), then to Rhodes (1309–1522) and finally to Malta, where its dominion lasted for over three centuries, until 12 June 1798, when Malta surrendered to the Napoleonic forces on board the French vessel L'Orient. During this long period (1530–1798) Cyprus, Rhodes and Malta provided the Order with a territorial basis, which was successively granted, as vassal States to the Order, in a political scheme akin to that of the princes or bishops of the time who were united under the Holy Roman Empire. Hence, for the five centuries during which the Order had a territorial basis, its sovereignty over such spaces was recognised by its contemporary international actors. Such historical foundations were so deeply rooted that even after the loss of Malta, its last territorial remnant, the Order was de facto treated as a *sui generis* non-State sovereign entity with international legal personality.

B Towards the Recognition of 'Water States'?

Drawing from these two historical examples, doctrinal studies started tentatively to suggest that the same positive fate could be endured by submerged island States. The examples of the Holy See and the Order of Malta are helpful in the sense that they point to the importance of historical contingency and

288 See Olivier Farran, 'The Sovereign Order of Malta in International Law', *International and Comparative Law Quarterly*, vol. 2 (1954), 217–234. For a more recent account, see Karol Karski, 'The International Legal Status of the Sovereign Military Hospitaller Order of St John of Jerusalem, of Rhodes and of Malta', *International Community Law Review*, vol. 14 (2012): 1, 19–32.

context as factors which may render possible the recognition of unusual forms of sovereignty not ascribed to a territorial State entity. To be sure, their peculiar status in today's international legal and political system is undoubtedly the result of the special bond of both institutions with the long process of formation of the modern State and of the early European Law of Nations. While in the 20th century international law has undergone a process of 'universalisation', its current status may be seen as a reminder of both the European origins of the international legal order and even as a reminder of Eurocentricism. Henceforth, simply recalling these two examples to support the cause of de-territorialised Pacific Island States and arguing that they may survive as non-territorial sovereign entities is not, as such, particularly useful nor enlightening, since the reasons why the Holy See and the Order of Malta are recognised today as special forms of non-State sovereign entities cannot be applied to the case of Pacific Island States.

Yet, they do help in one sense. In order to counter the historical weight represented by the two precedents of non-territorial sovereign entities and make of their example a useful tool for supporting the continuity of Pacific Island States, it is essential to emphasise that international law accepts and accommodates forms of functional non-territorial sovereignty. Stressing this point, Rosemary Rayfuse and Emily Crawford have considered that if it is possible to conceive of governments in exile, it may be possible to conceive of 'States-in-exile' – understood as a *theoretical construct* without dominion over their original territory, whose lost was the result of physical degradation (as opposed to political change).[289] Following similar lines, Maxine Burkett's thought-provoking article argues that Pacific Island States could survive as 'nations ex situ',[290] which essentially calls for putting greater emphasis on preserving as much as possible all the core or constitutive elements of the 'nation' composing, in particular, their cultural identity – so that it endures the challenge of de-territorialisation. Under this status, the newly-coined category of international actors would benefit from the continuity of their sovereign rights and benefits. In practice, it would require "the creation of a government framework that could exercise authority over a diffuse people."[291] This could be operated, for instance, through the revival of a system akin to the precedent of political trusteeship, although adapted as in the cases of ex-situ nations, Burkett stresses, elected citizens of the nation ex-situ would serve as trustees, so that

289 Emily Crawford and Rosemary Rayfuse, 'Climate Change and Statehood', 243–253, 213.

290 Maxine Burkett, 'The Nation Ex-Situ: on Climate Change, De-territorialised Nationhood and the Post-Climate Era', 345–374.

291 Ibid., 346.

the principles of self-governance and self-determination are preserved. This transition, she explains, could be overviewed by a United Nations office specifically established for this purpose, and complemented with dual citizenship for the displaced population – the nationality of their ex-situ nation, in addition to that of their new 'host' State. Another form of governmental framework, Rayfuse suggests, could be the development of co-operation and co-ordination of the States involved with regional fisheries management organisations, the International Seabed Authority, the International Maritime Organisation, and other relevant organisations. If a new authority were to be created, or the revival of the political trusteeship effected, it would be entitled to hold and administer the maritime resources, which could serve as 'resource rents' funding the relocation and livelihood of the displaced population in the new host State.[292]

The legitimacy and operation of 'nations ex-situ' or 'water States' and the resolution of the uncertain future will undoubtedly depend on the recognition by the international community, and its willingness to accept new possible categories of international actors. As Derek Wong recalls, recognition may be manifested either through the unilateral acceptance of 'great powers', or by the conservation of the de-territorialised State of their UN membership.[293] The development of a firmly grounded ethical and political argument would therefore be of crucial importance for gathering the necessary international political support. And so at this point, the interrelations between the territorial and the human dimensions of the State surface.

1.4 Conclusions

Just as the nature of the State – privileged habitat of the territory – lies outside the realm of law, the place of territory in international law is equally unsettled and its nature doubtful from a legal perspective. One way of explaining the different legal conceptions of territory is by approaching it from a historical perspective and drawing from such a context the various functions that the territory has had in the international legal order. In the context of the modern European State, the territory was understood as a property of the State (property theory), an element or integral part of the State itself (constitutive theory), or the space wherein the competences of the State were displayed (competence theory). A fourth understanding of the territory concept emerges from

292 Rayfuse, 'International Law and Disappearing States', 12.

293 Derek Wong, 'Sovereignty Sunk?, 381–383.

the role it plays in Pacific Island States for which the territory is an expression or manifestation of the right of Pacific Islanders to self-determination and political independence from former colonial domination. This particular meaning, which the concept of territory has for post-colonial Pacific Island States, proves that a solid normative ground may balance and even supersede the viability requirement, so that a very limited territory may be able to provide the necessary grounds for a political entity to acquire full-fledged statehood irrespective of its real capacities. Moreover, Pacific Island States' spatial dimension is not as small as it generally seems at first sight. In fact, the small extent of their land is largely compensated by the vast maritime spaces constituting, to the outer limit of the territorial sea, part of their state territory under the Law of the Sea – and which determines both their major spatial expression and their principal source of economic revenues. Paradoxically, for these 'Ocean States', the sea is as much the fundamental element of their identity and the primary source of their livelihoods – as it is also the origin of the threat to their continuity as States.

Henceforth, de-territorialisation due to the adverse impacts of climate change on Pacific Island States constitutes, first and foremost, a challenge to these States' maritime entitlements. As their shorelines progressively retreat as a result of sea-level rise, baselines – generally considered as 'ambulatory' under the present Law of the Sea – retreat accordingly, and thus so do the outer maritime limits of various maritime zones of the State concerned. Devices seeking to halt this effect would include the ratification of maritime delimitation agreements among Pacific Island States and, arguably, the possibility to fix their baselines and to claim archipelagic status by drawing archipelagic lines and then refraining from updating the official charts. Physical devices, such as the construction of sea walls, have also been taken into consideration.

And yet, as efforts seeking to freeze, preserve or protect the spatial dimension of Pacific Island States against adverse climate change impacts flourish and develop in the region, the prospects of a total de-territorialisation scenario cannot be ignored or overlooked. Alternative response measures to this extreme example of climate change impacts include making incursions into both re-territorialisation strategies, such as the purchase of land or the eventual merger with another State and, most prominently, the possibility to argue that completely submerged Pacific Island States may continue to be recognised as a form of a non-State sovereign entity in the Anthropocene epoch. The cases of the Holy See and the Order of Malta, which currently have an atypical status, prove that there is room in international life for some flexibility in the forms of subjects allowed by the international legal system, as long as strong historical and/or political reasons support this special treatment. It is also a reminder that although sovereignty is the main and principal characteristic of a State, it can also exist outside a State framework.

If the territory is a fundamental – yet not always necessary – condition for the recognition of sovereign entities, the next question that arises is: on which grounds may the continuity of Pacific Islands' statehood or international legal subjectivity be defended, in spite of their prospective total submergence? This question invites turning attention towards the role played by the second dimension of a State, namely, the population, as the disappearance of the State as a sovereign political entity with international legal personality may result first from a State's total or acute de-population prior to total submergence. The next chapter therefore seeks to show why the continuity of the State or, at least, of its international legal personality, even in extreme cases of total territorial loss, may be justified as a means of fulfilling a specific function: a device of comprehensive, albeit unique, protection offered to Pacific Islanders.

CHAPTER 2

The Challenge of De-Population

It is our belief that Tuvalu, as a nation, has a right to exist forever. It is our basic human right. We are not contemplating migration. We are a proud nation with a unique culture which cannot be relocated elsewhere. We want to survive as a people and as a nation. We will survive. It is our fundamental right.

APISAI LELEMIA, Tuvalu's Prime Minister, Poznan (Poland), 2008.

• • •

We don't want to lose our dignity. We're sacrificing much being displaced, in any case. We don't want to lose that, whatever dignity is left. So the last thing we want to be called is 'refugee'. We're going to be given as a matter of right something that we deserve, because they've taken away what we have.

ANOTE TONG, President of Kiribati, Tarawa (Kiribati), 2009.

2.0 Introduction

Equally important as the spatial dimension, the State is defined by the presence of a population within its territorial boundaries.[294] To be sure, scholarly positions regarding the importance of a population for the construction of the State as a political and a cultural entity vary widely. Preference for the population as the central criterion of statehood tends indeed to be linked to the wider reaction against what Charles Rousseau named the *préjugé étatique*, that is, the traditional conception of the international legal order tending to consider the State and not the individual as the primary reference object and subject of international law. In contrast, State practice – particularly that derived from the proliferation of micro-States of less than one million inhabitants following decolonisation – is consistent and clearly indicates that quantitative demographical weakness is not an obstacle to the acquisition of statehood by

294 Montevideo Convention, Article 4(1).

 | DOI 10.1163/9789004321618_004

a political entity,[295] *as long as* the presence of the population within its territory is continuous. Nowadays, as sea-level rise and coastal erosion gradually increase, the living conditions of Pacific Islanders are jeopardised by the socio-economic consequences of the de-territorialisation described in chapter 1. Conceptually, the correlation between the territorial transformation of Pacific Island States resulting from climate change impacts, on the one hand, and the socio-economic conditions of their inhabitants, on the other hand, can be apprehended through the concept of *habitability* – that is, the capacity of a given territory to sustain human habitation and the economic life necessary for such purpose.[296] Since the habitability of a given space informs population behaviour and movements, such a concept is key to addressing the current issue of climate-induced displacement and relocation, also referred to by some international legal scholars as 'climate refugees'.

As will be developed in section 2.2, definitional constraints have been encountered at the onset of the debate on this issue.[297] They are due, primarily, to the fact that climate change tends to operate as a factor stressing pre-existing socio-economic and environmental conditions composing the core of a population's vulnerability. As a result, climate-induced displacement is informed by multiple causes; it takes place in complex socio-economic and political frameworks which often lead to the analytical and practical difficulty – if not impossibility – to disentangle the 'real' subjective motivation behind the observed displacements in and among Pacific Island States.[298] Given this obstacle, the terms 'displacement' and 'relocation' will be used indifferently in this chapter, and will be understood as *the fact* of moving to a different place within a recognised context of environmental stress, irrespective of the possibility to trace a clear causal link between climate change impacts and the subjective will of the people involved to relocate elsewhere. If anything, preference will be given to the use of the term 'relocation', as it invites taking into account the process of displacement as a whole – that is, also encompassing the process of integration of the displaced people into their *new* habitat.

---

295 James Crawford, *'State', Max Planck Encyclopedia of Public International Law* (Oxford University Press, online resource), last entry January 2011, para. 21.

296 Such a definition of 'habitability' is drawn from Article 121(3) of the United Nations Convention on the Law of the Sea, which establishes a different legal regime for features considered as 'rocks' instead of 'islands'.

297 On the difficulties of conceptualising climate-induced displacement (referred to by McAdam as 'climate change-related movement'), see Jane McAdam, *Climate Change, Forced Migration, and International Law*, chapter 1, 15–38.

298 Walter Kälin, 'Conceptualising Climate-Induced Displacement', in Jane McAdam, ed., *Climate Change and Displacement: Multidisciplinary Perspectives* (Oxford: Hart Publishing, 2010).

Controversies over the use of the term 'refugee' might arise at a later stage, as a legal category dealing with the status of the person in the new relocation site and outside the borders of the country of origin.

While climate-induced displacement emerges as a striking and paradigmatic image of today's global environmental crisis,[299] international awareness of the impact of environmental conditions on population displacements is not an entirely new phenomenon. Indeed, following the exponential normative and institutional development that international environmental law underwent since the 1972 Stockholm Conference on the Human Environment, the United Nations Environment Programme (UNEP) initiated the first study of this issue almost three decades ago. The first attempt to create a legal category of 'environmental refugees' thus dates back to the 1985 Report commonly referred to after the name of the study's Chief Director, Essam El-Hinnawi.[300] The end of the Cold War brought about a development of some aspects of the El-Hinnawi Report, particularly after the publication of the 1994 Report of the United Nations Development Programme on human security. This Report considered, for the first time, that the need for adequate preservation of environmental conditions constitutes one of the pillars of human security.[301]

Finally, from the late 1990s onwards, political action headed by the Maldives' delegation to the United Nations Human Rights Office, led to the unanimous adoption in 2007 of a new agenda item titled 'Human Rights and Climate Change'. Although the object of study has been narrowed down to displacement situations connected to one concrete dimension of the global environmental crisis – namely, climate change – the term 'climate-induced displacement' can still cover a wide range of situations. These include not just slow-onset climate change impacts, such as severe desertification, sea-level rise, or coastal erosion, but may also result from extreme weather events or climate-related environmental disasters such as severe river flooding, cyclones, typhoons, or tsunamis.[302] In this sense, it is important to note that, while the timely framework for dealing with the challenge of de-territorialisation

299 See Arran Gare, *Postmodernism and the Environmental Crisis* (London: Routledge, 1995); and Leigh Glover, *Postmodern Climate Change* (London: Routledge, 2006).

300 Essam El-Hinnawi, *Environmental Refugees* (Nairobi: UNEP, 1985).

301 Nonetheless, the notion of human security dates back to Basket II of the OSCE, and came back to the forefront within the United Nations with the publication of the 1994 Human Development Report, which included 'environmental security' as one of the thematic lines composing human security.

302 This has become particularly noticeable in the international political arena since the adoption in 2014 of the Warsaw International Mechanism on Loss and Damage for Climate Change Impacts, discussed in section 2.1.1.A, below.

was centred on sea-level rise as a slow-onset process, the challenge of depopulation requires widening the temporal spectrum by incorporating also the possibility of climate-induced displacement taking place in contexts of sudden weather events. As a result of this wide range of climate impacts, and different possible time frameworks, the policy and doctrinal debate on climate-induced displacement seems at the moment scattered and heterogeneous.

The analysis undertaken in this chapter does not aim at covering all situations that have been included in present studies on climate-induced displacement, but will be restricted to national and transnational displacements in and among Pacific Island States. Besides, while all displacement experiences in the region are highly valuable to help understanding the complexities of these situations; displacement cases originating in any of the three Pacific Island States most affected by de-territorialisation bear a fundamental specificity: the fate of the population is inextricably bound to the continuity of the State itself. The study is therefore of particular interest when it comes to displacements originating in the Pacific Island States most at risk – namely, Tuvalu, Kiribati, and the Marshall Islands. It is in these cases that the study of climate-induced displacement supersedes the idea that, as Jane McAdam pointed out, "in some respects, discussions about climate-related movement simply represent a repackaging of the broader debate in 1990s about environmental displacement".[303] Which are the most appropriate legal and policy responses that may ensure the protection of all climate-displaced Pacific Islanders? How may the issue of the continuity of statehood raised for some Pacific Island States jeopardise the formulation of such responses? These are the questions raised in this chapter.

Most studies dealing with climate-induced displacement tend to respond to these and similar questions by preferring one legal regime to another (e.g. international refugee law versus international human rights law); suggesting the creation of a new international instrument specifically designed to respond to these new challenges (e.g. 'Climate Refugee Protocol' to either the UNFCCC or the Refugee Convention); or upholding a wide interpretation of the mandate of existing international organisations tasked with refugee and human rights issues, such as the UN High Commissioner for Refugees (UNHCR). Besides, an important number of these studies become trapped in a never-ending terminological *impasse* – for instance, when regarding the scope of the legal definition of 'refugee' covered by Article 1(A)(2) of the 1951 Refugee Convention.

Terminology is key in the realm of law, since the categorisation of facts of life into juridical categories serves the association of legal consequences with

303 McAdam, *Climate Change, Forced Migration and International Law*, 3.

each category.[304] Yet, the transformation of Pacific Islanders' displacements into legal categories ought to take into account at least the following three factors: (1) whether the displacement is national or transnational (*spatial variable*); (2) whether the continuity of the State itself may be at risk as a result of partial or total *en masse* transnational displacements (*contextual variable*); (3) whether either of the two previous factors is subject to change and progress over time, transforming the displacement situation itself, as well as its political consequences, into the continuity of the State (*temporal variable*). While the first of these three factors is generally presented as being the key to the analysis and the basis for the solution proposed, the other two tend to be by-passed.

Therefore, the present chapter suggests approaching climate-induced displacement of Pacific Islanders differently, through a methodology enabling to grasp the diversity of this process within the region, taking into account all three factors, with particular emphasis on the contextual one. It also argues that a multi-layered legal protection scheme could better adapt to the present and prospective conditions of population displacement induced by climate change in and among Pacific Island States. Section 2.1 begins with a comparative study of national and transnational law and policy approaches to relocation in and among Pacific Island States, used as factual grounds on which the proposal for a multi-layered legal protection scheme is based. Sections 2.2 and 2.3 explore this proposal through two scenarios divided by one central variable: whether the continuity of the State is or is not at risk as a result of population displacements.

2.1 The Facts: Climate-Induced Relocation in and among Pacific Island States

Migration is an inherent part of the history of Humankind and has always been practised as a form of adaptation to the changing conditions of the 'habitat'.[305]

304 This point is raised in Jane McAdam, ed., *Climate Change Displacement: Multidisciplinary Perspectives* (Oxford, Portland: Hart Publishing, 2010), 3.

305 The term 'habitat' generally refers to the natural home or environment of a living organism, be it an animal, a plant or an individual or group of human beings. For historical-anthropological accounts on 'environmental migration', see, for instance, Guy Richard, *Ailleurs, l'herbe est plus verte: histoire des migrations dans le monde* (Paris: Arléa-Cordet, 1996); Clive Ponting, *The New Green History of the World: the Environment and the Collapse of Great Civilisations* (New York: Penguin Books, 2007); or Jon Barnett and Michael Webber, 'Migration as Adaptation: Opportunities and Limits', in Jane McAdam, ed., Climate Change and Displacement: Multidisciplinary Perspectives, 37–55.

In a recent study, Volke Boege analysed the difference between the sea-faring Australasians who settled in South Pacific islands and used to relocate every time an atoll had become uninhabitable, centuries BC; and today's inhabitants of Pacific Island States threatened by the progressive degradation of their living conditions in the same islands.[306] Boege highlighted four main differences: (1) in the historical case, the issue was locally confined; (2) there was a direct relationship between cause and effect, for the people forced to resettle were also the ones responsible for the degradation; (3) at the time, they were not informed either of what was happening or of their implication in the degradation; and (4) today, "not only is climate change a global phenomenon, but also the discourse about it [which] frames the thinking and talking about climate change and migration".[307] While agreeing with all these points, at least one more fundamental difference must be noted: at the time of the Australasians, the State (as a socio-political reality, as well as a legal concept), was not the organising form of human societies. Nor had it spread and settled as the universal form of organisation of collective and social life, and as a fundamental element structuring contemporary international relations. Thus, while the phenomenon of migration may be familiar to our remote history, the tools through which Humankind responds to it are but a matter of our time and ought to be tackled with present-day instruments. The role that migration has played in the different stages of our evolution becomes historically contingent when we look into the present means to understand it and react to it.

Although migration may still be considered today as a form of adaptation, it is generally not thought of as a 'desirable' reaction policy with respect to climate change impacts. Two points support this appreciation. First of all, the South Pacific is often seen and depicted in western mainstream social studies as a rather simple and homogeneous regional unit characterised by the remoteness, underdevelopment and almost insignificant power of leverage of a 'bouquet of mini-States'. There is in addition a marked tendency to depict the whole region as being traditionally highly migratory, giving an inaccurate impression of homogeneity. Indeed, while the migration rate is very high in Tonga and high in Tuvalu and Fiji; on the other hand, Palau and the Marshall Islands, the Federated States of Micronesia and Kiribati have a medium-low rate, while Papua New Guinea's, the Solomon Islands' and Vanuatu's rate is

306 Volke Boege, 'Challenges and Pitfalls of Resettlement Measures: Experiences in the Pacific Region', COMCAD *Arbeitspapiere/Working Papers* (Series on Environmental Degradation and Migration), vol. 102 (2011), 1–40, 7.

307 Ibid.

low to very low.[308] Secondly, even Pacific Island States with a high or very high migration rate do not contemplate displacement (whether internal or external) as a natural and uncontroversial reaction policy. Rather, when officially taken into consideration in adaptation, development or disaster reduction plans, relocation is always considered as the solution of last resort or the 'ultimate' form of adaptation.[309] The following comparative analysis thus seeks to illustrate the common perspectives and fundamental differences of national and transnational climate-induced relocation strategies in the region, and will differentiate between preventive and reactive relocation actions.

2.1.1 *Regional Homogeneity of Preventive Relocation Actions*

A Common Vulnerabilities

Preventive relocation actions can be defined as the range of measures which primarily aim at *preserving* the habitability of human settlements affected by different forms of environmental stress (drought, cyclones, flooding, saltwater intrusion into agricultural lands, etc).[310] The recognition of preventive relocation actions as a distinct category is not incompatible with the acknowledgement that relocation as such may take place and be considered as a possible preventive adaptation solution. Relocations may be integrated into preventive relocation plans and still be considered under this preventive heading, given that what distinguishes preventive from reactive relocation action is the *principal focus* of the respective policy under study. While most preventive relocation policies are decided by national governments, provincial and

308 See Richard Bedford and Graeme Hugo, 'Population Movement in the Pacific: Future Prospects', Labor and Immigration Research Centre, Department of Labor of the New Zealand Government (with Collaboration of the Australian Government), February 2012, full report available at: http://www.dol.govt.nz/publications/research/population-movement-pacific-perspective-future-prospects/04.asp (25 November 2014).

309 See, for instance, the Statement of the Global Migration Group (GMG) on the Impacts of Climate Change on Migration, adopted in Paris on 11 November 2011 by the Principals of the GMG. Document available at: http://www.un.org/esa/population/migration/GMG%20statement_Paris_english.pdf. The GMG is an inter-agency group bringing together sixteen agencies (14 UN agencies, the World Bank and the International Organisation for Migration) to promote the application of relevant international instruments relating to migration and encourage the adoption of more coherent approaches to international migration generally.

310 See Edward Page, 'Theorising the Link between Environmental Change and Conflict', *Review of European Community and International Environmental Law*, vol. 9 (2000): 1, 33–43, 33.

local decision-making and participation also tend to be involved.[311] Moreover, this category includes actions seeking to secure access to basic needs (food, water, and acceptable conditions of sanitation and adequate housing), as well as measures seeking to protect human settlements from material and physical destruction by climate change impacts (construction of shoreline protection, such as sea walls, and protection of communication infrastructure, such as airports and roads, etc.). They are to be found in three different policy strands: climate change adaptation measures, disaster risk-reduction programmes, and development plans of the region – the latter having been transformed by the increased attention paid to the first two strands.

An overall view of the configuration of preventive relocation actions in Pacific Island States indicates that the protection of the habitability of human settlements in the region is being undertaken rather homogeneously, through similar forms and aiming at analogous purposes. This trend can be explained by factors which nurture the consolidation of Pacific Island States as constitutive parts of a regional unit economically headed by Australia and New Zealand.[312] Other than these two regional leaders, Pacific Island States are first of all characterised by their status of micro-States, which is reflected not only in their limited territorial extent, but also in their primary sector-based economies which rely heavily on coastal agriculture and fisheries as principal sources of income and survival of the population. The region's overall low level of economic development is in turn connected to the equally low level of resilience and general adaptive capacity to climate change impacts. Consequently, these States embody climate vulnerability, which is all the more reinforced when contrasted with the overwhelming environmental challenges they are bound to face. While climate change impacts and climate variability phenomena provoke a range of habitat disruptions that may vary from one Pacific Island State to another,[313] their geographical closeness implies that,

311 This may be the result of the recent tendency, widespread in international environmental governance policies, to favour bottom-up approaches for their higher level of integration of the local communities and the positive consequences that such involvement may in turn generate for the effective implementation of international environmental obligations. For instance, one of the areas in which the bottom-up approach and enhancement of local participation has been developed is the implementation of REDD+ projects.

312 For an overview of the regional context and the role played by Australia and New Zealand in it, see Richard Herr and Anthony Bergin, *Our Near Abroad: Australia and Pacific Islands Regionalism* (Barton: Australian Strategic Policy Institute, November 2011), 1–84.

313 Detailed study of the climate change impacts suffered by each Pacific Island State and territories has been conducted by the Applied Geo-science and Technology Division of the Secretariat of the Pacific Community (SOPAC) under the South Pacific Sea Level Rise

when a disruption occurs, most States of the region are, at least to some extent, affected. Given the geographical closeness, general economic weakness, and common high degree of exposure to climate change impacts, Pacific Island States have chosen similar policy strategies to prevent the further degradation or destruction of the habitability of their respective territories by climate challenges. However, the crucial factor that promotes the homogeneous or similar approach to preventive relocation actions is the important degree of intraregional co-operation on these matters which, as developed below, can be found in climate change adaptation actions, disaster risk-reduction programmes, and development plans of the countries at stake.

Adaptation was recognised as an important element of the international regime on climate change ever since its inception, although one that was always flawed with problems.[314] While the global nature of climate change invited a global solution and gave impulse to a process of international co-operation seeking to involve as many States as possible,[315] the inherently local (or, at best, regional) nature of adaptation measures became an important obstacle to securing adaptation funding in the earlier stages of existence of the regime. Besides, the confusion (deliberately sought by OPEC countries) between adaptation to climate change impacts and 'response measures' – namely, negative economic side-effects of mitigation actions on the oil-exporting industry – also contributed to the prevention of an exponential development of adaptation under the UNFCCC. Notwithstanding the earlier difficulties, the latest period of the climate change negotiations (particularly from the 2007 Bali Road Plan

---

and Climate Monitoring Project. Country reports can be found at: http://www.sopac.org/index.php/south-pacific-sea-level-a-climate change-monitoring (25 November 2014).

314 Farhana Yamin and Joanna Depledge, *The International Regime on Climate Change: A Guide to Rules, Procedures and Institutions* (Cambridge: Cambridge University Press, 2005), Chapter 8, 213.

315 On the move to wide multilateralism or even universalism in international environmental co-operation which particularly spread after the 1972 Stockholm Conference on Environment and Development, see Dan Sarooshi, *The Art and Craft of International Environmental Agreements* (Cambridge, Mass.: Harvard University Press, 2011). These voices against the use by developing countries of their power of leverage in international environmental agreements, such as the UNFCCC, to foster their developmental goals include Eric A. Posner and David Weisbach, *Climate Change Justice* (Princeton, N.J.: Princeton University Press, 2010). *Contra*, Friedrich Soltau, *Fairness in International Climate Change Law and Policy* (Cambridge: Cambridge University Press, 2009); and Karin Mickelson, 'South, North, International Environmental Law and International Environmental Lawyers?', *Yearbook of International Environmental Law*, vol. 11 (2000), 52–81.

onwards) has been marked by an increasing attention to the role and importance of adaptation, and by the acknowledgment that, in some cases, loss and damage for climate change impacts may not even be adequately resolved by adaptation measures. Besides, after the climate change negotiations for the second commitment period of the Kyoto Protocol for the post-2012 period went through a structural impasse – particularly after the failure of the 2009 Copenhagen Summit – the original focus on gathering concerted mitigation action shifted in favour of developing the operability of adaptation measures.

Following the establishment of the Adaptation Fund and the launch of the Nairobi work programme 'Understanding Vulnerability, Fostering Adaptation' the Conference of the Parties adopted, in 2010, the Climate Adaptation Framework in Cancun (Mexico). As the dynamic fostering adaptation gradually took off, the most vulnerable State Parties began setting the basis for common (regional) approaches to climate change adaptation and the understanding of vulnerability. The UNFCCC Expert Meeting on Climate Change Adaptation for Small Island Developing States, held in Rarotonga (Cook Islands) in February 2007, is a good illustration of the grounds on which a homogeneous approach to preventive relocation actions in Pacific Island States was based. As a result of the positive development of adaptation funding, coupled with the development of regionally concerted co-operation on this matter, all adaptation strategies of Pacific Island States are – unsurprisingly – linked to the UNFCCC process.

To begin with, approaches to preventive relocation actions are particularly similar in the least developed countries (LDCs) of the region that qualify for funding of National Adaptation Programmes of Action (NAPAs). Since the adoption of the UNFCCC, NAPAs were intended to be a way for least developed countries to identify their most urgent adaptation needs which would be funded and implemented as a priority. Today, the five LDCs of the Pacific region that qualify for this funding category have submitted their respective NAPAs – Samoa (2005), Kiribati (2007), Tuvalu (2007), Vanuatu (2007) and the Solomon Islands (2008) – and are now starting the implementation of projects in a variety of sectors ranging from, inter alia, disaster risk management, food and water security, and protection of coastal and marine ecosystems, to control climate sensitive diseases or the development of sustainable forestry and tourism. Overall, more than sixty-eight million USD have been allocated to the thirty-seven projects in these five Pacific Island States.[316] The greater level

316 Samoa, *National Adaptation Programme of Action* (Apia: Ministry of Natural Resources, Environment and Meteorology, 2005); see Republic of Tuvalu *National Adaptation Programme of Action* (Funafuti: Ministry of Resources, Environment, Agriculture and

of homogeneity in the approach to the configuration of preventive relocation actions through NAPAs is essentially derived from the common institutional framework in which these plans are developed. All qualified countries are advised by the same board of experts, which guides them through the same vulnerability assessment guidelines and encourages participation at the community level in the elaboration of NAPAs. The priority areas identified by the five LDCs of the region focus on securing access to basic needs (in particular, access to water) and improved sanitation (Kiribati). Most interestingly, only two of the Pacific counterparts (Samoa and Tuvalu) make reference to the protection of settlements themselves and the possibility of relocation as part of the set of adaptation strategies. Yet, neither of these two NAPAs addresses the issue of relocation as a key priority area in itself. On the one hand, Tuvalu's Plan simply mentions relocation/resettlement as a situation which should possibly be taken into account during the implementation of a key priority area.[317] On the other hand, Samoa's approach goes a bit further; while it also considers the protection of human settlements as one of the issues which may need to be dealt with in order to secure the effectiveness of another priority area, it nonetheless suggests that such protection may be achieved through, inter alia, the construction of artificial islands.[318] However, even Samoa's Plan falls short when compared with the place that relocation was given in the NAPA of the Maldives. This Plan as a whole is indeed centred on the 'Safer Island' project, which openly and directly seeks to address the issue of relocation of human settlements as a key priority area.[319]

Secondly, since the 2012 COP.18/MOP.8 held in Doha, the Conference of the Parties accepted that unavoidable loss and damage for climate change impacts in cases where these can no longer be managed through either mitigation or adaptation would be tackled, *within the UNFCCC*, through a newly established

Lands, Department of Environment, May 2007); Republic of Vanuatu, *National Adaptation Programme of Action* (Port Vila: National Advisory Committee on Climate Change, 2007); Solomon Islands, *National Adaptation Programme of Action* (Honiara: Ministry of Environment, Conservation and Meteorology, November 2008); and Republic of Kiribati, *National Adaptation Programme of Action* (Tarawa: Ministry of Environment and Social Development, 2007). Detailed information on the projects per country can be found at: http://unfccc.int/files/co-operation_support/least_developed_countries_portal/napa_project_database/application/pdf/napa_index_by_country.pdf (5 December 2014).

317 Tuvalu, NAPA.

318 Samoa, NAPA.

319 See Republic of Maldives, *National Adaptation Programme of Action* (Tarawa: Ministry of Environment, Energy and Water', December 2006). Among the actions planned in Maldives' NAPA, the so-called 'Safer Island Project' is particularly noteworthy, 40 ff.

institutional framework.[320] This decision sets a milestone in the long-standing struggle of the most vulnerable countries, launched by the Alliance of Small Island States (AOSIS) in 1991 during the negotiations on the adoption of the UNFCCC. It is a political success of these countries, not devoid of an emotional force, against the insistence of developed countries to leave loss and damage outside the UNFCCC structure and exclusively address it under two already existing frameworks, namely, the International Strategy for Disaster Reduction (ISDR), and the Hyogo Framework for Action (HFA). As M.J. Mace and Michiel Schaeffer explain, references by developed countries to these two frameworks are perceived by vulnerable countries "as a means to neutralise the politically sensitive discussions on State responsibility for transboundary harm".[321] Indeed, the fundamental differences between the risk-management and disaster-management frameworks, and the UNFCCC structure is such that, in the former, the causal link between emissions and impacts is diffused. Consequently, the ISDR and the HFA place responsibility on the vulnerable countries to find ways to reduce their own vulnerabilities, using their own resources. They thus conceive international financing for these issues as being voluntary. In other words, relying on the ISDR and the HFA to tackle loss and damage for climate change impacts was a strategy of developed countries to by-pass the fairness blueprint on which the UNFCCC is built, particularly the principles of historical responsibilities and respective capabilities from which the *legal obligations* to lead the reduction of greenhouse gas (GHG) emissions (Article 4.2 UNFCCC), and assist financially and technically (Article 4.3 UNFCCC), were established. A year after the adoption in Doha of the bench-

320 Decision 1/CP.16, adopted in 2010 in Cancun launched the Work Programme on Loss and Damage as part of the Cancun Adaptation Framework. The year after, in Decision 7/CP.17, three thematic areas for the Work Programme were approved during the COP in Durban, namely: 1/assessing the risk of loss and damage associated with the adverse effects of climate change and the current knowledge on it; 2/ approaches to address loss and damage associated with the adverse effects of climate change, including impacts related to extreme weather events and slow onset events, taking into consideration experience at all levels; and 3/ the role of the UNFCCC in enhancing the implementation of approaches to address loss and damage associated with the adverse effects of climate change. The institutional anchorage of loss and damage within the UNFCCC framework was then approved in Decision 3/CP.18, adopted in 2012 in Doha. See Laura Schäfer and Sönke Kreft, 'Loss and Damage: Roadmap to Relevance for the Warsaw International Mechanism', Briefing Paper (first version) (Bonn: Germanwatch, March 2014), 1–24.

321 M.J. Mace and Michiel Schaeffer, *Loss and Damage under the UNFCCC: What Relationship with the Hyogo Framework* (Bonn: Germanwatch/The Loss and Damage in Vulnerable Countries Initiative, October 2013), 1–16, 5.

mark Decision 3/CP.18, the COP set up a new institutional framework, known as the Warsaw International Mechanism for Loss and Damage Associated with Climate Change Impacts.[322] While the way in which this mechanism will produce concrete actions is yet to be seen, its existence is already important, for it recognises loss and damage resulting from both slow-onset and extreme events associated with climate change; covering a wide range of challenges – such as loss of ecosystems or livelihoods. But also, most interestingly, it embraces losses which cannot be quantitatively measured, nor adequately addressed by financing alone, such as the loss of statehood.

B Common Responses

Other than the special bond and similar approaches to preventive relocation actions which are found in the five State Parties to the UNFCCC that qualify for NAPA funding, all twelve Pacific Island States have a regional approach to climate change adaptation (into which the five NAPAS may be inserted). This common 'background' essentially results from the creation of the Pacific Adaptation to Climate Change Project (PACC), funded by the Global Environment Facility (GEF) along with the co-operation aid agencies of Australia (AusAid) and the United States of America (USAID) and the support of the C3D+ programme of the United Nations Institute for Training and Research (UNITAR), which aims at developing climate change adaptation and capacity-building.[323] The project covered activities from 2009 to 2013, and was structured along three key lines that build the resilience of Pacific Island States to climate change. Food production and food security is the focus of Fiji, Palau, Papua New Guinea and the Solomon Islands; coastal-management capacity is pursued by the Cook Islands, the Federated States of Micronesia, Samoa, Vanuatu and Tokelau (New Zealand's territory); and strengthening water-resource management is the area of development stressed by Nauru, the Marshall Islands, Tonga,

322 See Decision 3/CP.18, 'Approaches to Address Loss and Damage Associated with Climate Change Impacts in Developing Countries that are Particularly Vulnerable to the Adverse Effects to Climate Change to Enhance Adaptive Capacity', adopted during the CP.18/MP.8 of the UNFCCC held in Doha (Qatar) on 26 November till 8 December 2012, document reference: FCCC/CP/2012/8/Add.1, 21–24.

323 As explained in the official website of the Pacific Adaptation Climate Change Project, the C3D+ project is supported by a grant from the European Commission and is part of the larger "Capacity Development for Adaptation to Climate Change and GHG Mitigation" project. The C3D+ thus includes, as partner institutions collaborating with SPREP: the International Institute for Sustainable Development (IISD), the Stockholm Environmental Institute (SEI), Climate System Analysis Group (CSAG), Munasinghe Institute for Sustainable Development (MIND).

Tuvalu and Niue (New Zealand's territory). Additional funding provided by AusAid and USAID will enable Kiribati to become an additional participant of the PACC. One of the most important characteristics of the PACC is the clear objective of mainstreaming the adaptation measures to climate change and climate variability in key development sectors. This connection, in tune with the previously adopted 2008 Niue Declaration on Climate Change, is bluntly indicated in the main internet portal of PACC, which defines the Project as being "designed to promote climate change adaptation as a key prerequisite to sustainable development in Pacific Island Countries".[324] The United Nations Development Programme is thus one of the implementing agencies of the PACC, working in partnership with the Secretariat of the South Pacific Region Environment Programme (SPREP).

Moreover, the homogeneous approach to preventive relocation actions is also facilitated by the South Pacific Regional Environment Programme (SPREP) which, working in partnership with UNDP, was appointed by Pacific Islands' Heads of State as the leading co-ordination agency in the region's response to climate change impacts. To fulfil this function, SPREP works through three multi-stakeholder co-operation channels: the Pacific Climate Change Roundtable, hosted every two years; the Pacific Islands Framework for Action on Climate Change;[325] and the Council of Regional Organisations of the Pacific CEOs Working Group on Climate Change. The Secretariat conducts the development of guidelines and best practices to support Pacific Island States in planning and implementing national adaptation strategies, as well as to foster the mainstreaming process. While the forms or schemes of implementation may vary because the stakeholders involved in each country differ, the majority of national adaptation plans in Pacific Island States do not include relocation plans.

Finally, the way of dealing with preventive relocation actions in the Pacific is, in addition, characterised by the association of climate change adaptation and development programmes with Disaster Risk Reduction Plans, a third sector which has progressed a lot (and rather autonomously) during the past decade,

---

324 Information extracted from the official website of the Pacific Adaptation to Climate Change Project conducted through SPREP, available at: http://www.sprep.org/pacc-home (30 October 2014).

325 See the *Pacific Islands Framework for Action on Climate Change* (*2006–2015*) (Apia, Samoa: SPREP, 2nd ed. reprinted 2011), which timeframe builds on the previous *Pacific Islands Framework for Action on Climate Change, Climate Variability and Sea-Level Rise* (*2000–2004*).

particularly since the adoption of the 2005 Hyogo Framework for Action.[326] The recognition of the necessity to link these sectors in the Pacific was fostered after the 2007 tsunami hit the whole region. The traumatic and destructive force of this phenomenon dramatically reminded that many of the disasters experienced in this region are associated with climate-variability phenomena that will be exacerbated by: climate change, such as increase in storm number and frequency; changes in the paths of cyclones and an increase in their destructive force; longer periods of drought; and more acute and uncontrolled flooding. The link between climate change adaptation, development plans and disaster risk reduction is thus crystallised in the design of Joint National Action Plans (JNAPS) which result in inter-State co-operation among these three sectors and seek to reduce duplication of effort and achieve a more efficient use of already scarce resources.[327] To date, Tonga has already finished the design of its JNAP and is starting the implementation phase of its pilot projects;[328] the Marshall Islands has completed the preparation of the plan; whilst the governments of Tuvalu, Nauru, the Cook Islands and Niue (New Zealand's territory) are currently developing their JNAPS.

2.1.2 *Regional Heterogeneity of Reactive Relocation Actions*

A Differing Scenarios

Reactive relocation actions may be understood as the range of measures primarily dealing with relocations per se. Although some of these measures may

326 On the occasion of the World Conference on Disaster Reduction, held in Kobe, Hyogo (Japan), in January 2005, the 'Hyogo Declaration' and the 'Hyogo Framework for Action 2005–2015: Building the Resilience of Nations and Communities to Disasters' were adopted. The Hyogo Framework for Action tasked the United Nations Office for Disaster Risk Reduction – established in 1999 after the adoption by the General Assembly of the International Strategy for Disaster Reduction (UNISDR) – with support for the implementation of the Framework for Action.

327 See SPREP–JNAPS website, available at: http://www.sprep.org/Adaptation/current-programmes (25 November 2014). The current JNAP process has regional organisations working side by side with national governments to support them in this process. A joint partnership exists between SPREP, SPC (SOPAC Division) and UNDP, which has been supported through bilateral assistance from governments such as Australia's.

328 See Joint National Action Plan on Climate Change Adaptation and Disaster Risk Reduction (2011–2015), Second National Communication Project, July 2010, Ministry of Environment and Climate Change and National Emergency Management Office, Tonga. Tonga's JNAP was financed by the Global Environment Facility through the United Nations Development Programme, the ACP–EU Natural Disaster Facility through the South Pacific Applied Geoscience Commission (SOPAC) and the Secretariat of the South Pacific Regional Environment Programme (SPREP).

be launched before the relocation actually takes place, they are nonetheless considered in this section as *reactive* and not *preventive*, for, instead of being focused on preserving the habitability of human settlements generally, they operate within a context or plan which is centred on an ongoing and specific relocation of individuals or a community. The category of reactive relocation actions thus includes the preparation of a relocation plan, its implementation, as well as the *ex post facto* measures directed at the creation and maintenance of the habitability of the new relocation area.

Table 4 summarises reactive relocations in the Pacific. It aims, first of all, at identifying whether and when relocations have already taken place in and among Pacific Island States, the number of people involved, the places of origin and resettlement and, most importantly, whether the relocation involved autonomous individuals, family units or the displacement of entire villages at a time. The table then focuses on the policy and legal aspects of accomplished relocations. It thus begins by identifying the existence of a relocation 'plan' – or lack thereof – and (in the affirmative) indicates the actors involved in it. The existence of a 'plan' may include negotiations of national land tenure agreements or new overseas migration agreements that are capital to secure a right to housing in the relocation area. Table 4 also pays attention to the level of participation of the people involved in the preparation or implementation of the relocation, distinguishing three different scenarios: (1) where the plan emerged from national or regional authorities with little or no participation of local actors (top-down approach); (2) where the plan was produced by the local communities themselves, which then tried to convince regional or national authorities to support the plan politically and financially (bottom-up approach); and (3) where the plan resulted from co-operation amongst the different relevant levels of governance, which is generally headed by the national government (mixed approach). Finally, table 4 takes into account the range of relocation funding strategies, where the presence of international actors (States' external aid programmes, and foreign non-State institutions, such as NGOs, foundations, etc.) can be measured against those of governmental authorities (national, provincial or local), and those of community actors (Churches, Council of Elders, private donations, etc.).

Table 4 is not intended to be exhaustive; rather, it seeks to illustrate some of the population movements taking place in and among Pacific Island States in the context of environmental stress due to slow-onset events; and it shows how the region is not only aware of the prospects of climate-induced displacement, but has already experienced a fair number of precedents.

Some preliminary observations are due before commenting on the contents of Table 4. To begin with, the table is only concerned with climate-induced relocations, which can be understood as the ultimate form of adaptation to climate change impacts and to climate variation. Indeed, one can often find references to the relocation that followed the devastating environmental effects of nuclear testing undertaken by the USA in the northern part of the Marshall Islands. Kwalajei atoll was eventually fully evacuated and most of its inhabitants resettled in Majuro, the name of the main atoll and the capital city of the archipelagic State. The relocation of the Nauruans and of Banabans (Kiribati) to the island of Rabi (Fiji), who were forcibly expelled by former colonial and administering powers wishing to exploit phosphate resources, are the two other often cited examples.

Depicting Pacific Islanders as belonging to a region familiar with and traditionally open to migration, and supporting such a description by recalling these three past forced displacement episodes, can be criticised from two angles. First of all, using past abuses by foreign powers before Pacific Islands gained political independence to support the present appreciation of a regional migratory behaviour is controversial, if not totally ill-considered. While past forcible displacement and present climate-induced relocations share the same typology of 'victims' and are a consequence of the harmful actions of foreign State actors, each situation should nonetheless be correctly (re)contextualised. This being said, it also seems important to pay closer attention to the inequalities in the present regional migration rates and to introduce some qualifications of the generalised view that the Pacific is, still today, a highly permeable space of constant human circulation.[329]

The second preliminary observation is that it has only been possible to find documentation on relocation cases in nine out of twelve Pacific Island States. While the existence of climate-induced relocation episodes in Palau, Federated States of Micronesia and the Marshall Islands was mentioned in an official statement by the Micronesian President to the United Nations, no other documents on this subject have been accessible. Nonetheless, the information found about the remaining States displayed in table 4 provides an adequate illustration of the existence of fundamental factual and policy differences among Pacific Island States in respect of climate-induced relocation within the region.

329 See, Vera Martins Almeida, *Overview on South-South Migration and Development in the Pacific: Trends and Research Needs* (Geneva: International Organisation for Migration/ ACP Observatory on Migration, 2011), 1–40; Richard Bedford, *Population Movement in the Pacific: A Perspective on Future Prospects* (Wellington, New Zealand: Labour and Immigration Research Center, February 2012), 1–110; and Richard Bedford and Graeme Hugo, 'International Migration in a Sea of Islands: Challenges and Opportunities for Insular Pacific Spaces', *Population Studies Centre*, vol. 69 (July 2008), 1–40.

TABLE 4 *Climate-induced relocations in and among Pacific Island States*

	Relocation data					Relocation planning		
	Year	Type	Number of people	Place(s) of origin	Place(s) of relocation	Instrument(s) and/or actions	Approach and Actors involved	Funding
FSM				…				
Fiji	2007	National Planned Village Relocation	150	Shoreline Village of Vunidogoloa (Vanua Levu Island)	Higher Land Baptised 'Kenani' – The Promised Land – (Vanua Levu Island)	…	INTEGRATED – Local Communities – Ministry of Provincial Development – National Disaster Management Office – Ministry of Local Government, Urban Development, Housing and Environment	Governmental Assistance (water tanks)

	On-going	National Spontaneous Individual/Family Relocation	…	Outer islands (e.g. southernmost Betio Atoll) (Gilbert Islands)	Tarawa (capital city)	Kiribati Adaptation Program Pilot Investment Phase KAPIII, Lands Acquisition and Resettlement Policy Framework (2011)	TOP-DOWN – Ministry of Environment, Lands and Agricultural Development (MELAD) – Resettlement Committee	World Bank; GEF; UNDP; Japan; PHRD; Climate Change Fund
Kiribati	Since 2002	Overseas Planned Individual Relocation	75 citizens per year	Kiribati	New Zealand	New Zealand Pacific Access Category	TOP-DOWN – Government of Kiribati – Government of New Zealand	…
	Since 2007	Overseas Planned Individual Relocation	50 people currently enrolled	Kiribati	Australia	Kiribati-Australia Nursing Initiative	TOP-DOWN – Government of Kiribati – Government of Australia	…
	Forth.	Overseas Planned Families Relocation	500 families	Kiribati	Vanua Levu Island (Fiji)	No official Agreement. Purchase of 2000 hectares of land in 2014	TOP-DOWN – Government of Kiribati – Private Fijian land owner	Kiribati's Revenue Equalisation Reserve Fund

TABLE 4 *Climate-induced relocations in and among Pacific Island States* (cont.)

	Relocation data					Relocation planning		
	Year	Type	Number of people	Place(s) of origin	Place(s) of relocation	Instrument(s) and/or actions	Approach and Actors involved	Funding
Marshall Islands					…			
Nauru					…			
Palau					…			
Papua New Guinea	2001	National Planned Village Relocation	1000	Duke of York Islands	New Britain Island	…	…	East New Britain Provincial Government (provided emergency supplies; purchased resettlement land)

	2005	National Planned Village Relocation	1750	Carteret Islands	Bougainville main island (Tinputz, Tearouki and Mabiri host communities)	Carteret Integration Relocation Plan	BOTTOM UP – Tulele Peisa (local NGO) – Council of Elders – Autonomous Region of Bougainville Regional Government – PNG National Government	Catholic Church, NGOs, PNG National Government
Samoa	2010	National Spontaneous Village Relocation	…	Fagamalo Settlement	Higher land (unspecified location)	…	BOTTOM UP …	Christian Church of Samoa (water tanks donation)
Solomon Islands	2012	National Spontaneous Village Relocation	…	Ava village and North Vella Lavella	Giza Island	…	BOTTOM UP – Led by Chiefs of Communities	…
Tonga	2012	…	…	…	…	Joint Implementation Adaptation and Disaster Risk Reduction Plan, Pilot Project	MIXED …	…

TABLE 4 *Climate-induced relocations in and among Pacific Island States* (cont.)

	Relocation data					Relocation planning		
	Year	Type	Number of people	Place(s) of origin	Place(s) of relocation	Instrument(s) and/or actions	Approach and Actors involved	Funding
Tuvalu	Reg. Basis	National Spontaneous Individual/Family Relocation	…	Outer islands (unspecified location)	Funafuti (capital city)	…	…	…
	Since 2002	Overseas Planned Individual Migration	75 citizens per year	Tuvalu	New Zealand (Auckland)	New Zealand Pacific Access Category	TOP DOWN – Government of Tuvalu – Government of New Zealand	…
	Not yet	Overseas Planned Community Migration	…	Tuvalu	Australia	Formal Request to Australian Government (2008)	MIXED …	…

						– denied but followed by Informal Positive Response of Australia Tcrres Straits Inhabitants		
Vanuatu	2005	National Planned Village Relocation	300 approx.	Lateu Settlement (Tegua Island)	Higher land	Lateu Community Relocation Project Implementation Plan	INTEGRATED – South Pacific Regional Environment Programme (SPREP) – National Advisory Committee on Climate Change (NACCC) – Consultations with Local Communities	Canadian International Development Agency; NGOs; Church of Melanesia; NACCC

As table 4 shows, most climate-induced relocations which have been documented as such in Pacific Island States have taken place within State boundaries. Only Tuvalu, Kiribati and Tonga may be said to be officially embarking on projections of overseas climate-induced relocation of their population. The main instrument serving the purpose of permanent relocation of Tuvaluans, I-Kiribati and Tongan islanders is the Pacific Access Category. Under this immigration agreement concluded by these three countries with New Zealand in 2002, up to 650 people can annually be granted permanent residence and working permits, with special quotas being reserved for 75 Tuvaluans, 75 I-Kiribati and 250 Tongans. Although the quotas cover not only primary but also secondary applicants, such as spouses and dependent children under 25 years old, the demanding eligibility conditions for primary applicants initially reduced the impact of this new immigration provision. Apart from having a fair understanding of the English language, as well as a job offer in New Zealand, primary applicants were required to demonstrate good health and a sound character. Given that, initially, the quotas were not filled, these conditions were relaxed after 2005.[330] The Pacific Access Category (PAC) provides a stable framework for migration which undoubtedly facilitates the creation of islanders' communities in New Zealand which can promote their integration into this 'host' country.[331] Besides, the adoption of the PAC seemingly marked the beginning of a dynamic and engaging stage. Moreover, the two regional powers continued to open up to their Pacific island neighbours through the adoption of three new immigration agreements. In 2007, Australia and Kiribati agreed on the Nursing Initiative, an immigration and capacity-building framework "created to enhance Kiribati's ability to cope with overpopulation and climate change by educating Kiribati youth and gain Australian and international employment in the nursing sector".[332] The same year, New Zealand launched the Recognised Seasonal Employer Policy, allowing the temporary

330 Narayan Subramania, *Pacific Islands Immigration Provisions* (Center for Climate Change Law, 2011).

331 See François Gemenne and Shawn Shen, 'Environmental Change and Forced Migration Scenarios (EACH-FOR): Tuvalu and New Zealand Case Study', 2008, 1–32, (paper resulting from a research project financed by the European Commission, available at: http://www.each-for.eu (25 November 2014). See also Maryanne Loughry and Jane McAdam, 'Kiribati: Relocation and Adaptation', *Forced Migration Review*, vol. 31 (October 2008), 51–52.

332 This is a five-year pilot programme sponsored by AusAid (Australian official co-operation agency), which takes place at Griffith University. It requires an undergraduate student to attend a four-month preparation in Kiribati, followed by another four months of nursing-diploma preparation at Griffith University in Australia, and then to complete a

entry of workers into the horticulture and viticulture industries, with a preference for people from Pacific island countries (except Fiji).[333] This policy was reproduced two years later by Australia with the 2009 Pacific Seasonal Worker Scheme, created for the purpose of "encouraging Pacific island workers to contribute to the economic development of their home countries through remittances".[334]

While the unprecedented high level of overseas immigration from Kiribati, Tuvalu and Tonga to Australia and New Zealand is undeniable, the question of whether any of these new immigration schemes can be considered as responses to climate change impacts instead of regular economic migration is very controversial. Interestingly, Kiribati and Tuvalu are both direct beneficiaries of the permanent immigration scheme agreed with New Zealand, and yet each has shown radically opposed official approaches to this issue. On the one hand, the government of Kiribati openly considers it necessary to include overseas relocation as an important element of Kiribati's climate change adaptation programme. While the level of consciousness of the I-Kiribati population regarding the medium- and long-term effects of climate change impacts for their country is uneven and rather uncertain,[335] the opinion of President Anote Tong is quite straightforward:

nursing diploma before starting a bachelor degree in nursing. Eighty students are currently enrolled in this programme.

333 Ibid. This visa is valid for up to seven months in an eleven-month period and is only granted in case of a clear need for additional seasonal workers. It allows employers to recruit non-New Zealand citizens or resident workers. Since 2009, up to 8000 Pacific Islanders have worked under this policy, coming from Kiribati, Samoa, Tonga, Tuvalu, Vanuatu, the Marshall Islands and Palau.

334 Up to 2500 visas granted over a period of three years to workers from Kiribati, Papua New Guinea, Tonga and Vanuatu. For the horticultural industry, the need has to be proven by the employer. The visas were issued for periods of up to seven months in any twelve-month period, and did not allow dependents to benefit from it. The first phase, which lasted until February 2010, was used by fifty Tongan and six Vanuatu workers. Due to the lack of demand for workers, 44 of 100 visas were unused.

335 See Dev Nadkarni, 'An Ailing Island in the Sun', New Zealand Herald, 28 October 2008, which notes that, for instance, Harry Tong, brother of current Kiribati President Anote Tong and leader of the opposition, stated that the islands can never sink because his church believes that "God promised Noah there would not be another flood after the last one." Similarly, Michael McKenzie, Vicar-General of the Diocese of Tarawa and Nauru, qualified this position by responding that "God's promise is, of course, valid. But this sea-level rise has man-made causes and God has nothing to do with it."

> [W]e can never be too well prepared for the effects of climate change [...] if circumstances force the migration of our people to other nations at some future date, we want them to go there not as climate change refugees but as people who are equipped to contribute meaningfully to their host nations' economies.

Not only does he use the term 'climate change refugees', but even considers it a duty of the regional leaders to the people they govern to prepare them for worst-case scenarios. The most recent governmental external policy action was the purchase of 2000 hectares of land, in Fiji's Vanua Levu Island, from the Church of England.[336] The early conversations between the two Parties were facilitated by Fiji's Real Estate Agents Licensing Board, according to which, the purchase would be followed by the implementation of a development plan which would include the settlement of around 500 I-Kiribati families and the production of food products, as well as some materials for construction of shoreline protection and sea walls, to export to their home country.[337] Moreover, this openly embracing climate-induced relocation policy also had manifestations at the national level, as epitomised by the preparation, in 2011, of a Land Acquisition and Resettlement Policy Framework as part of the third phase of the Kiribati Adaptation Project. This policy framework – supported by international funding institutions World Bank, the Global Environmental Facility, the Japan PHRD Climate Change Fund and the UNDP – was launched with the acknowledgment that some of the activities involved in the implementation of Pilot Island Adaptations "might require land acquisition and/or resettlement of the vulnerable communities to climate change and sea-level rise".[338]

On the other hand, regardless of the real intentions of Tuvalu's governmental actions – such as the negotiation of the conditions of the Pacific Access Category – the official position of this country contrasts with that of Kiribati, being much more reluctant than Kiribati to acknowledge the need for an overseas climate-induced relocation planning. Ever since the possibility of

336 See chapter 1, section 1.3.1.A.

337 See, for instance, 'Kiribati Mulls Fiji Land Purchase in Battle against the Sea', BBC News, 12 March 2011, available at: http://www.bbc.co.uk/news/world-asia-17295862 (25 November 2014).

338 See Kiribati Land Acquisition and Resettlement Policy Framework, 2005 (amended 2011), 8. In particular, the Resettlement Committee was established by the Government of Kiribati to deal with the resettlement of the atolls of Teraina, Tabuaeran and Kiritimati, Ibid., 10.

relocating the whole population to the Fijian island of Kioa was raised in 2006 by a Tuvaluan-born Australian climate change campaigner,[339] the successive governments seemed to have maintained the position expressed by Prime Minister Maatia Toafa who was in power at the time. Back then, Toafa stated that relocation was "not a priority", and that his government did not regard rising sea levels as being such a threat that the entire population would need to be evacuated.[340] While Kiribati's approach may be more pragmatic, Tuvalu does not resign itself to the idea that the fate of future Tuvaluan generations may have to be found beyond the nine atolls that acquired political independence in the late 1970s. The citation, at the beginning of the present chapter, of the current Prime Minister of Tuvalu, Apisai Lelemia, clearly conveys this message:

> [I]t is our belief that Tuvalu, as a nation, has a right to exist forever. It is our basic human right. *We are not contemplating migration.* We are a proud nation with a unique culture which *cannot be relocated elsewhere.* We want to survive as a people and as a nation. *We will survive.* It is our fundamental right.[341]

Scholars, such as Jane McAdam, have shed light on the reasons behind this position, explaining how the 'refugee' label is rejected because it conveys the idea of powerlessness and victimisation that is shameful in Tuvaluan culture, and which, in contrast, is not found in the traditional label of 'economic migrants'.[342] Given this reluctant approach to climate-induced overseas relocation planning, it does not come as a surprise that Tuvalu does not count so far with an instrument akin to Kiribati's Land Acquisition and Resettlement Policy Framework directed at planning possible *internal* climate-induced

339 Don Kennedy, retired scientist born in Tuvalu, moved to Australia forty years ago and more recently stated in several conferences that Tuvaluans must move to preserve their culture and survive. It has been argued that he may have the influence to speak directly with the president of Tuvalu and convince him of the need to move Tuvaluans to the Fijian Island of Kiao. Several newspapers seem to follow his plea and have widely covered the news. See, for instance: 'Tuvalu Population Must Move to Preserve Culture', The Epoch Times, 18 February 2006, available at: http://www.theepochtimes.com/news/6-2-18/38332.html (25 November 2014).

340 The Statement by President Maatia Toafa was given as a response to Don Kennedy's contentions. It was reproduced by Tuvalu News on 21 February 2006, available at: http://www.tuvaluislands.com/news/archived/2006/2006-02-21.htm (25 November 2014).

341 [Emphasis added].

342 McAdam, *Climate Change, Forced Migration and International Law*, 41.

resettlements. This gap is all the more facilitated by the fact that both Tuvalu and Kiribati have experienced – particularly during the past decade – a high level of inter-island migration from outer islands and atolls to their respective capitals (Funafuti and Tarawa). Given that such population movement is due to a wide range of socio-economic reasons, the exact number of internal climate-induced relocations is difficult to determine. One thing they have in common is that the overpopulation of both capitals, coupled with sea-level rise, has become an important source of concern in Tuvalu, Kiribati, and the Marshall Islands. While Funafuti has lost one metre of land around its circumference, efforts are now being displayed to protect with sea walls the airports of Tarawa and Majuro (capital of the Marshall Islands) from being constantly flooded.

Contrary to the high rates of inter-island migration in Tuvalu and Kiribati previously mentioned, internal climate-induced relocations in other Pacific Island States have been reported by national and even by international media, and therefore tend not to pass unnoticed or become dissolved in the wider category of national economic-driven population movements. One of the first cases of climate-induced relocation that attracted widespread attention was the resettlement, in 2005, of the inhabitants of the Carteret Islands, located in the Autonomous Province of Bougainville in Papua New Guinea. Four years earlier, the same country had witnessed another previous relocation triggered by analogous causes. Yet, despite the fact that the precedent had affected over one thousand inhabitants of the Duke of York Islands, who were relocated to the province of New Britain, and implied the purchase, by the East New Britain Provincial Government, of resettlement lands and emergency supplies, the media remained quite silent about it and the situation passed almost unnoticed. The low coverage of the Duke of York relocation strikingly contrasts with the attention given to the 2005 Carteret Islands relocation, which was not only widely reproduced in the media, but also 'competed' against the parallel relocation of the Lateu settlement in Vanuatu for the title of first scenario of 'climate refugees' in history. Such a shift may be understood as a reflection of the progressive rise of public concern about climate change impacts during the first decade of the 21st century. What is most relevant about the Carteret and Lateu experiences, ever since they were labelled and identified as the first cases of 'climate-induced relocations', is that they show how relocation is far from being an easy or generally 'natural' experience in this region. Most importantly, they serve as a starting point from which it is possible to extract some of the variables that mark the differences between relocation experiences in and among Pacific Island States.

B Differential Factors

Relocations in and among Pacific Island States are first of all determined by a socio-cultural variable that either qualifies or enhances the traumatic aspect of most resettlement experiences. This variable is the existence of historical ties between the people affected and their land of origin – or lack thereof. Notwithstanding the observable deterioration of their living conditions due to climate change impacts, communities which have construed their identity in association with a specific and confined ancestral setting tend to be highly apprehensive of the idea of relocation. Evidence of this obstacle to socio-cultural relocation is for instance revealed by several interviews with Carteret islanders, and explains the tendency to delay as much as possible the beginning of the evacuation, even in cases where a relocation plan is ready for implementation. Besides, the relevance of this variable is all the more reinforced when relocation may directly affect indigenous peoples or an ethnic minority. In these cases, not only will the impact of the resettlement on the socio-cultural identity of the communities be more acute, it may also impinge upon their successful integration into a new place irrespective of whether the relocation takes place within the same State or close to the original lands. This concern has already been raised by Tulele Peisa – the local NGO of Papua New Guinea which is currently leading the relocation of Carteret families – with regard to the inhabitants of the Mortlock, Tasman and Nuguria Islands, who are expected to be relocated before the end of 2015. Unlike the Carteret islanders and the Bougainvillians, who are predominantly Melanesians, the Mortlock, Tasman and Nuguria Islands are mainly inhabited by a Polynesian minority. This fact impedes the reproduction of successful actions which have been promoted by Tuelele Peisa to ensure the social integration of the Carteret islanders with the Bougainvillians, such as the holding of inter-community meetings prior to relocation, and even the conclusion of marriages. It may also constitute an additional obstacle to the acquisition of new lands from customary landowners in Bougainville Island.

In addition to the socio-cultural and historical characteristics of the affected population, the variety of relocation experiences in and among Pacific Island States is highly determined by political factors. This variable takes into account: (1) whether the national government actually acknowledges the need to resettle part of its population or is, in contrast, more inclined to deny such a necessity (see the opposed official positions taken by Kiribati and Tuvalu explained above); and (2) in the affirmative, whether the presence of the national government in people's daily life is effectively rooted so as to ensure the preparation of a relocation plan, as well as the availability of the resources necessary to implement it. Besides, such a level of governmental 'presence' in communities affected by climate change impacts will determine not only the existence of a relocation plan – or lack thereof – but also the approach to the implementation of the relocation as a whole. Thus, when the community is isolated from the national government and functions autonomously, lead by local actors (Council of Elders, Heads of the Church, etc.), customary rules and bottom-up approaches will take the lead over top-down relocation planning.

Relocation experiences in Papua New Guinea, the Solomon Islands and Fiji provide a good illustration of the relevance of this variable. The Carteret Islands are located in the province of Papua New Guinea, known for its mineral resources and cradle of several years of civil strife that confronted the Papuan national government with the secessionist forces of the Autonomous Province of Bougainville (APB). Today, the future legal status of the province remains uncertain, as the organisation of a referendum has been delayed on several occasions ever since the war came to an end. As a result of this delicate political and historical context, the effective presence of the national government of Papua New Guinea in the region is rather limited. Thus, when the relocation of Carteret islanders began to be seen as inescapable, the Council of Elders of the community decided to create a local NGO which received the meaningful name of 'Tulele Peisa', a most poetical motto which means 'surfing the waves in our own way'. This community-based organisation elaborated a Carteret Integrated Relocation Plan which, after approval by the Council of Elders, was endorsed by the regional government of Bougainville. While the bottom-up approach to relocation planning has the advantage of being more responsive to the real needs of the population affected, as well as of better taking into account the practical post-relocation integration challenges, a limited presence of the national government is a potential source of financial struggle and limits the possibility to find appropriate relocation lands, as only the national government can conduct expropriations.[343] For instance, the 2 million kina (approximately 790,000 USD) that the national government of Papua New Guinea had earmarked for the Carteret relocation, and deposited with the provincial government, were never used. This contrasts with the increased governmental presence in the Solomon Islands in relocation areas, which shows how participation of national political actors and institutions facilitates access to international funding. Indeed, following the resettlement of the community of

343 For an early and thorough explanation of the land issues in Papua New Guinea, see Michael J. Trebilcock, 'Customary Land Reform in Papua New Guinea: Law, Economics and Property Rights in a Traditional Community', *The Adelaide Law Review*, vol. 9 (1983): 1, 191–228.

North Lavella, due to the 2007 tsunami that hit the region, the President of the Solomon Islands stated that the community relocations of the Ava Labella will be conducted in the near future with the financial support of the European Union under the EU Global Climate Alliance.[344] Likewise, the involvement of Vanuatu's national government permitted the involvement of the Canadian International Development Agency, which financed the preparation and implementation of the relocation of the Lateu settlement.

These cases of accomplished relocations in the Pacific also indicate how the denial by a national government of the need to conduct relocation, as well as its ineffective or weak presence in the locations concerned (whether for organisational, historical, cultural or political reasons) correlatively strengthens local stakeholders and enhances the development of tailored forms of organisation. In this sense, other than the paradigmatic example of the Council of Elders of the Papua New Guinean case, it is worth mentioning the capital presence of the Church in the Pacific. Sometimes, the importance of the Church is linked to its privileged position as a private land-owner, which makes it a particularly important stakeholder in relocations as a potential land donor. For instance, in Fiji, the Methodist Church played a fundamental role during the resettlement of the coastal village of Vunidogoloa. Not only did it help the villagers prepare spiritually for relocation and view it as a positive change, it was also fundamental in making them feel integrated into the new location of their village at a higher altitude. The new site was thus sanctified and received the biblical name of 'Kenani' – which means nothing less than 'the Promised Land' – and it was established that the first Friday of each month would be dedicated to fasting and praying in gratitude for the relocation. Besides, the Methodist Church acted in Vunidogoloa as a representative of the village before governmental authorities, and thus (quite successfully) claimed financial support on behalf of the village.

Finally, the third variable that determines the fundamental differences among relocation experiences in the South Pacific is geographical. Geography plays a role in conjunction with the two other socio-cultural and governmental factors described above, and addresses the question of whether the nature of the original location has, for instance, changed from an agricultural to an urban setting, as occurred in Tuvalu and Kiribati. In cases where the relocation occurs in an agricultural setting, access to new land is the central issue, particularly in a region which is characterised by the limited extent of public lands (generally not over two per cent). This feature seriously limits the

344 Relocation of other communities, such as those of Lord Howe, Sikaiana, Fanalei and Walande Islands have also been earmarked.

capacity of respective governments to design and implement appropriate relocation plans (in particular in the Marshall Islands, Fiji and Vanuatu), is a source of conflict between public and private actors, and gives great importance to land-tenure negotiations. Moreover, the geographical variable also marks a difference between cases in which the relocation site is in the same island at a higher altitude, close to the place of origin of the community – such as the Lateu settlement in Vanuatu – and cases in which the village is 'decomposed' into family units which are moved to an already inhabited and different island, as happened to the Carteret islanders, who got integrated into three existing communities of Bougainville Island and worked in plantations there.

All in all, in contrast to the overall homogeneous and regional approach to the preservation of habitable conditions, disparities among Pacific Island States surface quite quickly when it comes to reactive relocation actions. The important efforts displayed in preventive relocation actions described above strikingly contrast with the underdeveloped reactive relocation measures in the region. Not only is relocation not being dealt with from a common regional perspective, it has not yet even entered the policy agenda of some of the Pacific Island States. Moreover, in cases where reactive relocation actions can be identified, a country-to-country comparative study reveals a rather heterogeneous response to relocations per se, indicating that the cultural, historical, political and perhaps even psychological differences in each of the relocation scenarios play a crucial role. The heterogeneity of reactive relocation actions is thus driven by a set of socio-cultural, political and geographical variables that form the context in which relocations take place. Therefore, studying how reactions to the same climate-induced relocation challenge differ among Pacific Island States becomes an opportunity to pierce the 'outsider viewpoint' of this region and to explore its more complex and diverse nature.

2.2 The Proposal: A Multi-Layered Legal Scheme for Climate-Induced Relocation

The comparative study discussed above has shown how preventive and reactive relocation actions in and among Pacific Island States have been dealt with differently, receiving uneven attention from governmental authorities, regional and local actors of the States in question. This factual diversity strikingly contrasts with the 'one-dimensional' scholarly tendencies which elaborate legal responses to climate-induced displacement by focusing almost exclusively on one international legal regime or avenue. The present section argues that any legal scheme for climate-induced displacement based on international law should reflect the factual diversity described above. It therefore proposes to conceptualise the legal framework of climate-induced displacement in multiple and complementary layers of protection, so as to address the issue through an integrative and pluralistic scheme. Such a multi-layered legal scheme must not only reflect the fundamental connection between climate-induced displacement and the progressive de-territorialisation of Pacific Island States detailed in chapter 1 (the former being a reflection of the latter), but should also relocate the human and territorial pillars within the ultimate issue of this study, namely, the continuity of Pacific Islands' statehood. To achieve this purpose, the continuity of statehood is conceived as the central variable structuring the multi-layered legal scheme proposed.

2.2.1 *Assessment of the Existing Approaches to Climate-Induced Relocation*

A Relevant Diagnosis and Legal Regimes Analysed

Migration flows associated with the wide range of environmental impacts attributable to climate change is becoming an ever more recurrent object of analysis in international legal scholarship. As detailed below in this section, several studies have already reflected upon the statutory framework in which these situations might or should fall, identifying normative gaps in areas of international law applicable to these situations, and correlatively suggesting how these gaps could or should be filled. Studies have scrutinised international regimes of various kinds, ranging from international environmental law to human rights law or refugee law, and have also suggested a variety of solutions, from enhanced inter-regime synergies to institutional co-operation, among others. The existing literature is at the core of the proposal for a multi-layered protection scheme and forms the basis of its content. As already mentioned, the interest in climate-induced relocations can be traced back to the mid-1980s, following the publication, in 1985, of the El-Hinnawi Report in which a group of experts of the United Nations Development Programme dealt with how the effects of strong environmental disruptions could trigger human displacements. The specificity of these situations was recognised by coining the term 'environmental refugees', and even providing this widely referred definition of the concept:

> [E]nvironmental refugees are those people who have been forced to leave their traditional habitat, temporarily or permanently, because of a marked environmental disruption (natural and/or triggered by people) that jeopardised their existence and/or seriously affected the quality of their life (...). By 'environmental disruption' in this definition is meant

> any physical, chemical, and/or biological changes in the ecosystem (or resource base) that render it, temporarily or permanently, unsuitable to support human life.[345]

While this landmark definition introduced this issue from an international law and policy perspectives, it also gave rise to a definitional impasse which is revealed by what Jane McAdam identifies as the dependency of the legal realm on language and terminology.[346] This formula was primarily characterised by its wide personal, temporal and spatial scope. Firstly, the term 'environmental refugees' – referring, in plural, to the groups of people affected by environmental disruptions, contrasts with the individual approach of the Refugee Convention – defining the category of *refugee* (in the singular), as *any person* who is outside of his/her country of nationality owing to a well founded fear of being persecuted.[347] The use of the plural in the El Hinnawi Report seemingly acknowledged that, as a matter of principle, displacements induced by environmental disruptions are likely to generate migration *flows*. Although these tend to have a higher impact in communities with lower adaptive capacities and lesser resilience, environmental disruptions affect groups of population indiscriminately. Secondly, no mention is made of the boundaries of the State of origin of the displaced people; rather, the definition refers to the 'natural habitat' from which the affected people are forced to migrate. This somehow geographically 'neutral' concept seemingly indicates that the definition of 'environmental refugee' is not limited to cases of transnational displacement, but is also applicable to cases of internal migration. What matters is whether a move from a specific point of departure (the 'natural habitat') has taken place, irrespective of the point of arrival. Thirdly, this definition held that the condi-

345 El-Hinnawi, 1.

346 In McAdam, ed., *Climate Change and Displacement: Multidisciplinary Perspectives*, 3. See also Fabrice Renaud, Janos J. Bogardi, Olivia Dun and Koko Warner, 'Control, Adapt or Flee: How to Face Environmental Migration?' *InterSecTions Publication Series of United Nations University Institute for Human Security*, vol. 5 (2007), 1–48.

347 This narrowness of the main material scope of application of the two instruments of international refugee law contrasts with the broadness of the UNEP definition of 'environmental refugees'. See Article 1(A)(2) of the Refugee Convention, which defines the term 'refugee': "[A]s a result of events occurring before 1 January 1951 and owing to well founded fear of being persecuted for reasons of race, religion, nationality, membership of a particular social group or political opinion, is outside the country of his nationality and is unable or, owing to such fear, is unwilling to avail himself of the protection of that country; or who, not having a nationality and being outside the country of his former habitual residence as a result of such events, is unable or, owing to such fear, is unwilling to return to it".

tion of 'environmental refugee' was not dependent on whether the displacement was temporary or permanent – or, in other words, the irreversibility of the disruption was not an element of the definition of 'environmental refugee'. This is also an important point of difference from the protection offered to 'traditional' refugees, which is, by definition, temporary.[348]

The legal controversy generated by this UNEP definition arose with the need to deal with the determinant element for a migrant to be characterised as an 'environmental refugee': the triggering 'cause' of the displacement being "because of a marked environmental disruption".[349] This definition operates *ex post facto* – that is, once the disruption has occurred, as opposed to the 'well-founded fear of persecution' of the refugee regime, which looks into the future and does not require a material persecution to have taken place already. Most importantly, the use of the words 'because of' instead of any other softer expression, such as 'following a marked environmental disruption' or 'after a marked environmental disruption' point out that, in order to qualify as an 'environmental refugee', it was necessary to prove a *causal link* between the existence of an environmental disruption and the resulting specific human displacement. This has led to a definitional impasse over the term 'environmental refugees' (which *a fortiori* also affected the sub-category of 'climate refugees'), for, even assuming that a 'disruption' could be identified objectively, the place where such disruption occurs, in the mind of the migrant, is inescapably subjective. Some sectors in favour of recognising human migration caused by environmental disruption as a distinct category called for the introduction of the migrant's motivation as an element of the category. Christel Cournil and Pierre Mazzega explain that the notion of motivation refers to the numerous complex and interdependent factors converging in a person's psychology and which eventually lead the person to abandon the traditional habitat. This permits reintroduction of the notion of 'coercion', central to the characterisation of 'traditional' refugees fearing persecution.[350] Moreover, eclectic positions have sought to reconcile both outlooks. For instance, Diane Bates suggested a

348 This point, as well as a more detailed account of the characteristics of the refugee regime, is developed further in section 2.3, below.

349 For a general account of the issues raised by the definition provided in the UNEP report, see Jessica B. Cooper, 'Environmental Refugees: Meeting the Requirements of the Refugee Definition', *New York University Environmental Law Journal*, vol. 6 (1998), 480–529; and David Keane, 'The Environmental Causes and Consequences of Migration: a Search for the Meaning of "Environmental Refugees"', *Georgetown International Environmental Law Review*, vol. 16 (2004), 209–223.

350 See Christel Cournil and Pierre Mazzega, 'Catastrophes écologiques et flux migratoires: comment protéger les "réfugiés écologiques"?', *Revue Européenne de Droit de l'Environnement*, vol. 6 (Décembre 2006), 417–427.

distinction between three sub-categories of 'environmental refugees', according to the different degrees of free will left by the intensity and immediacy of the environmental disruption: (1) extreme cases leaving no space for free choice (natural disasters, close to 'the coercion of nature'); (2) mid-level cases with a high degree of coercion, but leaving some time for the person to elaborate an adaptation strategy; (3) low-level cases of simple environmental degradation where long-term socio-economic factors can be evaluated.[351]

Faced with the lack of objective data on the links between the existence of an environmental disruption and human migration flows, legal accounts reacted with the expression of a normative position. This shift from description to prescription resulted in polarised positions about which situations *should* be covered or excluded from the 'environmental refugee' category, or even whether such a category should be recognised at all. Susan Martin points out that apocalyptic scenarios of the issue generally come from environmentalists, whereas migration lawyers and security studies experts tend to be cautious about or even deny the existence of migration flows caused (exclusively) by environmental disruptions.[352] Proponents from the 'environmentalist sector' generally base their normative positions on the data provided by Norman Myers, who estimated that, at present, twenty-five million people could fall within the category, a number which could possibly increase up to two hundred million people.[353] In contrast, the group of opponents is more heterogeneous. While the migration law sector fears that an excessive extension of the refugee regime may diminish the protection of 'classical' refugees, opposition from the field of security studies pertains to a wider intra-disciplinary debate about the revision of the notion of security, which was launched after the end of the Cold War. Some branches of security studies (such as the Copenhagen School) then argued in favor of a holistic notion of security, which would not only encompass its traditional politico-legal dimension, but would also integrate economic, social and environmental sectors into it.[354]

---

351 See Diane Bates, 'Environmental Refugees: Classifying Human Migration Caused by Environmental Change', *Population and Environment*, vol. 23 (2002): 5, 465–477.

352 In Susan Martin, 'Climate Change, Migration and Governance', *Global Governance*, vol. 16 (2010), 397–414.

353 See, Norman Myers, 'Environmental Refugees', *Population and Environment*, vol. 19 (1997): 2, 167–182; and the widely cited alarmist account by Robert Kaplan, 'The Coming Anarchy', *The Atlantic Magazine*, (February 1994), available online at: http://www.theatlantic.com/magazine/archive/1994/02/the-coming-anarchy/304670/?single_page=true (5 December 2014).

354 See Barry Buzan, Ole Waever and Jaap De Wilde, *Security: A New Framework for Analysis* (Boulder, Colo.: Lynne Rienner Publishers, 1998).

More than twenty years after the release of the El-Hinnawi Report – first precedent in the international arena that indicated a trend towards the recognition of a category of 'environmental refugees' – the issue is back to the front line within the specific framework of climate change. Most recent scholarly developments on this issue have been undoubtedly connected with the increased concern about this issue shown by some international specialised institutions and agencies which have conducted workshops on this topic, and sometimes even incorporated it into their work programmes. As a result, migration flows induced by any of the environmental disruptions associated with climate change began being referred to as 'climate refugees' or 'climate-displaced people'. Although climate change has arguably become the paradigm of modern ecological challenge and is usually bound to images different from those in the mid-1980s, the structure of present works on the protection of 'climate refugees' generally follows the same pattern.

Legal accounts on the 'climate refugees' issues primarily focus on describing and evaluating the current state of the international legal framework within which such situations may fall. At least three international regimes have been scrutinised by different scholars engaged in this evaluation: international refugee law, international human rights law; and international climate change law. Most assessments consider that the real problem is either the lack of regulation or its inappropriateness to tackle the phenomenon in question. Thereby, the overall conclusion is generally formulated as a problem of regulatory scarcity.[355] The lack of or the inappropriateness of international regulation dealing with 'climate refugees' should not come as a surprise, for migration remains an issue primarily regulated at the national level (or regional level for States Parties to a regional integration organisation like the European Union). As policy decisions in this field tend to be closely associated with State sovereignty, bottom-up legal developments – from the national regulation to its eventual recognition in the international arena – are thus more likely to be found in relation to migration issues than in top-down law-making – from adoption of an international agreement to its implementation at the domestic level.

If most regulations of migration flows are in principle adopted at the domestic level, international refugee law is an exception.[356] Given the generalised

355 Sergio Marchi, 'Global Governance: Migration's Next', *Global Governance*, vol. 16 (2010), 323–329, 323.

356 The two main international instruments that will be referred to in this thesis are the 1951 Convention Relating to the Status of Refugees (adopted 22 July 1951; entered into force 22 April 1954) 189 UNTS 137; and its Protocol Relating to the Status of Refugees (adopted 31 January 1967; entered into force 4 October 1967) 606 UNTS 267.

use of the 'climate refugee' formula, attention turned towards this international regime when seeking to assess the level of response that the international legal order can provide in these situations. However, the results of this first exploration were quickly disappointing. International refugee law is a regime characterised by the specific historical context from which it emerged, and the correlative political objective embedded in the regulation. It originally sought to respond to the case of numerous Europeans who had been displaced by the Second World War.[357] This strong chronological and geographical specificity explains the inherent tendency of the regime to remain narrow in scope. The 'fear of persecution' by the authorities of the migrant's State of citizenship must be the cause of the displacement for the refugee status to be recognised. This central condition thus requires a persecution of a specific nature and a high degree of coercion attached to it. On the one hand, the difficulty of seeing how an environmental disruption induced by climate change could in any way be conceived as a persecution induced by political, ethnic, racial, religious or social motivations. On the other hand, most situations of acute environmental stress caused by disruptions potentially associated with climate change do not seem to fulfil the essential requirement of coercion either (apart from cases of unexpected and grave natural disasters). Finally, the inherently narrow scope of application of international refugee law is also seen as unsuited to cover the situation of 'climate refugees', for it requires the proof of an individualised or personalised persecution and does not grant refugee status to internally displaced people, two figures which contrast with the general acknowledgment that 'climate refugees' will take the form of mass flows[358] and may not necessarily lead to transboundary displacements. To some extent, the potential openness of the regime to cover these situations may be evaluated by pointing out the trend particularly followed by some EU Member States. When discussions about the evolution of the EU Temporary Protection Directive arose,[359]

357 See for instance, Jérôme Elie, 'The Historical Roots of Co-operation between the UN High Commissioner for Refugees and the International Organisation for Migration', *Global Governance*, vol. 12 (2010), 345–360.

358 See Satvinder Singh Juss, *International Migration and Global Justice* (Hampshire, UK/ Burlington, USA: Ashgate Publishing Ltd., 2006), 211 – highlighting that governmental oppression is very rarely the specific cause of environmental refugees' displacement, particularly because of the need to prove that the persecution is personal or individualised.

359 Council Directive 21/55/EC, of 20 July 2001, *on minimum standards for giving temporary protection in the event of mass influx of displaced persons and on measures promoting a balance of efforts between Member States in receiving such persons and bearing the consequences thereof*, Official Journal of the European Communities, 7 August 2001, L 212/12–23. Although this Directive is a regional instrument pertaining to the field of asylum law, it

Finland suggested introducing an explicit reference to 'climate refugees' as a specific category falling within the scope of Article 2, recognising that the European temporary protection could be granted to:

> [W]ho may fall within the scope of Article 1A of the Geneva Convention or other international or national instruments giving international protection, in particular (...) persons at serious risk of, or who have been the victims of, systematic or generalised violations of their human rights.'[360]

Although opposition from Spain and Belgium prevented the Finnish proposal from being adopted in the EU instrument, Member States remained free to widen the scope of application of the protection when transposing the objectives of the European regulation to their national legal orders. This possibility was followed by Finland and Sweden, whose respective national legislations currently provide for specific protection of 'environmental refugees'.[361]

In turn, scholars have addressed the possible solutions that the human rights regime may provide to protect climate-induced displaced people, either as an alternative or as a complementary regime to refugee law. Jane McAdam, for instance, is one of the scholars who have consistently argued against the characterisation of climate-induced displaced people as 'refugees'. Instead, she emphasises the importance of the human rights regime in the context of complementary protection for these cases.[362] According to that line of reasoning, the inapplicability of international refugee law for cases of 'climate

---

accepts a wider scope of applicability than the 1951 Refugee Convention and its Protocol, and seemingly becomes also an instrument of temporary human rights protection through Article 2.

360 Following this line, the United Nations High Commissioner for Refugees recommended in 2008 granting protection to people affected by climate change, natural disasters, and other forms of acute distress, UNHCR 'Climate Change, the Environment, Natural Disasters and Human Displacement: A UNHCR Perspective', (Policy and Evaluation Service, 29-01-08). See also the results of a seminar held by the International Organisation for Migration in the same year, International Organisation for Migration, 'Expert Seminar: Migration and the Environment', *International Dialogue on Migration*, vol. 10 (2008), 1–107.

361 See The Finnish Aliens Act (301/2004, amendments up to 973/2007 included), Section 109(1); and The Swedish Aliens Act (2005/716). See also Vikram Kolmannskog and Finn Myrstad, 'Environmental Displacement in European Asylum Law', *European Journal of Migration and Law*, vol. 11 (2009), 213–326, 316, highlighting that a press release from the United Kingdom Foreign Office seemingly indicated a possible evolution in this sense.

362 McAdam, *Climate Change, Forced Migration and International Law*, chapters 2 and 3, 39–98. This point is developed in section 2.3.2.A., below.

refugees' can be overcome by international human rights law, which is characterised by its wide material and personal scope of application as well as by its universal vocation. By definition, international human rights law is applicable to 'every human being'.[363] Besides, the fact that most international human rights law instruments only timidly (if ever) recognise an autonomous human right to the environment is not an obstacle to the claims of climate-induced displaced people. The severe consequences of acute environmental disruption enable basing their claims on 'core' human rights, rather than on the 'right to a healthy environment' (which has not received strong judicial recognition yet).[364] Arguably, the displaced people could claim the application of the principle of *non-refoulement* to situations where socio-economic forms of harm affect the enjoyment of certain rights – such as the right to health or shelter. The affected rights may be characterised independently, or in connection with civil and political rights such as the right to life, or the protection against 'inhuman or degrading treatment'.[365] Likewise, the protection of traditional knowledge may be at stake, the most paradigmatic case at present being that of the Inuit People.[366]

---

363 See, for instance, United Nations Declaration on Human Rights (adopted 10 December 1948), Preamble *in fine* ("The General Assembly proclaims this Universal Declaration on Human Rights as a common standard of achievement for *all peoples and all nations* (…)"); or the International Covenant on Civil and Political Rights (adopted 16 December 1966, entered into force 23 March 1976) 999 UNTS 171, Article 6 ('*every human being* has a right to life'); Article 7 ('*no one* should be subjected to torture'), Article 8 ('*no one* shall be held in slavery'), Article 9 ('*everyone* has a right to liberty and security'). [All emphasis added].

364 See for instance, Inter-American Court of Human Rights, *Petition to the Inter-American Court of Human Rights Seeking Relief from Violations Resulting from Global Warming Caused by Acts or Omissions of the United States*, submitted by Sheila Watt-Cloutier with the support of the Circumpolar Inuit Conference, on Behalf of All Inuit of the Arctic Regions of the United States and Canada, 7 December 2005. For more information, see the website of the Circumpolar Inuit Conference, available at: www.inuit.org (25 November 2014).

365 McAdam, *Climate Change, Forced Migration and International Law*, 10.

366 This matter has recently become an area of co-operation and study by UNESCO. See the publication launched by UNESCO with United Nations University: Douglas J. Nakashima et al., *Weathering Uncertainty: Traditional Knowledge for Climate Change Adaptation and Assessment* (Paris/Darwin: UNESCO/UNU, 2012), 1–120. In addition to this publication, the UNESCO Institute for Information Technologies in Education (UNESCO IITE) with the support of the UNESCO's Intersectoral Platform on Climate Change Mitigation and Adaptation has developed and launched the new website 'Climate Change Adaptation: Traditional Knowledge of Indigenous Peoples Inhabiting the Arctic and Far North', available at: http://iite.unesco.org/courses/climate_change/en/index.html (17 January 2015).

In this context, the report of the Inter-Agency Standing Committee (IASC) lists four main categories of human rights which, following an acute environmental disruption, are very likely to be curtailed: (1) protection of life, security and physical, mental and moral integrity; (2) protection of rights related to the basic needs of life; (3) protection of socio-economic and cultural rights; and (4) protection of other civil and political rights. These categories follow an order established on the basis of the 'urgent need of protection' criterion.[367] The work of this agency highlights the fact that the factual context in which 'climate refugees' emerge can affect the system of human rights protection just as exceptionally as armed conflict situations do.

However, contrary to armed conflict situations (which are covered by international humanitarian law), the discernible factual context which produces 'climate refugees' does not gives rise to the application of a specific set of rules operating as *lex specialis*. Aware of this gap, the work of the IASC points to a possible alternative path for future protection of 'climate refugees' under international human rights law. This would consist of the recognition of an emerging obligation of the State of origin to build up an effective system of risk prevention *prior to* the manifestation of the acute environmental disruption. Hence, under this perspective, the State's lack of preventive action would be tantamount to a human rights 'violation'. The *ex ante* character of the protection is central to this arguably controversial perspective, for once the environmental disruption has crystallised the State may suffer from a structural deficiency affecting its capacity to protect the population. The violation itself would therefore consist of the negligence of the State in not acting before the onset of an acute environmental disruption which diminished its capacity to protect and assist its population thereafter. The 'human rights violations' would therefore not stem from the environmental disruption as such, but from the incapacity of the State to continue guaranteeing the protection of its population's human rights faced with the disaster. In this sense, Vikram Kolmannskog

---

367 Inter-Agency Standing Committee, *Operational Guidelines for the Protection of Human Rights in Natural Disasters* (Brookings-Bern Project on Internal Displacement: Washington D.C., 2006), followed by a manual conceived to guide the action in situ, IASC, '*Human Rights and Natural Disasters, Operational Guidelines and Field Manual on Human Rights Protection in Situations of Natural Disasters* (Brooking-Bern Project on Internal Displacement: Washington D.C., pilot version, March 2008). See also, International Federation of Red Cross and Red Crescent Societies, *Preparedness for Climate Change: a Study to Assess the Future Impact of Climatic Changes upon the Frequency and Severity of Disasters and the Implications for Humanitarian Responses and Preparedness*, 2–6 December 2003, (Report Prepared for the 28th International Conference of the Red Cross and Red Crescent).

and Finn Myrstad have argued that the EU Temporary Protection Directive could be applicable to cases of 'mass flows' of 'climate refugees', since acute environmental disruptions can affect the institutional structure of the State in a similar way to armed conflicts.[368]

Finally, given that the central element of characterisation as a distinct category is primarily drawn from the cause of their displacement, namely, an acute *environmental* disruption, a look into the possibilities that international environmental law may offer for the protection and governance of 'climate refugees' was bound to be explored by international scholars. At the beginning of the 1990s, voices had already been raised in favour of constructing an international environmental regime into which the issue of 'climate refugees' would be integrated. When the 1992 Rio Summit on Environment and Development took place and the three Framework Conventions on climate change, on biological diversity and on desertification were adopted,[369] Gregory McCue argued that the increased international co-operation in the protection of the environment, on the one hand, and the issue of migration flows induced by environmental crisis, on the other, were bound by a cause and effect link.[370] Therefore, McCue argued, setting both issues apart would damage the effectiveness of any attempt to find solutions to the global environmental crisis.[371]

In sum, the findings of the studies on the possibilities offered by the international legal order for the protection and governance of 'climate refugees' can be summarised as follows: (1) international refugee law is too context-specific and closely bound to State sovereignty to enable the automatic inclusion of 'climate refugees' within its scope of application *ratione personae*; (2) despite international human rights law's wider scope of application, the regime is not effective enough either, for it has not so far recognised acute environmental disruptions as a specific context in which curtailment of human rights may occur. Thus, the regime leaves 'climate refugees' legally 'unarmed' both vis-à-vis the country of origin (by limiting their claim of human rights violation

368 Kolmannskog and Myrstad, 316–317.

369 United Nations Framework Convention on Climate Change (UNFCCC) (adopted 9 May 1992, entered into force 24 March 1994), 31 ILM 849 (1992); Convention to Combat Desertification in Those Countries Experiencing Serious Drought and/or Desertification, Particularly in Africa (UNCCD) (adopted 17 June 1994; entered into force 26 December 1996), ILM 1328 (1994); and Convention on Biological Diversity (CBD) (adopted 5 June 1992, entered into force 29 December 1993), 31 ILM 822 (1992).

370 See for instance, Gregory S. McCue, 'Environmental Refugees: Applying International Environmental Law to Involuntary Migration', *Georgetown International Environmental Law Review*, vol. 6 (1993–1994), 151–190.

371 Ibid.

against their own State) and also vis-à-vis the country of arrival. Finally, (3) the international regime on climate change centres on an inter-State agreement which is built up independently of non-State actors, so that any attempt to link the regime with the recognition of 'climate refugees' would be bound to meet this structural obstacle. All these findings lead to a shared and unique diagnosis which can be summed up as the lack of appropriate international regulation. Henceforth, proposals to respond to these lacunae followed this diagnosis. These have taken different forms and have been conceived at different levels of analysis, yet they all share a common ground – namely, that further normative and institutional development is needed.[372] From this standpoint, three main strands of development have been suggested in the literature on 'climate refugees': reform of the existing instruments; creation of a new instrument; and/or development of synergies among existing legal regimes.

B Limited Normative and Institutional Solutions Proposed

Proposals of normative reforms suggest that the scope of application of the 1951 Refugee Convention and the 1967 Refugee Protocol may be extended to cover 'climate refugees', either by amendment of the Convention itself, through wide interpretation of the instrument by the relevant international institutions, or through specific regional developments of the regime, such as that within the EU. Given the political difficulty associated with treaty amendment procedures and the context-specific rationale of the regime, proposals have generally focused on more indirect ways of obtaining the protection of the refugee regime, such as the use of the margin of interpretation that international institutions have regarding the instruments regulating their mandates. Thus, the UNHCR could, for instance, be asked to endorse an evolutive interpretation of the High Commissioner's mandate, as it was already called upon to do, soon after its creation, to permit its continuity beyond the three-year mandate that had originally been agreed. Following this line, it is worth noting that, in 2008, the High Commissioner released a recommendation to the European Union in favour of incorporating 'climate refugees' within the scope of application of the EU's Temporary Protection Directive.[373] Yet, notwithstanding this openness, the UNHCR has not changed its position when it comes to dealing with the scope of application of *its own* mandate. As Karen McNamara pointed out, by not recognising environmental disruptions as one of the possible 'forms of victimisation' which may be a ground for requesting the Commissioner's

372 See, for instance, Massimiliano Montini, 'Reshaping Climate Change Governance for Post-2012', *European Journal of Legal Studies*, vol. 4 (2011): 1, 8–24, 17–21.

373 Ibid.

assistance, the Commissioner maintains a very strict interpretation of both the terms of his Commission's mandate and of the conditions set up in the definition of 'refugee' in Article 1(2)(A) of the Refugee Convention.[374]

Proposals of normative reform not only encompass the revision of the international instruments of refugee law, but may also refer to regional regulations and, most particularly, those developed within a space potentially receiving 'climate refugees', such as the European Union.[375] In this sense, advocates of normative reform through regional arrangements have found support in the evolution of the 2004 EU Qualification Directive. Article 2(e) of the Directive indeed recognises that claimants who have been denied refugee status as defined in Article 1(2) of the Refugee Convention, may eventually benefit from a subsidiary protection regime, had they faced a "real threat of suffering a serious harm".[376] Although the Commissioner's final list of the acts and facts that can be considered as being likely to provoke 'serious harm' does not include grave environmental disruptions, the issue was discussed at the European Commission. This debate may find its roots in the 2002 petition from the European Parliament calling for the protection of transboundary migrants resulting from acute environmental disruptions.[377] Even more paradigmatic was the proposal for a *European Declaration on the Status of Environmental*

---

374 This conclusion has been reached in, for instance, a study based on 45 interviews of UN ambassadors and senior diplomats related to the institutions; see Karen E. McNamara, 'Conceptualising Discourses on Environmental Refugees at the United Nations', *Population and Environment*, vol. 29 (2007), 12–24. However, more recent accounts seem to point to a more open position of the UN since 2009. See, for instance, Experts Round Table, 'Climate Change and Displacement: Identifying Gaps and Responses. Discussion document available at: http://www.acnur.org/t3/que-hace/medio-ambiente/ (25 November 2014).

375 Christel Cournil, 'Les réfugiés écologiques: quelle(s) protection(s), quel(s) statut(s)?', *Revue de Droit Public*, vol. 4 (2006), 1035–1066.

376 Council Directive 2004/83/CE, 29 April 2004, *on minimum standards for the qualification and status of third-country nationals or stateless persons as refugees or as persons who otherwise need international protection and the content of the protection granted*, Official Journal of the European Union, 30 September 2004, L 304/12ff. This Directive was the result of a process initiated in 1999 by the 'Tampere Decision' to work towards the creation of a common European asylum system. European Council, Conclusions of the Presidency, adopted in Tampere on 15–16 October 1999, available at http://www.europarl.europa.eu/summits/tam_es.htm (25 November 2014).

377 European Parliament, *Report on the proposal for a directive of the European Parliament and of the Council on minimum standards for the qualification and status of third-country nationals or stateless persons as beneficiaries of international protection and the content of the protection granted*, COM (2001) 510-C5–0573/2001–2001/0207 (CNS), Committee on Civil Liberties, Justice and Home Affairs (Rapporteur Jean Lambert), 2002.

Refugees, of 6 April 2004, n°9/2004 (received 28 signatures from the European Green Party), which was not however adopted by the European Parliament. Moreover, it is worth recalling that the European Commission has explored the relationship between climate change and migration when discussing the establishment of a common European migration policy.[378] It has also been suggested that the protection of 'climate refugees' may be achieved through international or national judicial institutions, which could consider 'climate aggression' as falling within the evolutive interpretation of the notion of persecution; accept the 'right to environmental asylum'; or have recourse to alternative and less controversial notions, such as that of 'non-refugee survival migrants' suggested by Alexander Betts.[379] This perspective suggests extracting the common ground of all climate-induced migration, irrespective of the effect of the degree of coercion and the urgency of the environmental disruption on the migrant's psychology, so as to bind all these 'types' of migration under a single common triggering element: the search for survival.[380]

The second category of solutions proposed in reaction to the diagnosed problem of 'regulatory scarcity' suggests that the adoption of a new international convention for the protection of 'climate refugees' would be most suitable.[381] The advantage of this proposal, put forward by, for instance, Franck

378 See European Commission, *Communication on Strengthening the Global Approach to Migration: Increasing Co-ordination, Coherence and Synergies*, COM (2008) 611/3, 8 October 2008. On the recent increase in European concern about the issue of climate-induced displacement, see Nicole Itano, 'Europe Fears "Climate Migrants"', *Christian Science Monitor*, vol. 100 (2008): 76, 6.

379 Alexander Betts, 'Survival Migration: a New Protection Framework', *Global Governance*, vol. 16 (2010), 361–382. Betts' proposal has been backed by civil society actors; see Oxfam International, *The Right to Survive: the Humanitarian Challenge for the 21st Century* (Oxford: Oxfam International, 2009).

380 Ibid.

381 Frank Biermann and Ingrid Boas, 'Protecting Climate Refugees: The Case for a Global Protocol', *Environment Magazine*, vol. 50 (2008): 6, 10–16: and, by the same authors, 'Global Adaptation Governance: the Case of Protecting Climate Refugees', in Frank Biermann, Philipp Pattberg and Fariborz Zelli, *Global Climate Governance Beyond 2012: Architecture, Agency and Adaptation* (Cambridge: Cambridge University Press, 2010), 255–269. Following the same line, Mike Hulme, 'Climate Refugees, Cause for a New Agreement?', *Environment*, vol. 50 (November/December 2008): 6, 50–52; Bonnie Docherty and Tyler Giannini, 'Confronting a Rising Tide: a Proposal for a Convention on Climate Change Refugees', *Harvard Environmental Law Review*, vol. 53 (2009), 49–403; David Hodgkinson and Lucy Young, 'In the Face of a Looming Catastrophe: a Convention for Climate Change Displaced Persons', (2012), available at www.ccdpconvention.com (25 November 2014), based on previous works, such as David Hodgkinson et al., 'The Hour When the Ship

Biermann, Ingrid Boas and David Hodgkinson, is that it enables taking full account of the specific contextual dimension of climate-induced displaced people, and enables to construct a system of protection tailored to the specific characteristics of these situations. Besides, the proposal may even be mirrored at a national level through the creation of a right to environmental asylum. For instance, the Australian Labor Party called for an international coalition to accept climate-induced migrants from the Pacific, in response to the government's position, which was against the recognition of such a category, and the Australian Green Party (Senator Kerry Nettled) put forward a Migration (Climate Refugees) Amendment Act in 2007, which sought the revision of the Australian 1958 Migration Act by introducing a definition of 'climate change induced environmental disaster' and a clause on the granting of 'climate refugees visas'.[382] Nonetheless, co-operation on this matter at the international level remains highly controversial and, therefore, the adoption of a Climate Refugee Convention as a Protocol to the UNFCCC remains unlikely to crystallise in the short term.

Finally, alternative proposals stress the potential utility of developing inter-regime synergies to complement adaptation mechanisms of the international regime on climate change. Angela Williams has thus suggested that these synergies could be displayed through the development of the regional agreements operating under the UNFCCC.[383] This proposal remains close to the classical conception of migration as a form of adaptation to the changing conditions of the habitat, and bases the normative connection on the obligation of State Parties to the Convention and to the Kyoto Protocol to co-operate and help finance adaptation measures (Article 4(e) UNFCCC). Given that migration flows tend to take place within the same region for reasons of geographical – sometimes also cultural – proximity, the proposal to enhance regional co-operation in these matters has, in principle, various advantages: (1) it is practical and potentially more effective, since co-operation at this level helps tailoring the regulation to the specific reality of the region; (2) it avoids relying on the political will of a large number of States to adopt

Comes in: a Convention for Persons Displaced by Climate Change', *Monash University Law Review*, vol. 36 (2010), 69; David Hodgkinson et al., 'Copenhagen, Climate Change "Refugees" and the Need for a Global Agreement', *Public Policy*, vol. 4 (2009), 155–174; and David Hodgkinson et al., 'Towards a Convention for People Displaced by Climate Change', *New Critic*, vol. 8 (2008), 1–6, available at: http://www.ias.uwa.edu.au/new-critic (25 November 2014).

382 Amended text available at: http://www.austlii.edu.au/au/legis/cth/bill/mrab2007342/ (25 November 2014).

383 Angela Williams, 'Turning the Tide: Recognising Climate Change Refugees in International Law', *Law and Policy*, vol. 30 (October 2008): 4, 502–529.

an international regulation for 'climate refugees'; and (3) the preservation of the strict definition of refugee contained in the 1951 Refugee Convention, as understood so far, is ensured.

Moreover, another proposal advancing the benefits of inter-regime connections suggests approaching climate change from a human rights perspective.[384] This position stems from the fact that the adverse effects of the phenomenon are very likely to affect and even obstruct the object and purpose of the main international human rights instruments, previously elaborated for the protection of citizens from abuses by their own States.[385] Moreover, the international regime on climate change, built to decrease the aggravation of global warming (mitigation policies) and to halt its adverse effects once it becomes inevitable (adaptation policies), may create side-effects in the realisation of human rights. Mitigation policies establish the future distribution and usage of natural resources among States with no concern for the effects that such new redistribution may have on individuals. Finally, adaptation policies, which are by nature more suited to address the new climate change challenges, such as 'climate refugees', remain insufficiently financed and underdeveloped. However, the main difficulty of this proposal is to reconcile the two underlying rationales of each international regime, which are different in nature and structure, so as to regulate the issue by integrating both dimensions in a coherent and holistic way.[386]

Undoubtedly, the three international regimes, established *prima facie* to operate in cases of climate-induced displacement, which have already been dealt with in international legal scholarship, are seemingly appropriate to address different types and stages of relocation taking place in and

384 See Stephen Humphreys, ed., *Human Rights and Climate Change* (Cambridge: Cambridge University Press, 2010). This book is the result of a collective contribution and results from the 2008 programme initiated at the United Nations Human Rights Council by Resolution 7/23, 'Human Rights and Climate Change', adopted on 28 March 2008. Information about the programme currently in course is available at: http://www2.ohchr.org/english/issues/climatechange/study.htm (25 November 2014).

385 See, for instance, Margaux J. Hall and David C. Weiss, 'Avoiding Adaptation Apartheid: Climate Change Adaptation and Human rights Law', *Yale Journal of International Law*, vol. 37 (2012), 309–364; Marc Limon, 'Human Rights and Climate Change: Constructing a Case for Political Action', *Harvard Environmental Law Review*, vol. 33 (2009), 439–476; and Siobhan McInerney-Lankford, 'Climate Change and Human Rights: an Introduction to Legal Issues, *Harvard Environmental Law Journal*, vol. 33 (2009), 431–437.

386 The discourse on the protection of global commons, inclusive and communitarian, contrasts with the rationale of the human rights discourse, fundamentally individualistic and adversarial. See Stephen Humphreys, 'Competing Claims: Human Rights and Climate Harms', in Humphreys, ed., *Human Rights and Climate Change*, 37–69.

among Pacific Island States, described above. Firstly, as already pointed out in this section, international environmental law – particularly climate change adaptation law and policy – is already an indispensable tool for preventing the excessive degradation of the habitability conditions of settlements in the Pacific that may lead to international or transnational relocations. The regional co-operation among all Pacific Island States, conducted multilaterally through the SPREP Pacific Adaptation Project, as well as bilaterally through Joint National Action Plans, are currently vivid examples of the relevance of this regime to the issue of climate-induced displacement. Moreover, it is worth analysing whether transnational climate-induced displacements of I-Kiribati and Tuvaluans to Fiji, Australia or New Zealand may be covered by international refugee law, so as to eventually provide them with legal protection tools in 'host countries'.[387] Finally, in all situations indicated in this study, whether at a preventive or reactive stage of the relocation process, and regardless of whether the displacement is national or transnational, a wide range of Pacific Islanders' substantive and procedural human rights are fundamentally at stake.

Nonetheless, the pitfall of the approaches suggested lies in the fact that they tend to focus exclusively on one of the three legal avenues. As the issue of climate-induced displacement and relocation is analysed from one angle – mainly human rights, refugee law or international environmental law – solutions emerging from the identification of the normative and institutional deficiencies existing in each of the three regimes are also conceived from a unidirectional perspective. Although the proposals formulated so far are multiple and diverse – ranging from normative development and reform to inter-regime solution synergies, institutional co-operation and judicial activism – they are all based on an equally normative blueprint, which does not match the inherently heterogeneous – and therefore particularistic – condition of climate-induced relocations in and among Pacific Island States depicted above. Therefore, while acknowledging the relevance of the regimes analysed, as well as the solutions proposed so far, it is fundamental to consider that a legal framework must be able to adapt to the specificities of each relocation case, and to integrate the socio-cultural, geographical and political variables explained above. Rather than prioritising the use of problematic top-down theoretical approaches, which are then tested in pilot countries and often lead to definitional impasses, the reality displayed by this regional case study invites the elaboration of an inductive, bottom-up legal scheme.

387 On this point, see section 2.3.2.A., below.

2.2.2 *Proposal for a New Integrative and Pluralistic Approach*

A Widening the Spectrum of Applicable Legal Regimes

Vladimir Soloyov compellingly wrote of "a living man with his inseparable and essential national characteristics", who turns unto into "an empty abstract subject with all his determining peculiarities left out".[388] Taking stock of this ethical dictum, the present section suggests approaching the issue of climate-induced relocation through a scheme consisting of several layers of legal protection. Such a scheme contemplates and includes the protection that can be offered by the three legal regimes laid out above and already analysed in international legal scholarship, but not limited to them. While avoiding falling into an excessively casuistic approach to climate-induced relocation in the region, the multi-layered legal scheme is proposed to provide a legal translation of the factual heterogeneity previously highlighted. This approach is meant to acknowledge the variability of forms of vulnerability and the respective specific needs resulting from the variables previously identified. Such factual differences are thus the source from which the structure of the scheme emerges, which has the advantage of introducing a certain degree of flexibility into the comparison with the existing monolithic top-down approaches to this issue, while simultaneously pointing to new and unexplored legal avenues which may be relevant in cases of climate-induced relocation. These become all the more visible when the configuration of this multi-layered legal scheme (which will only consider treaty law) is based on the role the different States involved in climate-induced relocation cases are called on to play. To some extent, this proposal seeks to respond to what Jane McAdam has already defined as the *fragmentation of climate-displacement governance* by structuring such governance along one coherent axis. The study is intended to cover Pacific Islanders or the 'population' of Pacific Island States.[389] As a matter of principle, the category of 'population' will be understood in a wide sense covering all people subject to the (territorial) jurisdiction or control of a Pacific Island State, including – but not limited to – the citizens living in any of the countries under study, as well their permanent residents.

388 Vladimir S. Solovyov, *The Justification of the Good: An Essay on Moral Philosophy* (New York: Cosimo Inc. Publishers, 2010).

389 See Jane McAdam, *Environmental Displacement Governance*, UNSLRS Paper 1 (2010), 1–33.

TABLE 5 *Multi-layered legal scheme for climate-induced relocations in and among Pacific Island States*

RELEVANT LEGAL REGIME	SAFE STATEHOOD		ENDANGERED STATEHOOD			
	National displacement		Transnational displacement			
			Partial		*Total*	
	Prevention	*Relocation*	State of origin	Host State	State of origin	Host State
International Human Rights Law	Yes	Yes	Yes	Yes	Yes	Yes
National Human Rights Law (Constitutional Law)	Yes	Yes	Yes	Yes	Yes	Yes
International Environmental Law	Yes	Yes	No	No	No	No
National Environmental Law	Yes	Yes	No	No	No	No
International Migration Law	No	No	Yes	Yes	Yes	Yes
National Migration Law	No	No	Yes	Yes	No	Yes
Guiding Principles on Internally Displaced People	No	Yes	No	No	No	No
International Refugee Law	No	No	Yes	Yes	Yes	Yes
Statelessness Regime	No	No	Yes	Yes	Yes	Yes

B (Re-)Introducing the Role of the State

The fundamental place of State actions is generally and primarily associated with the State's role in preventive relocation measures – such as the development of climate change adaptation projects with neighbouring States. To a certain extent, this role is equally pinpointed with regard to reactive relocation actions – such as the elaboration of relocation plans, provision of funding, and organisation of its implementation phase – particularly in cases where relocation takes place at a national level and the State has full competence in all stages of the relocation process. The recognition of the presence of the State in these cases limits the traditional dichotomies confronting Humankind and the Environment, on the one hand, and the Individual against the State, on the other.[390] Instead, it invites the conceptualisation of a tri-dimensional relationship, in which the protection of the people against environmental disruptions of their habitat is ensured by the active role of the State in which such habitat lies. As table 4 showed, the role varies according to the specific needs of each case; whereas in bottom-up approaches to relocation planning, the State acts primarily as a co-ordinator of the actions elaborated and undertaken by provincial and local actors; top-down approaches are bound to endow the State with a more active role, in which it is called upon to design, finance and manage relocation plans, and even negotiate the international agreements that may be necessary for their implementation.[391]

Yet, there is a second and equally fundamental function of the State in the context of climate-induced relocation which has generally and consistently been neglected. While acting as an actor seeking to guarantee the basic living conditions of its population, the State *itself* becomes simultaneously the subject of the protection of *its own* continuity as a distinct independent political entity. This secondary subject of protection is particularly visible when taking into consideration the threat that large amounts of transnational displacements may represent for the continuity of Pacific Island States. In these cases, the protection of the State's continuity is a sort of by-product of the actions seeking to protect the population affected by relocation, since all contributions to the pre-emption of transnational climate-induced displacement may also be understood as being, in the long term, strategies for the preservation of Pacific Islands' statehood. Simply put, the fate of Pacific Islanders is bound to the fate of their respective Pacific Island State.

The multi-layered legal scheme proposed in table 5, above, is based on one main factor, namely, whether the continuity of the State is or is not at stake by

390 Binary divisions stem from enlightenment and French rationalism. They are also at the root of the anthropocentric view of environmental protection, which has been criticised extensively by, for instance, Catherine Redgwell, 'Life, the Universe and Everything: a Critique of Anthropocentric Rights', in Alan E. Boyle and Michael R. Anderson, eds., *Human Rights Approaches to Environmental Protection* (Oxford: Clarendon Paperbacks, 1996), 71–108.

391 On the specificities of the role of the State in environmental protection contexts, see Thilo Marauhn, 'The Changing Role of the State', in Daniel Bodansky, Jutta Brunée and Ellen Hey, eds., *The Oxford Handbook of International Environmental Law* (Oxford, New York: Oxford University Press, 2007), chapter 31, 727–747.

present or potential displacements of all or a significant part of its population. Table 5 shows that it is considered that the continuity of the State is not at risk when climate-induced displacements and relocations take place at a national level – cases in which there is only one State directly engaged in the relocation. The State's obligations and responsibilities, corollary of the rights of the population affected, may be divided into two groups according to whether they belong to a relocation effected in the preventive or in the reactive phase. In cases where climate-induced displacements and relocations take place at an international or transnational level, the continuity of the State may be at risk. Although these situations are, up to now, a minority in the Pacific, they cannot be disregarded in future projections, for they undoubtedly constitute the more challenging situations for international law and policy in various areas and a potential source of important developments and controversies in the field. This situation is therefore sub-divided into two groups – cases of partial de-population, and cases of total de-population. From a legal perspective, the web of responsibilities and obligations becomes more complex, for at least two States come into play – the State of origin and the host State. All in all, table 5 shows how different areas of national and international law may be relevant to the protection of people displaced as a result of the effects of climate change, depending on the scenario in which such relocation takes place. It serves as a road-map or frame of reference in which legal analysis of the extent of the rights and duties of both the States involved and the populations affected can be situated.

However, setting up the structure of a legal scheme is not enough. To make it meaningful, it is necessary to determine or identify in detail the specific typologies, boundaries and legal content of the most relevant situations indicated in table 5. The purpose of the following section is therefore to identify how the impacts of climate change on vulnerable communities can substantiate a claim for international protection under a *variety* of regimes. Both the aim and the scope of the following section are therefore limited, in so far as: 1) it does not attempt to draw up an exhaustive list of sources of protection obligations; 2) nor does it evaluate the feasibility or likelihood of the implementation of such protection obligations; and it is focused on protection obligations resulting from treaty law. Yet, in doing so, identifying the States involved in each situation is fundamental. The next section will thus reflect the divisions shown in table 5, primarily marked by the 'contextual variable' separating situations in which climate-induced displacement takes place within the boundaries of a State – and therefore does not risk jeopardising the continuity of the State – from overseas relocation situations potentially putting at stake the continuity of the State of origin of the population.

2.3 Possible Scenarios: Applying the Multi-Layered Legal Scheme to Pacific Island States

2.3.1 *'Safe' Statehood: National Relocations*

As shown in table 4, cases of climate-induced displacement so far encountered in the South Pacific have predominantly taken place at a national level. While relocation undoubtedly affects the population involved, it does not correlatively endanger the State itself nor does it raise the question of the continuity of the State itself, for no substantial or total 'de-population' is taking place. Therefore, this section focuses on the specific content to be given to the legal scheme for the people affected by relocation in cases where the contextual variable is not only steady but also revitalised, as the State has the active task of responding to the challenge that climate change disruptions represent for the habitability of the territory in which the displaced population is settled. One is thus led to scrutinise the responsibilities and obligations of the State towards its inhabitants, both at an individual and collective level. In this setting, the State can be approached as much the primary guarantor of the continuity of the bond between a population and their territory as the ultimate guarantor of the underlying conditions necessary for the creation of such a bond with the new site – in cases where the original location of the settlements has to be evacuated.

Following the distinction laid down in the preceding section, the two main sorts of strategies for response to climate-induced relocation that the State is expected to undertake are defined by a temporal variable: preventive relocation strategies – mainly focused on strengthening adaptation capacity and resilience with respect to the habitability of the original site of the human settlements; and reactive relocation strategies, centred on the evacuation and resettlement procedure, as well as integration into the new site. When determining the content of the legal framework applicable to internal policy strategies in response to an internal issue, national law will play a role as great as – if not greater than – international law.

A Legal Framework in a Preventive Relocation Phase

Situation n° 1: Human Rights Covenants

International human rights law is undoubtedly the first international legal regime to explore in order to identify and define the 'right-holders' and 'duty bearers' in a situation involving individuals and the State in a vertical relationship. The focus in this study on the 'classical' or 'original' forms of human rights law (operating between the State and its population) in no way constitutes a negation of the applicability of this regime to horizontal individual – individual relationships. Rather, the focus on vertical effects of human rights obligations stems from the explanation of the Human Rights Committee (charged with monitoring the implementation of the International Covenant on Economic, Social and Cultural Rights, hereinafter the ICESCR), which states that the obligations under Article 2(1) of the Covenant "are binding on States [Parties] and do not, as such, have a direct horizontal effect as a matter of international law".[392]

In preventive relocation situations – where the State is called upon taking measures to preserve the *habitability conditions* of its population – a range of fundamental rights, such as the right to health, the right to food, the right to access to water and the right to appropriate sanitation are involved, essentially corresponding to rights contained in the International Covenant on Economic, Social and Cultural Rights (ICESCR). In particular, these rights are recognised by the States Parties to the Covenant in Article 11 on "the right to an adequate standard of living, including adequate food, clothing and housing, and to the continuous improvement of living conditions"; as well as in Article 12, recognising the right to the "highest attainable standard of physical and mental health", and stating that steps necessary to the full realisation of this right include: (1) the provision for the reduction of the stillbirth-rate and of infant mortality and for the healthy development of the child; (2) the improvement of all aspects of environmental and industrial hygiene; (3) the prevention, treatment and control of epidemic, endemic, occupational and other diseases; (4) the creation of conditions which would assure to all the necessary medical service and medical attention in the event of sickness. As recalled by the World Bank in its 2011 Study on Human Rights and Climate Change, it is generally acknowledged that the recognition of such rights in the Covenant does not impose on State Parties an immediate obligation of result. Rather, the realisation of economic, social and cultural rights is conceived as a progressive evolution "in accordance with the capacities of each State".[393] As the Report explains, the language of the ICESCR "contemplates that, to some degree, the rights will be met through progressive realisation rather than immediately", although this level of flexibility in their compliance with their obligations does not prevent

392 UN Human Rights Committee, General Comment n° 31, 'Nature of the General Legal Obligation of State Parties to the Covenant', (2004) UN Doc. CCPR/C/21/Rev.1/Add.13.

393 Siobhan McInerney-Lankford, Mac Darrow and Lavanya Rajamani, *Human Rights and Climate Change: a Review of the International Legal Dimensions* (Washington D.C.: World Bank, May 2011), 6.

States from taking "immediate, concrete and targeted steps towards the realisation of such rights" and showing that "they are using the maximum extent of available resources towards the realisation of the rights".[394]

Moreover, the Commission on Economic Social and Cultural Rights (CESCR) has specified three different types of duties incumbent upon the State within the territory under its jurisdiction or effective control for the realisation of the rights contained in the Covenant. These are the duty to respect, the duty to protect, and the duty to fulfil treaty-based rights.[395] The 'duty to respect' refers to the duty of the State to refrain from actions that may interfere with, impair or violate the enjoyment of the rights of individuals under its jurisdiction; the 'duty to protect', in contrast, requires a positive and more active attitude of the States, which are required to preserve the enjoyment of the rights from interference by private action; finally, the 'duty to fulfil' requires States to take action seeking to facilitate, provide and promote the rights at stake. Although through such a tri-dimensional perspective on the duties of the State, the CESCR contributed to the construction of a comprehensive and global protection of the rights in question, its application in the South Pacific region has had so far rather limited beneficial effects, given that no Pacific Island State, other than Papua New Guinea and the Solomon Islands, has ratified the ICESCR.[396]

There are many reasons possibly explaining why this region has the lowest rate of ratification of human rights treaties in the world. While the Office of the High Commission on Human Rights (Regional Office for the Pacific) emphasises how the lack of resources impairs countries' capacities to comply with

394 Ibid.

395 Ibid. For an explanation of the meaning and content of these three duties, see United Nations, Commission on Economic, Social and Cultural Rights (CESCR), General Comment n°12, 'The Right to Adequate Food', 1999, E/C.12/1999/5, and CESCR, General Comment n°14, 'The Right to the Highest Attainable Standard of Health', 2000, E/C.12/2000/4.

396 The Solomon Islands succeeded the United Kingdom as a State Party to the ICESCR on 17 March 1982 (maintaining the reservations previously made by the United Kingdom). Besides, on 29 September 2009, the Solomon Islands signed the Optional Protocol to the International Covenant on Economic, Social and Cultural Rights (adopted in New York, 2008); Papua New Guinea acceded to the ICESCR on 21 July 2008 but has not yet signed the Optional Protocol. See table 6 below for the ratification of other relevant human rights treaties. Information available at the United Nations Treaty Collection website: http://treaties.un.org/Pages/ViewDetails.aspx?src=TREATY&mtdsg_no=IV-3&chapter=4&lang=en (8 November 2014), for status of ratification of the Convention; and http://treaties.un.org/Pages/ViewDetails.aspx?src=TREATY&mtdsg_no=IV-3-a&chapter=4&lang=en, (8 November 2014), for status of ratification of the Optional Protocol.

reporting obligations,[397] factors associated with the existence of local customs that clash with the western conception of economic, social and cultural rights, cannot be disregarded.[398] Cultural determinism can indeed preclude Pacific Islands' governments from being able to make their national legislation consistent with human rights instruments, particularly in areas that are commonly regulated by old customary rules and local traditions (such as family law, land rights, women's role in the community). For instance, the welfare of Samoan families very much depends on the capacity and willingness of each family's chief or *matai* to support and represent them, to the extent that the suitability to complement this ancestral and pre-colonial system of community organisation when the *matai* system fails, has already been pointed out by the Regional Office of the Pacific of the OHCHR.[399]

Interestingly, Pacific Island States have preferred avoiding ratification altogether, instead of overcoming the points of friction between the standards imposed by human rights instruments and their local customs, by accompanying the instrument of ratification with declarations of interpretation, or even reservations they might deem necessary. This choice might be motivated by the mistaken understanding that a State must comply with the human rights standards before it becomes a Party to the Treaty – a view which, according to the latest OHCHR report on this matter, is widespread in the region. However, it may also exemplify how States are sometimes reluctant to commit to obligations contained in an international treaty negotiated and perfected while the State in question had not yet acquired full political independence. This may partly explain why ratification rates of much more recent instruments, such as the Convention on the Rights of the Child and the Convention on the Elimination of All Forms of Discrimination against Women (CEDAW), are much higher.[400] The relatively recent attempts to set the basis of a regional human rights system – following the example of the European, the Inter-American and the African systems, although not yet so ambitious – can support this analytical interpretation of the behaviour of Pacific Island States. Already in 1989,

397 See United Nations Office of the High Commissioner on Human Rights (Regional Office of the Pacific), *Ratification of International Human Rights Treaties: Added Value for the Pacific*, Discussion Paper (Suva, Fiji: OHCHR Regional Office of the Pacific, July 2009), vii.

398 See Kelly Haines-Sutherland, 'Balancing Human Rights and Customs in the Pacific Region: a Pacific Charter of Human Rights', *ANU Undergraduate Research Journal*, vol. 2 (2010), 125–141.

399 OHCHR (Regional Office of the Pacific), *Protection of Human Rights of Internally Displaced People: Challenges of the Pacific*, Discussion Paper (Fiji: Davui Printery, 2011), 7.

400 See table 6, below.

the first attempt to elaborate a Pacific Charter on Human Rights had seen the light of day. More recently, the Pacific Islands Forum – today's principal form of regional institutional co-operation – places human rights as 'a fundamental component of the vision of the Pacific Island Leaders'. Initiative 12.4 of the Pacific Plan thereby supports the "appropriate ratification and implementation of international and regional human rights conventions, covenants and agreements, and support for reporting and other requirements", which is considered as an essential step to improve the institutional governance within and among Pacific Island States.[401]

Nonetheless, in spite of these regional developments, the level of ratification of international human rights treaties falls too short to formally consider international human rights treaty law (in particular, the ICESCR) as a possible conventional source of rights that a population can oppose to its State. This indirectly provokes a heavy reliance on constitutional law recognising fundamental rights, as well as on the judicial and non-judicial institutions tasked with their protection and enforcement. According to the Regional Office of the Pacific of the OHCHR, most Pacific Island States recognise civil and political rights in their constitutions, but only timidly – if ever – mention economic, social and cultural rights.[402] To be sure, an extended comparative study of Pacific Island States' constitutional legislation and case law on the matter would be needed in order to make a precise evaluation of the level of normative recognition of Pacific Islanders' economic, social and cultural rights opposable to their State of origin. While such a delicate task falls outside the scope of the present chapter, the context makes it worth mentioning that the creation of National Human Rights Institutions (NHRI) in the region seems to be in vogue. These Institutions apply the so-called 'Paris Principles'[403] and

401 More details on the background and the currently ongoing revision of the Pacific Plan can be found at the website of the Pacific Islands Forum, including specific actions derived from it (such as collaboration with the Regional Office of the Pacific of the OHCHR: http://www.forumsec.org/pages.cfm/political-governance-security/human rights/ (8 November 2014).

402 OHCHR, *Ratification of International Human Rights Treaties*, vii.

403 The 'Paris Principles Relating to the Status of National Institutions for the Promotion of Human Rights', were adopted in 1991 by the NHRIs at an international workshop held in Paris and marked the standardisation of this new form of international human rights co-operation. The Principles were then endorsed by UN General Assembly Resolution A/RES/48/134, 'National Institutions for the Promotion and Protection of Human Rights' (with the Paris Principles Annexed), 48th session, 4 March 1994; as well as by the UN Commission on Human Rights Resolution 2005/74, 'National Institutions for the Promotion and Protection of Human Rights', 59th meeting, 20 April 2005. See UN Office of

operate as interlocutors between international human rights treaty bodies and national institutions, seeking the promotion of human rights through national institutions' own systems of assimilation. Most recently, the first NHRI of a Pacific Island State was set up in Suva (Fiji), with the financial support of the European Union, and has just released an important survey of the level of Fijian recognition and respect of human rights, based on Fijian case law and national legislation.[404] If other NHRIs were to be set up in Pacific Island States, access to information by the Pacific Islanders of each State would not only be facilitated, but would also help in drawing up a comparative survey of the state of recognition of economic, social, and cultural rights in the region.[405]

Situation nº 2: A Human Rights 'Approach' to Climate Change Law

The fact that the main international human rights treaty relevant in a preventive relocation scenario – namely, the ICESCR – is not applicable in eleven out of the twelve Pacific Island States does not, however, imply that the region has completely neglected the realisation of the rights the treaty contains. Indeed, of the three duties incumbent upon the State for the realisation of economic, social and cultural rights, the duty to protect has been particularly implemented, albeit through indirect means associated with international environmental law obligations. In contrast to the low rate of ratification of international human rights treaties, all Pacific Island States are Parties to the United Nations Framework Convention on Climate Change, as well as to the Kyoto Protocol. Under the umbrella of the Association of Small Island States (AOSIS), they have been playing a very active political role ever since the start of the regime,[406] certifying the top level of concern that climate change has

the High Commissioner for Human Rights, *Assessing the Effectiveness of National Human Rights Institutions* (Versoix, Switzerland: International Council for Human Rights Policy and OHCHR, 2005). More detailed information on NHRIs can be found at the OHCHR website at: http://www.ohchr.org/en/countries/nhri/pages/nhrimain.aspx (8 November 2014).

404 See Fiji Human Rights Commission, Bill of Rights: International Legal Analysis, updated 2009, available at the website of Fiji's Human Rights Commission at: http://www.fhrc.org.fj/index.php?option=com_docman&Itemid=12 (8 November 2014).

405 For a previous thorough analysis of this topic, see Brian Burdekin, *National Human Rights Institutions in the Asia-Pacific Region* (Leiden: Brill Nijhoff, 2007).

406 Stephen Krasner, ed., *International Regimes* (Ithaca, NY: Cornell University Press, 1983), who defined regimes as "a set of explicit or implicit principles, norms, rules, and decision making procedures around which actor expectations converge in a given issue-area".

been given in the agenda of Pacific Islands' governmental institutions.[407] To be sure, as the World Bank Report recalls, the international regime on climate change is not tailored as a "construction of a legal framework of rights/obligations between the individual and the State".[408] Rather, its structure responds to a "global goods perspective"[409] and is more explicitly built upon the principle of reciprocity, as clearly revealed by the primary purpose of the regime – namely, the reduction of man-made greenhouse-gas emissions. Yet, this structural characteristic does not preclude the regime from indirectly serving the realisation of economic, social and cultural rights. To begin with, one of the difficulties raised by the international human rights legal framework is that, the World Bank report explains:

> [I]t appears somewhat *reactive* in its design, geared more towards redressing past or imminent harms than the speculative business of scientific predictions pertinent to climate change.[410]

Such an *ex post facto* perspective, inherent in human rights law, strikingly contrasts with the *ex ante* or preventive approach inherent in climate mitigation within the UNFCCC. Henceforth, from a structural perspective, the climate change regime as a whole may be better suited to filling some of the normative lacunae in preventive relocation situations. In addition, recent developments within the regime complement this *prima facie* beneficial characteristic, as the scientific- and economy-driven rationales that originally grounded the regime are now being progressively diluted in favour of a perspective more inclusive of the 'human face' of climate change. This slow and delicate shift is becoming particularly visible through the changing evaluation and quantification conducted by the relevant institutions of the regime in the context of climate change adaptation measures. The most prominent example of such a 'humanisation move' can thus be found in the criteria used by the IPCC when measuring climate change impacts in the most vulnerable countries. Particularly since the publication of the IPCC's 2007 Fourth Assessment Report, and more

407 For an overall view on the influence exerted by small island States in different international fora where their interests were involved, see Jenny Grote Stoutenburg, 'The Changing Tides of Small Island States Discourse: a Historical Overview of the Appearance of Small Island States in the International Arena', *Law and Politics in Africa, Asia and Latin America*, vol. 2 (2010), pp. 164–191.

408 McInerney-Lankford, Darrow and Rajamani, 53.

409 Ibid.

410 Ibid., 47. [Emphasis added].

recently confirmed in the 2014 Fifth Assessment Report, comprehensive social and community-based approaches have an increasingly prominent place in what the IPCC actually considers to be 'affected' by climate change impacts. All in all, the idea that environmental degradation may interfere with the enjoyment of human rights – or become a great obstacle to the realisation of such rights – is progressively being acknowledged in the international regime on climate change. This move will be referred to as the incorporation of a *human rights approach* into international climate change co-operation, taking the lead over the alternative strict approaches that argue in favour of inter-regime synergies between human rights and climate change *law* – understood in a formal sense.

As a corollary to the acknowledgement that environmental degradation may interfere with human rights, measures seeking to halt or diminish the impacts of climate change are increasingly considered as factors facilitating the realisation and enjoyment of such rights. Specific application of the human rights approach to climate change adaptation can thus be found, as the articulation of the relationship between adaptation and human development or human welfare is being promoted by the institutions of the human rights regime. The five Pacific Island States listed as least-developed countries have conducted such articulation mainly by mainstreaming human rights, human welfare and human development all together during the elaboration of their National Appropriate Plans of Action (NAPAS). From a substantive perspective, one of the main manifestations of the link between adaptation and human development is that their NAPAS are primarily directed towards the preservation of the populations' basic survival *needs*. It has also resulted in the incorporation by Pacific Island States, other than the five LDCS, of climate disaster management in both development and adaptation programmes, such as the Joint National Action Plans previously mentioned in this chapter. From a procedural perspective, the mainstreaming of these three human-centred pillars and spreading the application of common procedural standards are desirable – particularly the right of local communities to be informed and actively participate in climate change adaptation planning. Bottom-up approaches are therefore favoured, so as to reinforce the legitimacy of NAPAS, while also potentially improving the effective implementation at the local level of adaptation actions financed by the institutions of the international regime on climate change.

The mention of local communities suggests that climate change law does not just allow articulating a protection to the socio-economic rights of individuals. Group rights can also find some level of normative protection in preventive relocation situations through the climate change regime, which may not even be found in international human rights treaties. Indeed, consideration of the rights of future generations is contemplated by the UNFCCC, whereas they are not mentioned in human rights instruments. Regardless of the procedural difficulty of protecting and implementing the rights of future generations this point is nonetheless relevant from the standpoint of principle for the next situations – of reactive relocation actions – given that the descendants of the present generation are likely to suffer irreversible cultural harm derived from the relocation of their families from their original or traditional lands.[411]

All in all, we can conclude that, in preventive relocation situations, international human rights treaty law (in particular, the ICESCR) is of very limited applicability in the Pacific and therefore ill-suited to provide a normative basis for Pacific Islanders to claim specific actions from their respective governments directed towards the preservation of their habitability conditions. Yet, the preservation of their basic survival needs, in danger of being harmed by climate change impacts, can be ensured through compliance with legal obligations under the UNFCCC and the Kyoto Protocol. The progressive incorporation of a human rights approach into the institutional practices of the international regime on climate change implies that some socio-economic rights of Pacific Islanders are positively affected. To be sure, this by-product or 'positive externality' could be criticised for not being grounded in a concrete legal provision of the international regime on climate change. It is, however, compensated by the fact that the non-ratification of the most relevant international human rights instrument by Pacific Island States does not erase the obligations of all other States Parties to the UNFCCC to co-operate in the promotion of human rights. Although such obligation is limited, it could arguably constitute (along with Article 4(3) of the UNFCCC) a basis for Pacific Islanders to claim a right to get climate change adaptation plans financed.

411 International law protects specific groups, particularly indigenous peoples, after the adoption of General Assembly Resolution 61/295, United Nations Declaration on the Rights of Indigenous Peoples, 13 September 2007; and the ILO Convention nº 169 Concerning Indigenous and Tribal Peoples in Independent Countries, 27 June 1989, which do not necessarily constitute the affected communities of the Pacific region described in table 5. Hence, specific protection that international human rights law could have eventually given is not really relevant to the cases so far studied. Only the specific protection of the Convention against Racial Discrimination could be relevant to the Polynesian minority of the Papuan islanders, who may soon have to suffer the same fate as their neighbouring Carteret Islanders and will most probably find even more obstacles to access new lands than the Carteret Islanders.

B Legal Framework of the Reactive Phase

Situation n° 3: Specialised Human Rights Treaties

Efforts to protect basic survival needs of human settlements vulnerable to climate change impacts may not be sufficient to prevent individuals or groups of persons from displacement. In such circumstances, referred to in this chapter as the reactive stage, Pacific island governments are called upon to take measures directed at the relocation of part of their respective populations to a different locality falling under the government's jurisdiction or effective control. When seeking to determine the content of an international legal framework for national reactive relocation actions, the set of rights of Pacific Islanders to housing, land and property (HLP rights) is among the first deserving closer examination.

As Scott Leckie and Chris Huggins recall, there is no single international legal instrument specifically dealing with housing, land and property rights.[412] Thus, rather than considering them as constituting an area of specialisation in international law, it is a crosscutting issue, dealt with in numerous international sources of law. Some instruments recognise them as 'self-standing' human rights. Others do not single them out autonomously, but regard them as an element of the more general right to an 'adequate standard of living' (which also encompasses most of the economic, social and cultural rights referred to in the 'preventive' stage). In this context, the violation of HLP rights may thus be considered as a continuous violation of basic rights, such as the right to water, which, according to the Committee on Economic, Social and Cultural Rights' General Comment n° 15, is "inextricably related to the right to the highest attainable standard of health (Article 12, paragraph 1) and the rights to adequate housing and adequate food (Article 11, paragraph 1)".[413]

Interestingly, focusing on the level of normative recognition of HLP rights in the South Pacific implies approaching the territorial dimension of the State, which was the object of chapter 1 – albeit from a different angle. HLP rights address the issue of 'de-territorialisation', but from the perspective of the individual, which generally falls under the domain of private law in

412 See Scott Leckie and Chris Huggins, *Conflict and Housing, Land and Property Rights: a Handbook on Issues, Frameworks and Solutions* (New York: Cambridge University Press, 2011).

413 CESCR, General Comment n° 15 on the Right to Water 2003, comment n° 7.

western systems. This is perhaps the point where the connection between the two challenges inspiring the titles of chapters 1 and 2 is more visibly evident: de-territorialisation affects the territorial pillar of the State and, by the same token, impinges upon the HLP rights of its population. While the right to adequate housing is rather uncontroversial and can be commonly found in international instruments, land and property rights tend to remain unaffected by international law, for the land tenure and property systems of a State or government may be a reflection of its ideological disposition. In addition to this characteristic, successive waves of colonisation and administration by different foreign powers in Pacific Island States resulted in very complex housing, land and property systems. These are the real sources and causes of the heterogeneity of reactive relocation actions depicted in section 1 of the present chapter, impeding any sort of generalised legal scheme for the region.

Notwithstanding that the rights and obligations involved differ in each relocation phase, both the preventive and the reactive relocation phases share the same relevant regime of international law. Thus, there are five international human rights instruments relevant to reactive relocation situations: two general instruments – the International Covenant for Civil and Political Rights (ICCPR), and the previously mentioned International Covenant of Economic, Social and Cultural Rights (ICESCR); and three instruments specifically designed to protect certain groups of people – the Convention for the Elimination of All Forms of Racial Discrimination (CERD);[414] the Convention for the Elimination of Discrimination against Women (CEDAW);[415] and the Convention on the Rights of the Child (CRC).[416] Despite the fact that the reactive phase is better suited to the design of the human rights legal framework than the prevention phase, its applicability remains, once again, subject to the ratification of the instruments in question, as summarised in table 6, below.

414 Convention on the Elimination of All Forms of Racial Discrimination, adopted on 21 December 1965, entered into force on 9 January 1969, 660 U.N.T.S. 195.

415 Convention on the Elimination of all Forms of Discrimination against Women, adopted on 18 December 1989, entered into force on 3 September 1981, 1249 U.N.T.S. 13.

416 Convention on the Rights of the Child, adopted on 20 November 1989, entered into force on 2 September 1990, 1577 U.N.T.S. 3.

TABLE 6 *Ratification of human rights treaties by Pacific Island States*

Treaty	ICCPR 1966		ICESCR 1966		CERD 1965	CEDAW 1979		CRC 1989
Country	CV	OP	CV	OP		CV	OP	
FSM	No	No	No	No	No	Yes	No	Yes
Fiji	No	No	No	No	Yes	Yes (**R**)	No	Yes
Kiribati	No	No	No	No	No	Yes	No	Yes
Nauru	No	Signed	No	No	No	No	No	Yes
Marshall Islands	No	No	No	No	No	Yes	No	Yes
Palau	No	No	No	No	No	No	No	Yes
PNG	Yes	No	Yes	No	Yes	Yes	No	Yes
Samoa	Yes	No	No	No	No	Yes	No	Yes
Solomon Islands	No	No	Yes (**R**)	Signed	Yes	Yes	Yes	Yes
Tonga	No	No	No	No	Yes	No	No	Yes
Tuvalu	No	No	No	No	No	Yes	No	Yes
Vanuatu	Yes	No	No	No	No	Yes	Yes	Yes

CV = Convention; OP = Optional Protocol; R = ratified with reservations.

As explained by Scott Leckie and Chris Huggins, the most significant articulation of the right to housing – considered as part of the larger right to an adequate standard of living, but also as an independent and freestanding right,[417] can be found in Article 11(1) of the ICESCR, which provides that State Parties to the Convention "recognise the right of everyone to an adequate standard of living for himself and his family, including adequate food, clothing and housing, and to the continuous improvement of living conditions".[418] In 1991, the CESCR enumerated the seven components of such a right (including 'cultural

417 Leckie and Huggins, 60.

418 See CESCR, General Comment n° 4, 'The Right to Adequate Housing (Article 11(1) of the Covenant)', 1991, paragraph 3, which states that. "[A]lthough a wide variety of international instruments address the different dimensions of the right to adequate housing Article 11(1) of the Covenant *is the most comprehensive and perhaps the most important of the relevant provisions*" [Emphasis added].

adequacy').[419] It also stated that the right to housing should not be interpreted in a narrow or restrictive sense, but should be read as the right to "live somewhere in security, peace and dignity".[420] Besides, although the nature of HLP rights implies that they are in principle classified in the economic, social and cultural rights basket, Leckie and Huggins also note that the ICCPR – in particular Articles 4 and 17 – is increasingly being used to enforce housing rights, as homelessness is increasingly recognised as a threat factor in – or even a violation of – the right to life, while forced evictions are considered as contravening the right to be free from arbitrary or unlawful interference within the home.[421] Despite the fact that these formal sources of law could be a useful starting point for the fulfilment of a legal scheme in reactive relocation situations of Pacific Islanders, their applicability remains rather limited, given that only two States – Papua New Guinea and the Solomon Islands – have ratified the ICESCR, while three – Papua New Guinea, Vanuatu and Samoa – have ratified the ICCPR.

Nonetheless, it should be noted that HLP rights are also contained in other human rights treaties aimed at the protection of particularly vulnerable groups (minorities, women and children). Article 5(e)(iii) of the CERD expressly prohibits discrimination in the enjoyment of economic, social and cultural rights, and in particular of the right to housing, and compliance with such a provision is monitored by the United Nations Committee on the Elimination of Racial Discrimination.[422] Article 14 of the CEDAW also protects the rights of rural

419 Ibid. paragraph 7: "[T]his [the inherent dignity of the human person] from which the rights in the Covenant are said to derive requires that the term 'housing' be interpreted so as to take account of a variety of other considerations, most importantly that the right to housing should be ensured to all persons irrespective of income or access to economic resources. Secondly, the reference in Article 11(1) must be read as referring not just to housing but to adequate housing. As both the Commission on Human Settlements and the Global Strategy for Shelter to the Year 2000 have stated: 'Adequate shelter means (…) *adequate privacy, adequate space, adequate security, adequate lighting and ventilation, adequate basic infrastructure and adequate location with regard to work and basic facilities – all at a reasonable cost*'" [Emphasis added].

420 Ibid.

421 Leckie and Huggins, 61.

422 Ibid. Article 5(e)(iii) of the CERD reads: "[I]n compliance with the fundamental obligations laid down in Article 2 of this Convention, States Parties undertake to *prohibit* and to *eliminate* racial discrimination in all its forms and to *guarantee* the right of everyone, without distinction as to race, colour, or national or ethnic origin, to equality before the law, notably in the enjoyment of the following rights: […] (e) economic, social and cultural rights, in particular: […] (iii) the right to housing". [Emphasis added].

women to adequate housing – along with sanitation, electricity, water supply, transport and communications – and compliance with such provision is monitored by the CEDAW Committee. Similarly, Article 27(3) of the CRC, recognising the right of every child to a standard of living adequate for its development, calls upon States to assist the parents and other persons responsible for the child by providing material assistance and support programmes, in particular with regard to nutrition, clothing, and housing.

Although the level of ratification of the CERD is as low as the previous 'general' international human rights treaties (three ratifications, by Fiji, Papua New Guinea and Tonga); the CEDAW counts with eight Pacific Island States as Parties to the Convention – Cook Islands, Fiji, Kiribati, Marshall Islands, Federated States of Micronesia, Papua New Guinea, Samoa and Tuvalu – while the CRC enjoys ratification by all Pacific Island States. To be sure, the CEDAW and the CRC do not cover the population of Pacific Island States *as a whole*. Yet, given that the proportion of children and women amongst the people most affected by climate-induced relocations in Pacific Island States, the extended applicability of the CEDAW and CRC Conventions is very important and may actually provide the core of the legal scheme for reactive relocation actions in the Pacific.

Situation nº 4: Complementary Role of the Guiding Principles of Internally Displaced People

These meagre – yet existing – formal sources of law protecting the HLP rights of Pacific Islanders, which serve as the basis for governmental policies, may be complemented by non-binding human rights standards. Other than Articles 17 and 25(1) of the United Nations Declaration on Human Rights – which include the right to housing and property exercised individually or in association with others – most international legal literature concerned with climate-induced displacement pays special attention to the 2005 Guiding Principles on Internally Displaced People (hereafter, the Guiding Principles). This instrument sets standards analogous to those of refugee law, for cases where the displacement and relocation have taken place within the boundaries of the State. The standards contained therein serve as a guide to governments, international organisations and all other relevant actors, in providing assistance and protection to internally displaced people through identification of the rights and guarantees relevant to the different phases of displacement (that is, assistance during displacement as well as guarantees for resettlement).

Although the Guiding Principles essentially apply in armed-conflict contexts, they also encompass cases of internal displacement generated by 'natural disasters'.[423] Perhaps, by having an overoptimistic and exaggerated idea of the scope of the Guiding Principles, some scholars have seen in them an important road-map from which a legal scheme (through formal or informal means) for climate-induced displaced people could and should be elaborated. Roger Zetter (former Director of the Centre for Refugee Studies at the University of Oxford), for instance, considered that "pending further research and consensus-building towards specific new norms or institutions", the Guiding Principles might be seen as:

> [N]ot just a starting point, but also a model for the process of aggregating and adapting the norms and principles from a wide range of international instruments to protect the rights of the 'environmentally displaced'.[424]

Yet, the limits of this instrument, as well as the essential differences between displacements fostered by internal armed conflict and those generated by natural disasters or any sort of environmental stress and degradation, should not be underestimated. Leckie and Huggins raise this point and defend the need "for a separate set of guidelines on land and natural disasters".[425] Indeed, concerns over HLP rights in cases of climate-induced displacement do not match post-conflict situations. More precisely, issues of repatriation, return or eviction of second occupants are not relevant in climate-induced relocation situations. Rather, the focus in these situations is on issues pertaining to the means and possibilities of acquisition of *new* land, as well as the safety and sustainability and cultural adequacy of, and community participation in, the relocation to such lands, along with the need to avoid discrimination of minority and gender groups in their access to them.

In contrast, two other existing 'soft law' instruments may be better suited to serve as a model for protection of climate-induced displaced people (although the 'urgency' present in these reports does not always match the progressive nature of the deterioration of the habitability conditions due to climate change

423 See Walter Kälin, *Guiding Principles on Internally Displaced People: Annotations, Studies in Transnational Legal Policy* (American Society of International Law), vol. 38 (2008), 1–75. As the Annotations of Paragraph 2 of the Guiding Principles explain: "Paragraph 2 is a descriptive identification of the category of persons whose needs are the concern of the Guiding Principles. [T]his paragraph provides some examples of how internal displacement may be brought about – situations of generalised violence, violations of human rights *or natural* or human made disasters." [Emphasis added].

424 McInerney-Lankford, Darrow and Rajamani, 64.

425 Leckie and Huggins, 144.

impacts): the 2006 Operational Guidelines on Human Rights and Natural Disasters of the Inter-Agency Standing Committee,[426] as well as the UN-HABITAT Scoping Report on 'Addressing Land Issues after Natural Disasters' released in 2007.[427] This path has been recently taken by the Regional Office for the Pacific of the OHCHR, which organised the same year a regional workshop on 'Integrating Human Rights in National Disaster Management'[428] and more recently published a report on 'Protecting HRs of Internally Displaced People by Natural Disasters: Challenges in the Pacific'.[429] The purpose of the workshop was to increase the awareness and skills of UN and other staff involved in disaster-relief and recovery operations on the relevance of human rights issues in disaster management (in particular, the UN Disaster Management Team and Pacific UN Disaster Assessment & Co-ordination) and develop a human rights check-list, specific to the Pacific, to be used by all concerned actors in situations of humanitarian emergencies to ensure that human rights concerns are effectively incorporated into natural-disaster preparedness, response and recovery.

As opposed to the preventive relocation situation, whereby the UNFCCC was considered as relevant to the preservation of the habitability conditions of Pacific Islanders, no international environmental legal instrument is relevant as a formal source of HLP rights. Only Agenda 21, a non-binding instrument adopted at the 1992 United Nations Conference on Environment and Development, held in Rio de Janeiro, makes reference to housing, land and property rights. Chapter 7.6 of the Agenda recalls that the right to adequate housing is a basic human right enshrined in the Universal Declaration of Human Rights and in the ICESCR and considers that access to a safe and healthy shelter is essential to a person's physical, psychological, social and economic wellbeing and should be a fundamental part of national and international action. Moreover, chapter 7.9(b) states that countries should adopt and/or strengthen national shelter strategies, while chapter 7.30(f) provides that

---

426 UN Inter-Agency Standing Committee, *Protecting Persons Affected by Natural Disasters: Operational Guidelines on Human Rights and Natural Disasters*, Brookings-Bern Project (Washington, D.C.: IASC, June 2006); followed by the *Operational Guidelines and Field Manual on Human Rights Protection in Situations of Natural Disaster*, Brookings-Bern Project (Washington, D.C.: IASC, Pilot version March 2008).

427 Daniel Fitzpatrick, *Scoping Report: Addressing Land Issues after Natural Disasters* (UN-HABITAT; United Nations Settlement Programme, 2007).

428 See the Regional Workshop on 'Integrating Human Rights into National Disaster Management' (9–11 May 2007), in Suva, Fiji.

429 See the OHCHR, *Integrating Human Rights in Natural Disaster Management*, Workshop Report (Suva: UNDP Pacific Center), 1–12.

countries should consider developing national land-resource management plans to guide land-resources development and utilisation and, to that end, establish appropriate forms of land tenure. However, instead of considering it as a formal source of rights and obligations, the UNFCCC could be useful in reactive relocation situation as a means to enforce HLP rights and associated basic rights, particularly as an effective and reliable source of funding. This would constitute the reverse situation to the one described for the preventive relocation phase, whereby the main formal source of law was international environmental law (due to the low level of ratification of the ICESCR) and human rights approaches serve to inform the best forms of carrying out climate change obligations.

2.3.2 *Endangered Statehood: Transnational Relocations*

Most climate-induced relocations currently take place within the boundaries of one single State. Yet, as the deterioration of the habitability conditions of Pacific Island States becomes more stringent and the sustainability of the places within the country less exposed to climate change impacts (such as higher lands or urban centres) reaches its limit, the prospects of transnational climate-induced displacements will undoubtedly be confirmed. In fact, migration of islanders (individuals or families rather than whole communities) from Tuvalu and Kiribati – the two States with a highly reduced margin of internal relocation, due to their geographical configuration – to other island States, as well as to Australia and New Zealand, have already been documented.

Recent developments fostered by the adoption, in December 2010, of the Cancun Adaptation Framework, point to the possible creation in the near future of new norms on climate change, natural disasters, and displacement.[430] Paragraph 14(f) of the Adaptation Framework indeed recognises that Parties to the UNFCCC shall address "climate change induced displacement, migration and planned relocation" within the framework of climate change adaptation.[431]

430 A detailed account of these developments can be found in Jane McAdam, 'Creating New Norms on Climate Change, Natural Disasters, and Displacement: International Developments 2010–2013', *Refuge*, vol. 29 (2014): 2, 11–26.

431 Article 14(f) of the Cancun Adaptation Framework reads: "14. [I]nvites all Parties to enhance action on adaptation under the Cancun Adaptation Framework, taking into account their common but differentiated responsibilities and respective capabilities, and specific national and regional development priorities, objectives and circumstances, by undertaking, inter alia, the following: [...] (f) Measures to enhance understanding, co-ordination and co-operation with regard to climate change induced displacement, migration and planned relocation, where appropriate, at the national, regional and international levels".

While the provision limits itself to inviting States to undertake measures that help understanding, co-operation, and co-ordination on these issues – hence not imposing any binding obligation upon States to protect 'climate refugees' – it does open the door for national adaptation plans that include this matter to be funded by the Green Climate Fund.[432] On the basis of this provision, a year later – coinciding with the 'political momentum' represented by the 60th anniversary of the Refugee Convention and the 50th anniversary of the Convention on the Reduction of Statelessness – the UNHCR encouraged States to agree on a 'global guiding framework' on displacement in the context of climate change and natural disasters.[433] Negotiations within this framework eventually failed, given that, as Jane McAdam explains, States showed reluctance to being led – if not pushed – by an agency with no clear mandate to deal with climate-induced displacement.[434]

Nonetheless, the move of the UNHCR in pushing forward the introduction of this issue into the international agenda had an important impact. First, it prompted the celebration in June 2011 of an inter-State Conference, convened in Nansen by the Government of Norway with the Centre for International Climate and Environmental Research and the Norwegian Refugee Council. This event closed with the adoption of the ten Nansen Principles "designed to shape and inform further action on addressing the linkages between climate change and mobility, both normatively and practically".[435] These principles then led to the creation, in 2012, of the State-led Nansen Initiative on Disaster-Induced Cross-Border Displacement, a governmental process promoted by the Norwegian and Swiss governments, and in operation from 2013 to 2015. Through a bottom-up approach and regional consultations through

For an account of the context in which this provision was adopted, and its legal and political implications, see Koko Warner, *Climate Change Induced Displacement: Adaptation Policy in the Context of the UNFCCC Climate Negotiations* (Geneva: UNHCR, 2011). This background paper was prepared for UNHCR's Expert Roundtable on Climate Change and Displacement, convened from 22 to 25 February 2011 in Bellagio, Italy.

432 McAdam, 'Creating New Norms', 12.

433 Ibid. See also the UNHCR, Intergovernmental Event at the Ministerial Level of Member States of the United Nations on the Occasion of the 60th Anniversary of the 1951 *Convention Relating to the Status of Refugees* and the 50th Anniversary of the 1961 *Convention on the Reduction of Statelessness* (7–8 December 2011), Background Note for the Roundtables, document reference: UN Doc HCR/MINCOMMS/2011/08, 18 November 2011.

434 McAdam, 'Creating New Norms', 12.

435 Ibid., 15. The text of the principles can be found in the report of the Conference, see Christian Ghare, *The Nansen Conference: Climate Change and Displacement in the 21st Century, Oslo-Norway, 7–11 June 2011* (Oslo: Norwegian Refugee Council, 2011), 5.

which the specific challenge for each region can be earmarked, the Nansen Initiative seeks to develop a non-binding Protection Agenda. Noticeably, the Initiative avoids references to climate change, and instead uses the term *disaster*-induced cross-border displacement. It is also focused on the consequences of the disaster, rather than the causes of it. As already mentioned in the context of the loss and damage platform, where the same political move was also operated in order to gather developed States' co-operation, such strategy seeks to dilute the 'climate justice' perspective and circumvents the causation issue underpinning climate-induced displacement.[436] Nonetheless, as part of the Nansen Initiative, a Pacific Sub-Regional Consultation was held in the Cook Islands in May 2013. The focus of the consultation was on the long-term processes affecting Pacific Island States, including the need to secure migration agreements or solutions, and well as appropriate locations for resettlement.[437]

While bearing in mind these developments, it is worth exploring the relevance and regional applicability of existing international instruments to situations of transboundary climate-induced relocations. Table 7 below shows the level of ratification of the core instruments composing the three special regimes potentially applicable in cases of transnational climate-induced relocations. Firstly, the relevance and applicability of refugee law – consisting of the 1951 Convention on the Status of Refugees[438] and its 1967 Protocol[439] – and international migration law – essentially consisting of the Convention on the Protection of the Rights of All Migrant Workers and Members of Their Families[440] – will be dealt with in situations of partial de-population. In these cases, the possible role to be played by human rights law in the context of 'complementary protection' will also be mentioned. Then, as the possible extinction of some Pacific Islands' statehood following total de-population

436 McAdam, 'Creating New Norms', 19.

437 See the outcome of the consultation, The Nansen Initiative, 'Human Mobility, Natural Disasters, and Climate Change in the Pacific', *Report from the Nansen Initiative Pacific Regional Consultation* (Rarotonga: Nansen Initiative, 2013), 1–32.

438 Convention Relating to the Status of Refugees (adopted 22 July 1951; entered into force 22 April 1954) 189 U.N.T.S. 137.

439 Protocol Relating to the Status of Refugees, (adopted 31 January 1967; entered into force 4 October 1967) 606 U.N.T.S. 267.

440 International Convention on the Protection of the Rights of All Migrant Workers and Members of Their Families, adopted 18 December 1990, entered into force 1 July 2003, 2220 U.N.T.S. 3.

and de-territorialisation is anticipated, the international legal regime on statelessness may become relevant.[441]

Transnational climate-induced relocations bear two additional elements of complexity. Firstly, by definition, transnational relocations imply a plurality of possible 'duty bearers', as the territoriality and jurisdiction of at least two States – the State of origin and the State of arrival (hereafter referred to as the 'host State') – are involved. The movement of the individual leaving his/her country of citizenship to enter another State and become part of the shadowy category of 'alien' is viewed by international law as a transition towards a situation of increased weakness and potential defencelessness. Thus, the focus of international legal regimes relevant to international migration cases is primarily on the treatment that asylum-seekers (whether migrant workers or statutory refugees) receive on entering and eventually settling in a territory subject to the jurisdiction of a State that is new to the migrant. The State of origin is somehow left behind by international law, which accompanies the migrant from the moment of departure, closely follows his/her steps until the point of arrival, and then remains vigilant in respect of the treatment received thereafter under the jurisdiction or control of the host State.[442] Whereas such a 'behavioural characteristic' of the international legal order does not conflict with partial de-population situations, it is very ill-fitted for this purpose, calling for a thorough re-examination as the prospects of total de-population situations are anticipated. The second source of complexity lies in the characterisation of the overall context in which climate-induced displacements and relocations take place, since these are so much akin to situations of conflict that an extension of the regulation could be justified. Such debate informs the applicability of many of the international legal regimes relevant to cases of transnational climate-induced relocations, which depends not only on the ratification by the countries concerned, but also on the conceptual definitions that raise controversy both from a legal and a policy perspective.

---

441 Convention on the Status of Stateless Persons, adopted on 28 September 1954, entered into force on 6 June 1960, 360 U.N.T.S. 177; and Convention on Convention on the Reduction of Statelessness, adopted on 30 August 1961, entered into force on 13 December 1975, 189 U.N.T.S. 175.

442 The idea that the State of origin is ignored by international law once the refugee has crossed the border is nevertheless subject to the exception of the right of return. See Emma Haddad, 'The Refugee: the Individual between Sovereigns', *Global Society*, vol. 17 (2003), pp. 297–232.

TABLE 7 *Ratification in the South Pacific of the refugee, statelessness and migration treaties*

Treaty	Refugee Status		Statelessness		Migrant Workers
Country	Convention	Protocol	Status Convention	Reduction Convention	Convention
Australia	Yes	Yes (D)	Yes	Yes	No
Cook Islands	No	No	No	No	No
FSM	No	No	No	No	No
Fiji	Yes (R)	Yes	Yes (R)	No	No
Kiribati	No	No	Yes (R)	Yes	No
New Zealand	Yes (R)	Yes	No	Yes	No
Nauru	Yes	Yes	No	No	No
Marshall Islands	No	No	No	No	No
Palau	No	No	No	No	Signed
PNG	Yes (R)	Yes	No	No	No
Samoa	Yes	Yes	No	No	No
Solomon Islands	Yes	Yes	No	No	No
Tonga	No	No	No	No	No
Tuvalu	Yes (R)	Yes	No	No	No
Vanuatu	No	No	No	No	No

R = Ratified with reservations; D = Ratified with a declaration.

A Legal Framework for Ongoing Partial De-Population

Situation n° 5: International Refugee Law and International Migration Law

Migration issues have traditionally been subject to the discretionary power of sovereign entities, as States are in principle free to decide on the conditions of entry and residence of aliens in their territory. Nonetheless, under certain circumstances informed by humanitarian purposes, international law limits this discretional margin of appreciation. Refugee law constitutes the most stringent curtailment of a State's sovereign power over migration issues, and offers to its beneficiaries a comprehensive protection in terms of the certitude, scope and level of rights afforded.[443] Being aware of this fact, international

443 Thomas Gammeltoft-Hansen, *Access to Asylum: International Refugee Law and the Globalisation of State Control* (Cambridge: Cambridge University Press, 2012), 30 – referring

scholars quickly examined its applicability to situations of climate-induced relocations.[444] Its relevance for Pacific Islanders thus merits examination.

Refugee law is designed to benefit the refugee, a category defined by Article 1A(2) of the 1951 Refugee Convention as "a person with a well-founded fear of being persecuted on grounds of race, religion, nationality, membership of a particular social group, or political opinion". Therefore, the regime comes into play when the State of origin is unable or unwilling to offer protection normally due to its population on specific grounds,[445] and entails that another State may endorse the protective function of the State of origin by granting refugee status to the affected persons. Such functional replacement is designed to be provisional (limited to the period ranging from the departure of the refugee to his/her integration into a new community or return to the State of origin), and shall not be interpreted as an act of enmity towards the State of origin. As explained by Vera Gowlland-Debbas, originally the applicability of the refugee regime was only conditional upon the crossing of an international border, after which, a set of rights and obligations concerning exclusively the host State came into play. The focus of the 1951 Convention and its 1967 Protocol, as well as the approach taken by the UNHCR, was thus focused on the responsibility and the obligations of the host State, rather than the State of origin, vis-à-vis asylum-seekers likely to request refugee status. Only when, in the 1970s, the problem of mass flows arose, did the attention also turn towards the development of the responsibility of the State of origin under refugee law, as well as under human rights and

to the existence of a "protection lite" – intending to mean 'protection *élite*' – different from other standards with 'the presence of formal protection but with a lower degree of certainty about the scope and/or level of rights afforded'.

444 For a detailed and critical review of the arguments in favour of applying the refugee regime to these cases, see McAdam, *Climate Change, Forced Migration and International Law*, chapter 2, 39–50.

445 See Denis Alland, 'Le Dispositif international du droit de l'asile', in Société Française de Droit International, Colloque de Caen, *Droit d'Asile et des Réfugiés* (Paris: Pedone, 1997), 13–83, 49. As Alland explains, there are two possible ways of interpreting the duties of the State of origin that can possibly fill the criteria of 'persecution' contained in the definition of Article 1 of the 1951 Refugee Convention. According to a subjective conception, the State has an obligation *of means* (not of result) towards its citizens; therefore the absence of protection has to be 'wrongful' in order to create a fear of persecution fitting into the definition of the Convention. In contrast, under an objective conception, the obligation of the State of origin towards its citizens is one *of result*; therefore, the absence of protection derives directly from a factual situation. Only in these cases can *the inability* of the State to offer protection be admitted as a cause of persecution.

international humanitarian law.[446] Interestingly, Gowlland-Debbas explains, certain causes of mass displacement, such as environmental disruptions or extreme weather events, can be attributed to the State of origin when it is proved that it has breached an obligation of due diligence, of non-discrimination (for instance by refusing to help groups of its population on racial or ethical grounds), or any other human rights violations.[447] It is thus worth highlighting the fact that the set of actions undertaken by Pacific Island States, such as the purchase of land abroad by Kiribati, the negotiations with neighbouring countries aiming at the acceptance of migration quotas for Pacific Islanders threatened by sea-level rise, and the implementation of disaster-risk reduction strategies, development and climate change adaptation plans, may all be considered as State actions necessary to avoid international responsibility by complying with the principle of due diligence – if not also with a view to preserving the controversial 'right to remain' of Pacific Islanders.[448]

The primary attention remains, nonetheless, on the responsibility of the host State. This is also reinforced by the fact that, while the refugee regime contains definitions of the category of people who can benefit from its protection, the procedural norms establishing how such a status is to be determined by national law offers a certain margin of discretion to the host State.[449] The refugee is indeed, as Thomas Gammeltoft-Hansen puts it, "poised squarely between these two conceptions of sovereignty: sovereignty as freedom and sovereignty as responsibility".[450] At the heart of the regime and embodying the concept of sovereignty as responsibility lies the principle of *non-refoulement*, enshrined in Article 33 of the Refugee Convention. This principle entails that

---

446 On this point, see Vera Gowlland-Debbas, 'La responsabilité de l'État d'origine pour des flux de réfugiés', in Société Française de Droit International, Colloque de Caen, *Droit d'Asile et des Réfugiés* (Paris: Pedone, 1997), 93–131.

447 Ibid., 106.

448 The rationale behind the right to return has been eloquently defined by Guy S. Goodwin-Gill, 'The Right to Leave, the Right to Return, and the Question of a Right to Remain', in Vera Gowlland-Debbas, ed., *The Problem of Refugees in the Light of Contemporary International Law Issues* (The Hague, Boston, London: Martinus Nijhoff Publishers, 1995), 82–93. Cited in Gowlland-Debbas, 111. "[T]he right to remain comprises the common or garden sense of not having to become a refugee, not having to flee, not being displaced by force or want, together with the felt security that comes with being protected. It is another way of expressing, in concrete terms, the connection between individual, community and territory, but its effective realisation depends upon human rights and development considerations that are staggering in their breadth".

449 Alland, 42.

450 Gammeltoft-Hansen, 13.

the host State has the obligation not to return the asylum-seeker to his/her State of origin once the refugee status is granted, as well as – and consistent with the view that the refugee status is declaratory and not constitutive – while the application is still pending.[451] Guy Goodwin-Gill indeed remarks that although *non-refoulement* "is not synonymous with a right to admission, the principle of non-rejection at the frontier implies at least temporary admission to determine an individual's status".[452] The principle of *non-refoulement* would undoubtedly offer an important advantage to climate-induced displaced people, as their permanence in the territory of the host State would be protected from any eventual changes in economic migration policy, while their right to return to the State of origin – presuming that it is still habitable – remains formally preserved.[453]

Moreover, in addition to the protection granted by the principle of *non-refoulement*, regarding the living conditions of refugees in the host State, the refugee regime offers a wider set of socio-economic, civil and political rights. Such a set of rights is granted en bloc along with the refugee status, but its enjoyment by the beneficiaries depends on the level of attachment a refugee demonstrates to a given host country. This incremental approach, Gammeltof-Hansen thus explains, "reflects a seemingly sensible concern of the drafters not to extend the full scope of rights immediately in situations where refugees may arrive spontaneously in large numbers".[454] Thus, 'durable residents' enjoy more sophisticated civil and political rights and may have better access to social benefits than others. When it comes, in particular, to the right to adequate housing, Article 21 of the Refugee Convention establishes the obligation of host States Parties to the Convention to accord refugees lawfully staying within their territories a treatment "as favourable as possible". Such treatment is also protected by a non-discrimination rule (also enshrined in Article 3), as refugees shall not be treated less favourably than other aliens generally in the same circumstances.

As all these potential benefits offered by the refugee regime are laid down, the question that arises is whether the regime is applicable in the Pacific, both

451 For a thorough explanation of the particular issues arising from the application of the principle of *non-refoulement* upon admission at the border, see Guy S. Goodwin-Gill and Jane McAdam, *The Refugee in International Law* (Oxford: Oxford University Press, 3rd ed. 2007), chapter 5, 211–284, 206–208.

452 Ibid., 215.

453 On the right to return, see Goodwill-Gin, 'The Right to Leave, the Right to Return, and the Question of the Right to Remain'.

454 Gammeltoft-Hansen, 28.

from a geographical and a material perspective. Table 7 above shows that both the Refugee Convention and its Protocol have been ratified by six Pacific Island States – Fiji, Papua New Guinea, Tuvalu, Nauru, Samoa and the Solomon Islands – which is a rather high ratification rate when compared with the level of ratification of human rights treaties in general (ICESCR and the ICCPR) dealt with above. However, given that the refugee regime is more focused on the host State than on the State of origin, the only relevant ratification to be considered is that of potential host States. Rather than looking at the Pacific Island States of origin, such as Tuvalu or Kiribati,[455] the crucial point is that the two regional powers and main potential host States of climate-induced displaced islanders, Australia and New Zealand, have also ratified the Refugee Convention and its Protocol.

Yet, the unproblematic geographical applicability of international refugee law in the most relevant countries of the Pacific contrasts with the highly controversial extension of the Convention's material scope to cover cases of climate-induced displacement. To be sure, the traditional paradigm of refugee law is currently challenged by new factual realities that have transformed the 'asylum-seeking experience'. Gammeltoft-Hansen has, for instance, dedicated a thorough study to the measures of migration control set up before arrival in the host State and often carried out by private actors, and questions whether such a shift in location and actors for the encounter between the refugee and the host State is accompanied by an equal *de jure* shift in the scope of international refugee law.[456] Climate-induced displacement equally challenges the refugee regime – by, for instance, promoting rather forced re-examinations of the boundaries of the persecution criterion and questioning the relevance of the regime in cases of mass influxes – to such an extent that it leads to the rather useless definitional impasse already detailed in section 2.2.1 of this chapter. It is thus not surprising that Australian and New Zealand courts have consistently denied the refugee status to Tuvaluans and I-Kiribati who claimed to be 'climate refugees'. For instance, in 2013, the New Zealand Immigration and Protection Tribunal ruled against the claims of a citizen of Kiribati to have an entitlement to be recognised as a refugee on the basis of changes to his environment in Kiribati due to sea-level rise and associated with climate change.[457]

---

455 Ibid., 25, explaining that refugee law is 'international' in so far as "refugee protection is not guaranteed in a global homogeneous juridical space, but materialises as a patchwork of commitments undertaken by individual states tied together by multilateral treaty agreements".

456 Ibid., at 231, cases referred to by this author as 'offshore and outsourcing migration'.

457 Refugee Appeal, AF v. New Zealand (I-Kiribati citizen), [2013] NZIPT 800413.

On the one hand, the Tribunal admitted that the need for persecution to rest on some form of human agency "does not mean that environmental degradation, whether associated with climate change or not, can never create pathways into the Refugee Convention or protected person".[458] Acknowledging that "the relationship between natural disasters, environmental degradation, and human vulnerability to those disasters and degradation is complex", the Tribunal openly admitted that "pathways can, in some circumstances, be created into international protection regimes, including Convention-based recognition."[459] However, the situations considered to be likely to fall within these 'pathways' – cases of post-disaster relief following a natural disaster, and situations of armed conflict linked to environmental degradation – did not help the claimant's case. Therefore, the Tribunal did not grant the appeal, and considered that:

> [M]igration is generally regarded as existing somewhere along a voluntary-forced spectrum. While a substantial grey area exists between the ends and no bright-line demarcation can be made, displacement typically occurs towards the 'forced' end of this spectrum. In this case, it is clear to the Tribunal that this appellant has undertaken what may be termed a voluntary adaptive migration – that is, to adapt to changes in the environment in South Tarawa detailed in the 2007 NAPA, by migrating to avoid the worst effects of those environmental changes. While there is some degree of compulsion in his decision to migrate, his migration cannot be considered as being 'forced' and was to another country, New Zealand, and not to another place in Kiribati.[460]

Besides, the existence of a special category of 'environmentally displaced people' has not so far been acknowledged by national judges of the two regional powers.[461] Thus, the direct applicability of the Refugee Convention does not

458 Ibid., para. 55.

459 Ibid., para. 57.

460 Ibid., para. 49.

461 See, for instance, New Zealand and Australian cases where claimant's demand to receive refugee protection for climate change impacts was rejected. New Zealand cases: Refugee Appeal 72719/2001, RSAA (17 September 2001) – (Tuvaluan claimant); Refugee Appeal 72313/2000, RSAA (19 October 2000) – (Tuvaluant claimant); Refugee Appeal 72314/2000, RSAA (19 October 2000) – (Tuvaluan claimant); Australian cases: 10004726[2010] RRTA 845 (30 September 2010) – (Tongan claimant); 0907346 [2009] RRTA 1168 (10 December 2009) – (I-Kiribati claimant); N00/34089 [2000] RRTA 1052 (17 November 2000) – (Tuvaluan claimant). All cited in McAdam, *Climate Change, Forced Displacement and International Law*, 47.

seem possible – at least in partial de-population situations– although, McAdam suggests, some of the 'conceptual constructs' of the refugee regime may be usefully transposed to a future regulation of climate-related displacement.

Considering the difficult applicability of the refugee regime to cases of climate-induced displacement in the Pacific – particularly due to the lack of fulfilment of the 'persecution criterion' – international migration law appears to be the second alternative international regime limiting the freedom of States to decide migration issues within their territory, and possibly applicable to climate-induced relocation of Pacific Islanders. The first advantage of relying on this regime is that it seemingly matches the will of Pacific Islanders and of their authorities to avoid the victimisation and degrading effect of being labelled as 'refugees'. President Anote Tong's declaration, reproduced in the chapeau to the present chapter, is a striking and clear expression of this apprehension. Moreover, it is also consistent with the policy already practiced in the region, whereby several worker migration schemes already detailed above – such as the New Zealand Pacific Access Category or the Australian Pacific Immigration Scheme – have been agreed between the regional powers and Tuvalu, Kiribati and Tonga.[462]

The general preference of Pacific Islanders themselves to be considered under the general category of economic migrant workers, coupled with the existing regional State practice, raises the question of whether they could benefit from the formal protection offered by the 1990 International Convention on the Protection of the Rights of All Migrant Workers and Members of Their Families. While this instrument does not facilitate the access to the territory of host States, as the Refugee Convention does, it nonetheless sets minimum standards on the treatment that immigrant workers have the right to receive once they have been granted working and residence permits (whether temporary or permanent) by the host State. Article 43(1) of the Migrant Workers Convention, for instance, declares the right of all migrant workers to equality of treatment with nationals of the State of employment in relation to basic socio-economic rights, such as access to housing – including social schemes – and protection against exploitation in respect of rents. Unfortunately, the Convention does not serve as a formal source of such rights, for it has not been ratified by any State of the Pacific, including the two regional powers.[463] Therefore, other than the eventual possible applicability of specialised Conventions of the International Labour Organisation, which generally have a low level

462 See section 2 of this Chapter, above.

463 As shown in table 6, only Palau has signed the Migrant Workers Convention.

of ratification,[464] the treatment and conditions of entry of Pacific Islanders and workers into Australia and New Zealand (as well as into other Pacific Island States less threatened by climate change impacts) largely depends on these State's public statutory and case law.

Situation nº 6: Human Rights Law as Minimum Standard and Complementary Protection

Noting the limited applicability of international refugee law from a material perspective (*ratione materiae*), and of international migration law from a geographical perspective (*ratione loci*), scholars such as Jane McAdam have also explored the possible role of human rights law which not only sets a minimum standard of treatment of the displaced Pacific Islanders by their host States but may also, in some cases, provide for complementary protection to that afforded by the Refugee Convention.

Firstly, as shown in Table 7, both the International Covenant on Civil and Political Rights, and the International Covenant on Economic, Social and Cultural Rights, although still very poorly ratified by Pacific Island States, are yet in force in Australia and New Zealand. In these two major regional powers, and primary destinations of cross-border climate-induced relocation in the region, the two Covenants operate, as Jane McAdam points out, as a formal source of rights setting the minimum standards of treatment to the population under their control or jurisdiction, including aliens with or without refugee status.[465] General civil, political, cultural and socio-economic rights for Pacific Islanders and their families settled and/or working in these two regional powers, such as the right to family reunification, which is also recognised by the Rights of the Child Convention.[466] Moreover, although human rights law primarily serves as a source setting minimum standards of treatment in the host

464 The only example of a relevant ILO convention could be the 1958 ILO Convention nº 110, Concerning Plantations, Article 88(1) of which establishes that, where housing is provided by the employer, the conditions under which plantation workers are entitled to occupancy shall be not less favorable than those established by national custom or national legislation.

465 McAdam, *Climate Change, Forced Migration, and International Law*, 52.

466 Article 10 of the Rights of the Child Convention recognised the right of children and their parents to leave any country and to enter their own for purposes of reunion or the maintenance of the child-parent relationship. In addition, Article 22 provides that special protection shall be granted to children having or seeking refugee status, including through co-operation between States and the international organisations that provide such protection and assistance.

State, it may also display effects upon the State of origin and on the international community. Regarding the former, in addition to the obligations of due diligence and non-discrimination – aiming at preventing or limiting mass influxes due to environmental disasters – the State of origin must ensure the right of free movement of its population, which includes the right to travel and to receive appropriate travel documents. This right is also bound by a non-discrimination rule and can still be limited on grounds of national security or public order.

In addition to the possible role to be played by human rights as a source of minimum standard of treatment to aliens under the control and jurisdiction of the host State, the human rights regime may also play a role in a second context. The term 'complementary protection' refers to a State's protection obligations arising from international human rights treaties and customary law which are complementary to the protection provided by the 1951 Refugee Convention.[467] Some scholars even go a step further and call for reconsidering refugee law as a specialised area falling within the realm of human rights law.[468] In theory, the complementary protection of human rights law has the potential to widen the scope of the principle of *non-refoulement* under international law to meet the protection needs of people falling outside the definition of refugee provided in Article 1A(2) of the Refugee Convention. Nonetheless, in practice it is virtually impossible for an applicant to benefit from such protection in cases where an economic, social or cultural right has been breached. Not only, as McAdam reminds us, has the UN Committee on Economic, Social and Cultural Rights never established that any right of the ICESCR contains a *non-refoulement* obligation, but also, the provisions of such an instrument "permit a balancing test between the interests of the individual and the State, thus placing protection from *refoulement* out of reach in all but the most exceptional cases".[469] Indeed, although the violation of socio-economic rights can be characterised as a form of inhuman treatment triggering complementary protection, courts have been careful not to use its applicability as a means to respond to poverty, unemployment, or general lack of resources.[470] Considering that, as already mentioned, climate-induced displacement is,

467 See for instance, Goodwin-Gill, *The Refugee in International Law*, chapter 6, 285–354, 285; and McAdam, *Climate Change, Forced Migration and International Law*, chapter 3, 52–98, 53.

468 See for instance James C. Hathaway, *The Rights of Refugees under International Law* (Cambridge, UK: Cambridge University Press, 2005).

469 McAdam, *Climate Change, Forced Migration and International Law*, 53.

470 Ibid., 54.

for the time being, a multi-causal phenomenon where acute environmental stress mixes with and enhances pre-existing low levels of socio-economic development, climate change adaptation and resilience, the system of complementary protection will not be profitable *for the early stages* of climate-induced displacement. Following this line, McAdam notes that jurisprudence would indeed need to be substantially developed for climate change impacts to be accepted as a source of harm clearly falling within the concept of inhuman and degrading treatment.[471]

Only the right to life, along with the right not to be subjected to torture or cruel, inhuman or degrading treatment, clearly entail in existing jurisprudence an obligation of *non-refoulement*.[472] For human rights complementary protection to be applicable in this situation on the basis of a threat to the right to life, two main conditions should be met: the risk to life must be imminent or actual, and must personally affect the claimant. While it may still be premature to consider that a violation of such rights arises in present slow-onset climate-induced displacement cases, through time the increased gravity of the situation of Pacific Island States is likely to make such appreciation evolve; the looming environmental stress derived from climate change impacts pushes to the edge the habitability of Pacific Islands' salty shrinking territory, producing acute and inhuman distress for the population. As we project the evolution in time of the situations at stake and grasp the legal developments that may develop from such extreme cases, we shall explore the legal framework that may be relevant for situations of total de-population.

B Legal Framework for Prospective Total De-Population

The scenario in which the whole population of a State affected by climate change impacts relocates elsewhere remains, for the time being, a prospective challenge, yet one that low-lying developing island States will be facing in a relatively short time span. Since, in 2008, Mohammed Nasheed announced the project of setting up a public fund to finance the purchase of lands in neighbouring countries where all the Maldivian population would eventually resettle, the 'sinking island paradigm' has slowly progressed to become a reality. The purchase of 2,000 hectares of Fijian lands by Kiribati in 2014 for that ultimate purpose indicates that the total de-population is not only imagined, but clearly and publicly anticipated by some of the governments concerned.

471 Ibid.

472 Ibid., 55.

The situation of total de-population refers to the relocation outside the territorial borders of a State of all or at least a very significant part of its population, which previously fell under its control or jurisdiction. Projecting the total de-population of a State entails, first of all, turning the attention to the State of origin. In international refugee law, international migration law and international human rights law, the existence of the State of origin is always presumed, even when the focus of such regimes is primarily on the responsibilities and obligations of the host State, as is the case in refugee law. In fact, the very concept of 'host State' – a country under whose jurisdiction or control the asylum-seeker arrives in – may be said to be constructed in opposition to the concept of 'State of origin' – a country from whose jurisdiction or control the asylum-seeker originally escaped.

In contrast, the prospect of total de-population entails one all-embracing consequence, namely, that the very existence of the State of origin can no longer be taken for granted. And yet, as alarmingly threatening as total de-population may be, the disappearance of the State of origin or the extinction of its international legal personality cannot be automatically deduced from the mere observation of such facts. In other words, the total de-population of a State *may or may not* result in the loss of its statehood. As previously noted, the answer to such a question partly transcends the realm of law and will be highly dependent on the ethical grounds on which the international community will base pragmatic solutions seeking to accommodate these unprecedented cases. The fate of the State in both cases remains an outstanding issue that is, after all, likely to be resolved by political and informal – rather than legal and formal – mechanisms, including the capacity of the State concerned to defend its situation in international fora. However, in the meantime, some of the more salient legal consequences arising in each of these two possible scenarios can already be earmarked.

Situation n° 1: Total De-Population Not Resulting in Loss of Statehood

While total de-population may not necessarily result in the State of origin's loss of statehood, the intensified gravity of the situation is likely to have an impact on some of the core rights and obligations derived from the refugee regime, as well as of human rights law in the context of complementary protection. Firstly, when the level of environmental degradation is so acute that the country of origin becomes indisputably uninhabitable, the right to life as well as the right not to be subject to cruel, inhuman or degrading treatment may trigger the protection of the obligation of *non-refoulement* by the host State. It can be presumed that, at such a stage, official evidence of the magnitude of the degradation – including the possible declaration of a state of emergency in the country of origin – and the threat such a declaration may constitute for their lives and physical integrity will be available to asylum-seekers. Yet the most important challenge will arise at a later stage, when the 'habitability' of the country of origin degrades into a plain 'impossibility' of return, which will occur in any case no later than after total de-territorialisation occurs. The irreversibility of the situation that provoked the displacement, and the consequent need of the asylum-seekers for a permanent solution, clashes with the provisional nature of the refugee regime and of the complementary protection of human rights law.

However, as long as the extinction of the international legal personality of the State of origin is not internationally sanctioned, the situation of the asylum-seekers will not be of *de iure* statelessness, that is, of being without citizenship or nationality.[473] As a result, while the initial obligation of *non-refoulement* may be indefinitely prolonged in time, and the right to return becomes void of meaning, the displaced are likely to find themselves in a situation of juridical limbo or *de facto* statelessness. On the one hand, their State of nationality is unable to protect them adequately; on the other hand, the host State is not yet obliged to give them permanent protection either, be it through naturalisation or long-term leave to remain. Therefore, the protection of the 1954 Convention on the Status of Stateless Persons, which entitles stateless persons lawfully present in the territory of a State Party to the Convention to receive the same treatment accorded to aliens, as well as travel documents and identity papers, would not apply to these cases. Only the Final Act of the Convention, which refers to the issue of *de facto* statelessness, makes a non-binding recommendation upon State Parties to "consider sympathetically the possibility of according to that person the treatment which the Convention accords to a stateless person".[474] Similarly, the 1961 Convention on the Reduction of Statelessness – devoted to the *elimination* of statelessness rather than to the *protection* of stateless persons – obliges State Parties to grant nationality to persons born in their territories and who would otherwise be stateless.[475] While one can see the benefit that such an entitlement could constitute for the children of the climate-induced displaced people and the subsequent generations born on foreign soil, the application of such provisions is again only binding for cases

473 For a succinct introduction to the concept of stateless persons, see Katja Göcke, 'Stateless Persons', *Max Planck Encyclopedia of Public International Law* (New York: Oxford University Press, 2008–), internet resource.

474 Convention on the Status of Stateless Persons, Final Act, para. III.

475 Convention on Convention on the Reduction of Statelessness, Article 1.

of *de iure* statelessness, leaving to the good will of the host State the possibility to extend such protection to *de facto* stateless people or perhaps, as the Commission on Human Rights suggested, to grant a new form of status such as 'environmental asylum'.[476]

Yet the responsibility of the State of origin is also greatly affected, given that, as part of its obligation to protect its population, it may have to declare a state of emergency and order the forcible transfer of the remaining people abroad. This issue was already stressed in 2009 by the UNHCR in a submission which noted that, on the one hand, "each State has the duty to take life-saving measures to protect the life of its people", including the need to temporarily evacuate people away from danger zones "and prohibiting them from returning home, if necessary on a permanent basis".[477] On the other hand, the submission also acknowledged that the right to freedom of movement also applies to persons displaced by natural disasters.

Kiribati could for instance order such a transfer to the 2,000 hectares of land purchased on Viti Levu Island in Fiji. Given that, in principle, the forcible transfer of a population is illegal under international law, a clear depiction of the context becomes imperative so that it does not qualify as an international wrongful act engaging the responsibility of the State of origin.[478] On the contrary, in these cases, the situation could in fact be reversed, as not ordering the evacuation could be equated to a violation of the State of origin's obligation of due diligence towards its population. Notwithstanding this fact, in

476 On this point, see Françoise Hampson, 'Human Rights Implications of Indigenous Peoples in States and Territories Threatened with Extinction for Environmental Reasons', Commission on Human Rights, Working Group on Indigenous Peoples, 13 July 2004, document reference: E/CN.4/Sub.2/AC.4/2004/CRP.1., 1–8, 5, para. 12, stating: "[W]hat rights, if any, do the affected populations have vis-à-vis other States? A status analogous to refugee status would not appear to be adequate. The people affected need more than a right to *environmental asylum*. They need a right to be granted it. [Emphasis in the original]".

477 UNHCR, 'Forced Displacement in the Context of Climate Change: Challenges for States under International Law', Paper submitted by the Office of the UNHCR in co-operation with the Norwegian Refugee Council, the Representative of the Secretary-General on the Human Rights of Internally Displaced Persons, and the United Nations University, 6th session of the Ad-hoc Working Group on Long-Term Co-operative Action under the UNFCCC Convention (AWG-LCA 6), held in Bonn (Germany), 1–12 June 2009.

478 See Emily Haslan, 'Population, Expulsion and Transfer', *Max Planck Encyclopaedia of Public International Law* (New York: Oxford University Press, 2008-), last entry updated April 2011, internet resource, available at: http://o-opil.ouplaw.com.catalogue.ulrls.lon.ac.uk/view/10.1093/law:epil/9780199231690/law-9780199231690-e861?rskey=em3IKU&result=1&prd=EPIL (5 May 2015).

order to be lawful, the transfer ought to comply with a set of conditions. It must, first of all, be consented to by the population, who also ought to be duly informed of the causes of the displacement, the conditions, a timely framework and the place of relocation. Considering the close attachment that many Pacific Islanders have to their lands, their reluctance to submit to or their outright opposition to forcible transfer cannot be excluded. This matter becomes all the more delicate when it comes to indigenous peoples, given that their forcible relocation from ancestral lands can constitute an act of ethnocide unless it is undertaken in accordance with Article 16(2) of the ILO Convention n° 169 and Article 10 of the 2007 UN General Assembly Declaration on the Rights on Indigenous Peoples, and with the free prior consent of the indigenous population.[479] If the condition to proceed with the transfer without the free prior and informed consent of the population is not met, the State ought to rely on a legitimate cause – such as the protection of its population –, operate the relocation in a non-discriminatory way, and secure a place of resettlement that is adequate for living. Likewise, the consent of the host State is also necessary. In such cases, one could ask whether the welcoming words of the President of Fiji could be considered as a valid form of consent from the host population, expressed through its principal representative.

Situation n°2: Total De-Population Resulting in Loss of Statehood

Under the second alternative, the extinction of the State's international legal personality due to its total de-population could occur at two different moments (at least).

The first moment would be when the population leaves an inhabitable territory falling under the jurisdiction or control of the State in question, yet the influx takes place before total submergence of the territory occurs. In this case, the two 'material elements of the State' still exist and are tangible. It is the *link* or *connection* between the population of a State and its territory that gets discontinued. Such discontinuity is geographical, in so far as the population no longer inhabits the territory of the State; but it is also functional, to the extent that the primary role of the State as a form of organisation of a community of people becomes *prima facie* meaningless once such people fall outside its jurisdiction and control. Yet, such a presumption could be overturned if the international community resorts to innovative conceptual figures through which the functional discontinuity may resist or even supersede the effects of the geographical discontinuity. For instance, it could be recognised that

479 This is the main issue addressed by Hampson.

total de-population of some Pacific Island States induced by climate change impacts constitutes a case of 'population in exile' (including its government), a 'Diaspora People' or even, as Maxine Burkett already suggested, an 'ex situ nation'.[480] These figures aim at preserving the national and cultural distinctiveness of the population displaced abroad; as well as its unity, even in cases where no significant part of the population remains together in one place, but is scattered over several host countries. Any additional acknowledgement of a special form of governance for de-populated Pacific Island States, seeking to revive – if not reinvent– the functional dimension of the State, would first necessarily require the preservation of some form of distinctiveness and unity of the population.

The second moment when the loss of statehood could presumably occur is at a later stage; that is, when the total de-population of a State is followed by its total de-territorialisation as a result of the submergence of its territory. In this case, the geographical and functional discontinuity of the link between population and territory would *prima facie* result from the disappearance of one of the 'material dimensions' of the State; namely, its territory. Again, such a presumption could be overturned through international recognition of innovative figures, such as the concept of 'water States' or a new form of a non-State sovereign entity akin to the Holy See or the Order of Malta.[481]

In any case, regardless of the exact moment when statehood may be lost following its total de-population, what is certain is that the fate of the State of origin is the key to the determination of the legal status that the displaced population may uphold: the total de-population of a State leads to its loss of statehood, which in turn results in rendering its population stateless. Curiously and sadly enough, it would appear that a paradoxical situation arises, in which the displacement of the population, triggered by survival needs, becomes the cause of the change into a vulnerable legal status. A scenario in which the loss of statehood of the State of origin in sanctioned would in principle allow the displaced – now become *de iure* stateless people – to benefit from the legal and institutional protection of the Convention on the Status of Stateless People as well as of the Convention on the Reduction of Statelessness, including the possibility to resort to the UNHCR whose mandate extends not only to refugees but also to stateless people.[482]

---

480 On the proposal by Maxine Burkett, see chapter 1, section 1.3.4.B.

481 On the different possibilities, see chapter 1, section 1.3.4.

482 The UNHCR could also co-operate with other institutions specifically created for these cases, such as a special 'Authority' tasked with managing and distributing the economic

Yet, given that extinction of the international legal personality of the State of origin would not in these cases be followed by succession, resulting in an unprecedented form of statelessness, certain obligations derived from these instruments will need to be adapted.[483] As the Commission on Human Rights already noted in 2004:

> [I]n recent cases of State succession, citizenship and other status issues have been addressed by putting pressure on the successor State to make it possible, in practice, for long-term residents to acquire citizenship or equivalent rights (…). In the case of those States which are likely to disappear for environmental reasons, there would appear to be no successor State on whom obligations can be imposed and there will no longer be in existence the predecessor State.

In the absence of both a successor and a predecessor State, both of which are normally targeted by the international regime on statelessness, attention would thus turn towards the 'host State' or country where the displaced people may have found temporary asylum. The question that arises is whether the affected people could invoke their human right to receive the nationality of their host States, in view of their *de iure* statelessness outside a context of State succession.[484] The matter acquires particular relevance for vulnerable groups, such as children, minorities, and women, which tend to be the most affected by statelessness, and towards which the host State has specific protection obligations derived from their respective international instruments.

2.4 Conclusions

Closely connected with the looming de-territorialisation described in chapter 1, the primary manifestation of climate change impacts on the human

---

revenues derived from the exploitation of the natural resources of the maritime spaces of the submerged States. See chapter 1, section 1.3.4.B.

483 Hampson, 3, para. 6.

484 On this point, see for instance David Weisbrodt and Clay Collins, 'The Human Rights of Stateless Persons', *Human Rights Quarterly*, vol. 28 (2006), 245–276, who consider that "the right not to be stateless, or the right to a nationality, is widely recognised as a fundamental human right", 246. Citing US case law, the authors express the view that "the right to be a citizen of a State has been called 'man's basic right' for it is nothing less than the right to have rights", 248.

dimension of Pacific Island States is the decrease of the living conditions of Pacific Islanders. The paradigm reaction to the acute environmental stress suffered by the populations of Pacific Island States is their displacement and consequent relocation to a different place where their basic survival needs may be met. Although such reaction seems, *prima facie*, the most common ultimate form of adaptation, with several previous examples in world history, the closer look at cases of accomplished relocation in the South Pacific taken in the present chapter has unveiled both the fundamental differences as well as the underlying complexities accompanying such a process in this particular region.

First, actions conducted in the region seeking to preserve the living conditions of Pacific Islanders, so as to *prevent* their resettlement, are rather similar and homogeneous among Pacific Island States, given that these governmental actions are organised, co-ordinated and even financed at a regional level and are closely linked to implementation measures stemming from the international regime on climate change and from international action on disaster risk-reduction. In contrast, when the display of such a range of preventive relocation actions cannot guarantee the maintenance of basic survival needs of the population, and Pacific Islanders are forced to move, actions seeking to *react* to relocation scenarios differ widely from country to country and offer a highly heterogeneous landscape. These differences flourish as a result of socio-cultural, political and geographical factors encompassing, inter alia, the divergent levels of effective presence of the national government in their respective populations' daily life, the complexity of land-tenure systems, or the differences between relocations that remain within an agricultural setting, and relocations implying the move to an urban (and often already overpopulated) setting.

As these root differences are unveiled, a critical approach to how climate-induced displacement has so far been dealt with in international legal scholarship becomes necessary. Although, in general, existing studies rightly diagnose the legal challenges – namely, the scarcity and inappropriateness of the existing legal instruments and institutions – the solutions proposed to overcome such challenges – creation of new legal instruments, development of existing instruments and institutions or inter-regime linkages – are either ill-suited or insufficient to address the inherent complexities of climate-induced displacement in the South Pacific previously revealed. These are somehow top-down responses that do not respond to the heterogeneity existing in reality. Besides, and most importantly, the proposals made so far do not take into account the wider contextual variable in which relocations take place, and thus neglect or disregard the issue of the continuity of the State and how such an unresolved matter may greatly affect the responses to the problem of climate-induced displacements.

As a result, this chapter has suggested a new framework for understanding and tackling the issue of climate-induced displacement, one that encompasses all stages of a relocation (both the prevention and reactive phases), distinguishes between relocation taking place at a national level and cross-border resettlements. In doing so, it relocates the issue within its wider context by associating it with the unresolved question of whether the continuity of the State in question may or may not be at stake. This question is in fact the real axis structuring the proposal to deal with climate-induced displacement through a multi-layered legal scheme, and essentially highlights the fact that the fate of Pacific Islanders and the eventual legal protection they may have cannot be detached from the fate of their State of origin.

Then, when called upon to fill each stage of the multi-layered legal protection scheme with concrete rights, some of the proposals most recurrently upheld in international legal scholarship also appear to be useless in this region. In cases of national displacements – which do not jeopardise the continuity of the State – the applicability of human rights treaty law is very limited, due to the low level of ratification of the ICESCR in this region. Yet, the development of a human rights approach may positively inform the implementation of international environmental legal instruments – in particular of the UNFCCC, which has been ratified by all Pacific Island States. National relocation in the reactive phase fundamentally alludes to the level of guarantee or protection of housing, land and property rights which – apart from being quite poorly recognised in international human rights treaty law – are mostly ruled by traditional customary systems that may generate significant obstacles to climate-induced displaced people seeking access to new lands. On the one hand, in these cases, the housing, land and property rights of some minority groups (ethnic minorities, women and children) may be covered by specialised human rights treaties which count with a higher level of ratification than the general instruments. On the other hand, informal instruments, such as the Guiding Principles on Internally Displaced Peoples, despite serving as an appropriate road-map for future normative developments, are nevertheless not well suited to respond to displacements due to environmental – as opposed to conflict – causes.

In cases of transnational relocations – which constitute the real looming threat to the continuity of Pacific Islands' statehood – difficulties arise out of the multiplicity of relevant States involved (at least including the State of origin and the host State). Given that the moment in which the population of a State and its territory becomes formally discontinued is not settled, this chapter has differentiated between two main situations. In cases of partial de-population, the applicability of refugee law to climate-induced displaced people has not been so far accepted by the national courts of the two potential host countries of the region – Australia and New Zealand; international migration treaty law, whose material scope of application is less specialised than that of the refugee regime, is not yet applicable in the region, given the lack of ratification. Finally, the prospects of having to deal with total de-population of Pacific Island States (before total de-territorialisation actually takes place) has shown that, ultimately, the content of this multi-layered legal scheme cannot, in these cases, be fixed or defined as long as the contextual issue – namely, whether and when the State becomes extinct – is unresolved. The determination of the fate of the State challenged by high rates of, or total, de-population is a *prerequisite* of the determination of the eventual rights that climate-displaced people may have. Yet, the answer to this question lies outside the realm of law and forces us to come back to the weakly defined boundaries between international law and international politics.

Despite the fact that answers to this fundamental and structural question remain outstanding, one thing can be made clear. In all situations covered by the multi-layered legal scheme, the action and presence of the political dimension of the State is fundamental, for it is the government that organises and finances preventive and, in some cases, reactive relocation actions; and it is also the government that is tasked with negotiating migration agreements with other countries, providing its citizens with travel documents, purchasing land in other countries and, ultimately, raising its voice and defending its cause before the international community. Hence, as the protection of the human dimension of the State is necessarily ensured by its political dimension, chapter 3 will be devoted to the effects of climate change impacts on the governmental capacity of Pacific Island States and the correlative power this dimension may have for the defence of the continuity of their statehood.

CHAPTER 3

The Challenge of Government Failure

The theoretical possibility of [the State's] de-population or the total loss of its territory is hardly even of academic interest, but the loss of an effective government is altogether a different matter.

TOMAS BATY, *International Law in Twilight*, 1933.

• • •

Throughout its history, the development of international law has been influenced by the requirements of international life.

INTERNATIONAL COURT OF JUSTICE, *Reparations case*, 1949.

∴

3.0 Introduction

Both the geographical and human dimensions of the State which have been made the object of study of the two preceding chapters may be seen as the paramount raison d'être of the State. As the metaphysical or immaterial State has not (yet) seen the light of day, territory and population clearly and prominently reveal how the State, as a political concept, nonetheless requires the existence of a physical basis through which such a concept can be materialised.[485] The political dimension of the State, embodied in its government, is generally referred to as the third dimension of statehood. It is also one with the potential to eclipse all other dimensions given that, as Rosalyn Higgins considered, "if it is States who are leading actors, they find their physical manifestation in governments".[486] And indeed, there is a determinant quality of governments derived from the fact that States, as juridical persons, must

485 This argument does not amount to an organic theory of the State (generally found in geopolitical studies), but simply seeks to point out at the fact that the metaphysical State has not (yet) appeared as one of the possible expressions of statehood.

486 Rosalyn Higgins, *Problems and Process: International Law and How We Use It* (Oxford: Clarendon Press, 1994), 42.

 | DOI 10.1163/9789004321618_005

act through organs.[487] On the one hand, in the international sphere, governments are the agencies with exclusive and plenary competence to represent their respective States vis-à-vis other States or international organisations, irrespective of whether they are governments *in situ* or in exile.[488] On the other hand, when it comes to a State's action in the national realm, governments are the engine and agency through which the State's ruling power over its population and territory, or its competences therein, are displayed.

Yet, the political dimension of the State does not simply refer to governments from the standpoint of the functional role they play; an additional element is encompassed in this dimension, which marks its distinctiveness from the geographical and human dimensions. Indeed, while serving as the material agency through which a State operates, both at a national and international levels, governments are also the depositories of the distinctive attribute of the State, namely, its sovereignty. Therefore, the political dimension of the State plays a fundamental role vis-à-vis the geographical and human dimensions.

From the fact that the political dimension of the State encompasses not only the organ through which the State acts, but also the quality enabling its differentiation from other forms of political organisation,[489] it follows that this dimension conflicts with, and necessarily leads to, incursions into the realm of power. Indeed, to begin with, sovereignty as a legal concept can be commonly defined as the 'supreme power' which, as Jean Salmon explains, is distinct from independence or political and economic power, and applies equally to all States from the moment they acquire such a *statut étatique*, irrespective of the scope of their territory, the importance of their population, the form and nature of their government, or their autonomy or independence from third States.[490] Norberto Bobbio further specifies that, although sovereignty does not exhaust power, it does capture the most significant and potentially

487 *German Settlers in Poland*, Advisory Opinion, Permanent Court of International Justice, 1923 Series B, nº 6: 1, 22.

488 Ibid.

489 Jean Salmon, 'Quelle place pour l'État dans le Droit international d'aujourd'hui?', *Recueil des Cours de l'Académie de Droit international de La Haye*, vol. 347 (2010), 9–78, 21.

490 Ibid., 55 [my own translation]. The original version in French defines sovereignty as: "[U]n concept juridique distinct de l'indépendance ou du pouvoir politique et économique. C'est un concept juridique qui s'applique de manière égale à tous les États à partir du moment où ils accèdent au statut étatique, quelle que soit la superficie de leur territoire, l'importance de leur population, la forme et la nature de leur gouvernement, leur autonomie ou leur indépendance économique ou politique à l'égard d'États tiers".

dangerous form of concentration of power.[491] Among the countless definitions of sovereignty that may be found in international legal scholarship, political theory and international relations, the one that best crystallises this idea may be Carl Schmitt's. In the words of this German philosopher, "sovereign is he who decides on the exception" – understanding 'the exception' as a general concept belonging to the theory of State, and not merely as a construct applied to states of emergency or siege.[492] The exception, Schmitt continues, "is not codified in the existing legal order",[493] and can at best be characterised as a situation of "extreme peril, a danger to the *existence* of the State",[494] or the like. It is thus in the essence of sovereignty to be a borderline (as opposed to a vague) concept, although such a condition is precisely what "makes relevant the whole question of sovereignty".[495]

Furthermore, by linking the definition of sovereignty with the extreme situations in which the continuity of the State is at stake, Carl Schmitt's definition is also a reminder of how sovereignty operated as a fundamental conceptual basis of the modern State. Indeed, it is nowadays generally admitted that, since the birth of the modern European State, the concept of sovereignty cannot be dissociated from the concept of State (notwithstanding the special cases of the Holy See and the Order of Malta previously mentioned).[496] Correlatively, the idea of sovereignty was nurtured by the internal sphere of the State, where it finds its conceptual roots and origin, while its external dimension developed as a limit to the scope of the respective sovereign powers of other States.[497] Both the internal and external aspects of sovereignty – inextricably bound to the State's national and international political dimensions – are referred in Article 1 of the Montevideo Convention, which establishes that, for a

491 Norberto Bobbio, Nicola Matteuci y Gianfranco Pasquino, eds., *Diccionario de Política* (Méjico DF: Siglo XXI, 2008), vol. 2, 1215–1225, 1218.

492 Carl Schmitt, *Political Theology: Four Chapters on the Concept of Sovereignty* (Chicago: University of Chicago Press, translation by George Schwab, 1985), 5.

493 Ibid., 6.

494 Ibid. [Emphasis added].

495 Ibid., 6.

496 See for instance Jean Salmon, ed., *Dictionnaire de Droit international public* (Bruxelles: Bruylant, 2001), 1045–46: "la souveraineté expire la puissance suprême (suprema potestas) de gouverner, de commander et de décider et qui, *liée à l'apparition de l'État moderne, est inseparable de celui-ci*". [Emphasis added].

497 Bobbio et al., 1221.

political entity to acquire statehood, it must have a "government"[498] as well as "the capacity to enter into relations with other States".[499]

This final chapter considers the impacts of climate change on the political dimension of Pacific Island States, a complex task, for two main reasons. First of all, such a consideration entails a high risk of getting trapped in a somewhat circular logic. Indeed, in order to explore the impacts of climate change on the capacity of Pacific island governments to fulfil their roles, as organs of the State, governments by definition require the existence of a State.[500] Yet, it is difficult to evaluate whether and, eventually, when the adverse effects of climate change on the governmental capacity of Pacific Island States may truly endanger their statehood. Secondly, associating the concept of sovereignty, understood as a supreme power, with the politically weak, economically vulnerable, and environmentally threatened Pacific Island States may seem a bit far-fetched or disconnected from their reality in the international sphere.

The question raised in this chapter is two-fold: how may the adverse impacts of climate change challenge the governmental capacities of Pacific Island States? But also, how may the active participation of Pacific Island States in the international sphere operate as a determining variable of the question of their continuity as States? Section 3.1 begins by considering the challenge that climate change impacts may have on the capacity of Pacific Islands' governments to operate as *effective* organs of the State. It therefore examines how they may be challenged by the 'failed States' doctrine, or by the possibility of having to evacuate the governments from their respective territories. Section 3.2 explores the role of Pacific Island States as players in the international community, with particular emphasis on their position regarding the threat to their survival as States, and argues that such a position is not irrelevant to the solution to their situation. Relatedly, this section recalls the involvement that the international community has had in the acquisition and sustainability of Pacific Islands' statehood, and reveals that the question of the continuity of Pacific Islands' statehood cannot be resolved from a legal standpoint alone, but depends on the political decision of the international community to maintain their support of those States. Finally, section 3.4 argues that the continuity of Pacific Islands' statehood even beyond its 'material life', through for instance the temporary maintenance of international recognition, can be based on a set of ethical grounds.

498 Montevideo Convention, Article 1(c).

499 Ibid., Article 1(d).

500 Stefan Talmon, *Recognition of Governments in International Law – with Particular Reference to Governments in Exile* (Oxford: Clarendon Press, 1998), 15.

3.1 The National Challenge: Impacts on Pacific Islands' Domestic Governmental Capacity

3.1.1 *Partial De-Territorialisaton and De-Population: From 'Vulnerable' to 'Failed' States?*

A Controversial Labels: 'Failing', 'Failed', or 'Collapsed' States

Just as sovereignty became the indispensable and distinctive attribute of the State since the birth of the modern European State, formal equality among States is regarded as integral to sovereignty. The resulting principle of 'sovereign equality', "indispensable foundation of international society",[501] conveys the idea of an international legal order where all States are legally equal, while it also enables State diversity – also referred to as pluralism – to be tolerated.[502] And yet, while the fundamental character of this principle in the construction of the modern international legal order remains uncontested, its relevance and operability in the context of contemporary international relations has been open to question. First, as a result of decolonisation, strong political input – for instance found in the ideology for a New International Economic Order – focused on the need to centre on substantive and not only formal sovereignty.[503] Then, as globalisation spread at the end of the Cold War and interdependence reached unprecedented peaks, some legal scholars considered, somewhat apocalyptically, that the end of the State could be closer than ever.[504]

---

501 Lassa Oppenheim, 'The Future of International Law', *Carnegie Endowment for International Peace*, Pamphlet n° 39 (1921), 1–68, 20. Unsurprisingly, literature on such a grounding principle of the international legal system, such as sovereign equality, is very extensive. Some examples include, Carlos Espósito, 'Soberanía e Igualdad en el Derecho Internacional', *Estudios internacionales* (*Universidad de Chile*), vol. 165 (2010), 171–196, 172. Pieter H. Kooijmans, *The Doctrine of Legal Equality of States: an Enquiry into the Foundations of International Law* (Leiden: Nijhoff, 1964); and Christian Tomuschat, 'International Law: Ensuring the Survival of Mankind on the Eve of the New Century', General Course on Public International Law, *Recueil des Cours de l'Académie de Droit international de La Haye*, vol. 281 (1999), 9–438, 161.

502 Gerry Simpson, *Great Powers and Outlaw States: Unequal Sovereigns in the International Legal Order* (Cambridge: Cambridge University Press, 2004), xii.

503 See UN General Assembly, 'Declaration on the Establishment of a New International Economic Order', adopted at the 2229th plenary meeting, 1 May 1974, UN doc.: A/RES/S-6/3201. Doctrinal accounts of this move include, for instance, Mohammed Bedjaoui, *Towards a New International Economic Order* (Paris: UNESCO, 1979).

504 See for instance wordlist approaches, such as Ali Khan, *The Extinction of Nation-States: a World without Borders* (Michigan: Kluwer Law International, 1996). See also, Christoph Schreuer, 'The Waning of the Sovereign State: Toward a New Paradigm of International Law', *European Journal of International Law*, vol. 4 (1993): 1, 447–471.

However, the real challenge that the principle of sovereign equality has recently been confronted with does not simply rely on the factual inequality of a State's capacities (albeit connected to the historical phenomena of decolonisation and globalisation). Such a challenge is related to the spread of new labels that seemingly sought to establish 'categories' of States – be it 'outlaw'/'rogue' States, on the one hand, or 'failed'/'failing'/ 'collapsed' States, on the other.

These terms are misleading in so far as most literature on State failure or collapse generally uses such a term to refer to what in practice corresponds to *government* failure or collapse. On the one hand, the first category has been generally used to refer to 'criminal States' (i.g. States ruled by "tyrans" with no political or legal legitimacy) and seemingly introduces a Manichean view of international relations and has been particularly used after the 9/11 terrorist attack. On the other hand, second category of 'failed', 'failing' or 'collapsed' States has been particularly applied to post-colonial developing States, in view of their inherent political instability and economic underdevelopment that could potentially lead to violent national and regional conflict.[505] Both categories were born as highly controversial political constructs deployed after the end of the Cold War,[506] and result from the ideological homogenisation that followed such a period. They served to pierce the veil of formal sovereign equality and marked the launch of a doctrinal and political move towards the *grading* of sovereignty – which arguably amounted to an updated re-introduction of the modernist doctrine of the 'standard of civilisation'.[507] As these

505 For a critical explanation on how the social sciences developed representations of post-colonial States that arose as an adjunct to the hegemonic pretensions of the USA in the making of the 'Third World', see Pinar Bilgin and Adam D. Morton, 'Historicising Representations of "Failed States": beyond Cold War Annexation of the Social Sciences?' *Third World Quarterly*, vol. 23 (2002): 1, 23–50.

506 Voices opposing the concept of failed States from the policy and legal standpoints thus include, inter alia: Ralf Wilde, 'The Skewed Responsibility Narrative of the "Failed States" Concept', *ILSA Journal of International and Comparative Law*, vol. 9 (2003): 2, 425–429; Hans-Joachim Spanger, 'Failed State or Failed Concept? Objections and Suggestions', *Peace Research Institute of Frankfurt*, Paper presented at the Conference on Failed States III: Globalisation and Failed States, held in Florence (Italy) on 7–10 April 2000. For more recent works, see, for instance, Alison J. Ayers, 'The Illusion of an Epoch: Critiquing the Ideology of Failed States', *International Politics*, vol. 49 (2012): 5, 568–590; Neyire Akpinarli, *The Fragility of the "Failed State" Paradigm: a Different International Law Perception of the Absence of Effective Government* (The Hague and Leiden: Martinus Nijhoff Publishers, 2010).

507 Early and clear accounts in favor of the 'grading of sovereignty' by American political scientists include: Robert H. Jackson, *Quasi-States: Sovereignty, International Relations and the Third World* (Cambridge: Cambridge University Press, 1990); Stephen Krasner, 'Compromising Westphalia', *International Security*, vol. 20 (1995): 3, 115–151; Janice E.

concepts were essentially produced by policy-makers of hegemonic countries, no formal institutional definitions by the UN or regional organisations can be found; rather, a diverse range of definitions and attempts to establish formal criteria or thresholds are mainly found in national security strategies of hegemonic powers and doctrinal analyses of international relations scholars and political scientists.[508]

The category of 'rogue' or 'outlaw' States is left out of the present analysis, since it is not relevant to our study on the impact of climate change on the governmental capacities of Pacific Island States. The concepts of 'failed' or 'failing' States rely on the idea that, as sovereignty empowers the State, correlative responsibilities attached to such power ensue. States are, in particular, expected to perform through their governments certain functions directed at the protection and wellbeing of their respective populations, on which the State's political legitimacy is ultimately based. Thereby, political scientists have tended to define this category loosely as those States which have been unwilling or unable to fulfil the so-called core 'Weberian functions of the State'.[509] Some authors, such as Robert Rotberg, have established a detailed

---

Thomson, 'State Sovereignty and International Relations: Bridging the Gap between Theory and Empirical Research', *International Studies Quarterly*, vol. 39 (1995), 213–233; and Cristopher Clapham, 'Degrees of Statehood', *Review of International Studies*, vol. 24 (1998), 143–157. For a general explanation of the legal consequences of this movement, see Petra Minnerop, 'The Classification of States and the Creation of Status within the International Community', *Max Planck Yearbook of United Nations Law*, vol. 7 (2003), 79–182.

508 See, for instance, Stanley Hoffman, *Chaos and Violence: What Globalisation, Failed States and Terrorism Mean for US Foreign Policy* (Lanham, MD: Rowman and Littlefield, 2006); Robert I. Rotberg, 'Failed States, Collapsed States, Weak States: Causes and Indicators', in Robert I. Rotberg, ed., *State Failure and State Weakness in a Time of Terror* (Washington, D.C.: Brookings Institution, 2006) and by the same author 'The New Nature of Nation-State Failure', *The Washington Quarterly*, vol. 25 (2002): 3, 85–96. One of the main sources of controversy arises from the fact that these newly invented categories have the potential to serve justifying – at least from a political perspective – foreign military interventions in breach of the general prohibition of the use of force and of non-intervention in the internal affairs of other States- respectively prescribed in Articles 2(4) and 2(7) of the UN Charter. For a synthetic explanation of the legal implications of the spread of the concept of failed States for international security, see, for instance, Robin Geiss, 'Failed States: Legal Aspects and Security Implications', *German Yearbook of International Law*, vol. 47 (2005), 457–501; Claus-Dieter Classen, '"Failed States' and the Prohibition to Use Force,' in *Les nouvelles menaces contre la paix et la sécurité internationales: Journée franco-allemande*, Société Française pour le Droit international (Paris: Pedone, 2004), 129–140.

509 Jonathan Di John, 'Conceptualising the Causes and Consequences of Failed States: a Critical Review of the Literature', *School of Oriental and African Studies, Crisis States Working Paper Series*, vol. 2 (January 2008): paper nº 25, 1–54, 4.

list of the functions of the State, the fulfilment of which should be scrutinised when considering whether a State should be labelled as 'failed' or 'failing'.[510] These functions include, first of all, the security of the population (measured by the absence of enduring political violence as well as by the ability to control State borders); coupled with political participation, economic development, infrastructures and social services.[511]

Two points are worth noting in this context. First of all, State weakness or fragility is generally differentiated from failure; while the former refers to a reduced or limited *effectiveness* of the State to fulfil one or more of the functions laid down above, the latter implies that the specific functions at stake are not fulfilled at all; in other words, they are *absent*. Secondly, this gradation goes on to differentiate between failed States and collapsed States, another category which tends to be even more loosely defined, but generally refers to the failure of the State to fulfil through its government the set of Weberian functions *to the extent that* the conditions necessary for the State to "sustain itself as a member of the international community" are no longer met.[512] Therefore, in William Zartman words, State collapse can be defined as occurring when the governmental capacities of a State are so diminished that it "can no longer reproduce the conditions for its own existence".[513]

B Controversial Assumption: From 'Failed State' to 'Extinct State'?

Considering that the definition of failing, failed and collapsed States is not only controversial, but also variable, unsettled and multiform, one is faced with the question of how should concrete cases, such as those of certain Pacific Island States, be approached. Also, more importantly, how can it be possible to disentangle the concrete legal elements so that we can consider whether Pacific Island States will eventually become 'failed' or even 'collapsed' States as a result of the adverse impacts of climate change?

Given that State capacities are not uniform across functions and that, as Jonathan Di John explains, "it is imperative for any definition of "failure" to be explicit in which dimension a State fails",[514] the first question that arises is whether, when, and how may the *precise tipping points*, whereby State

510 Robert I. Rotberg, 'The New Nature of Nation-State Failure', *Washington Quarterly*, vol. XXV (2002): 3, 85–96, who considered that the severity of State failure could be ranked according to the specific function that the State was not fulfilling.

511 Ibid.

512 Gerard B. Helman and Steven R. Ratner, 'Saving Failed States', *Foreign Policy*, vol. 89 (1993), 3–20.

513 William Zartman, ed., *Collapsed States: the Disintegration and Restoration of Political Authority* (Boulder, Colo.: Lynne Rienner, 1995), 1.

514 Di John, 4.

weakness becomes failure or collapse, be identified.[515] To be sure, one could make a projection and estimate that the effects of climate change on Pacific Islands' territories and populations will gradually exert an increased pressure on the already limited capacities of Pacific Islands' governments. As already seen in chapters 1 and 2, sea-level rise and coastal erosion are already forcing these States to take active measures to stabilise and fix their maritime entitlements and enhance regional co-operation to develop infrastructures that may increase their resilience, whilst population relocations have forced some Pacific Island States to produce relocation plans, negotiate with the relevant stakeholders and even seek international funding. Yet, setting the criteria for establishing when the limited effectiveness of Pacific island governments to fulfil the Weberian functions of the State in the context of acute climate change impacts may deteriorate so as to amount to 'failure' of the State is undefined and potentially highly controversial.

If State failure of Pacific Island States was to be confirmed internationally – for instance, by a relevant United Nations body, or self-acknowledged by the States concerned – the second question that arises is whether State failure or collapse necessarily implies loss of statehood. As previously remarked, 'State failure' or 'State collapse' are terms generally used to refer to what in practice corresponds to *government* failure or collapse. This terminological equation is to some extent understandable considering that, as already pointed out above, States are juridical persons which necessarily act through an organ, namely, the government. The fundamental importance of an effective government at the time when a political entity seeks to acquire statehood is undeniable; its existence is also generally considered and balanced along with other considerations, such as the political legitimacy of such a government and its connection with the people's right to self-determination. And certainly, the subsequent loss of an effective government, as Tomas Baty's words opening this chapter illustrate, is no little matter. However, irrespective of the legal consequences that the lack of an effective government may be, State practice indicates that the failure of the State to fulfil its functions does not imply that its statehood is challenged. In the words of Rosalyn Higgins "what is absolutely clear is that a loss of 'stable and effective government' does not remove the attribute of statehood, once acknowledged".[516] Paul Williams for instance notes that one of the reasons supporting this may be that the failure of the governance structures

515 Ibid.

516 Higgins, 40.

of a State are not even generally homogeneous in all the territory.[517] Rather, different "degrees of success and failure can exist within a single State", for the failed zones are usually made up of "numerous (often interconnected zones) where different sources of authority may dominate the local governance structures".[518] Focusing on the example of Somalia, the author for instance notes that the collapse of the Somali central government "did not automatically exclude the possibility that zones of alternative forms of governance and authority existed within Somalia's officially recognised international borders".[519]

Yet one point deserves attention in this context, which relates to the specificity of Pacific Island States. Although government failure as such is not, according to consistent State practice recently exemplified with the cases of Somalia or Afghanistan, a circumstance that may challenge the continuity of the affected State, one may wonder whether such continuity could be at risk when the failure of the government is coupled with the partial loss of territory and the partial but significant de-population of the State itself. This question is challenging in so far as it directly touches on the already mentioned inherent bias of the international legal order to maintain the status quo for the sake of the stability of international relations, even if sometimes such stability may require the maintenance of legal fictions. Therefore, when and why the extinction of statehood takes place is ultimately a matter that will probably be resolved by the weight of the facts; it will invite scrutiny of the future adaptation of State practice and theoretical aspects of statehood in this matter.

The next question that arises when considering whether Pacific Island States may become 'failed' or 'collapsed' States therefore regards the obligations and responsibilities of other States in such a situation. Literature on State failure and collapse has sometimes developed the idea that the observed existence of 'failed States' engages the responsibility of the 'Great Powers' to act in the country and participate in the prevention of collapse or in the reconstruction of the State at stake. Robert Jackson for instance equates today's Great Powers' responsibility in the face of State failure to the earlier trusteeship function performed by the 'benevolent international society' during the decolonisation process.[520] This line of reasoning is seemingly compatible

517 Paul Williams, 'State Failure in Africa: Causes, Consequences and Responses', in Africa South of the Sahara (Europa World Yearbook, 3rd ed. 2010), 21.

518 Ibid.

519 Ibid., 21–28.

520 See, for instance, Robert H. Jackson, 'Surrogate Sovereignty? Great Power's Responsibility and "Failed States"', *Institute of International Relations of the University of British Colombia*, Working Paper n° 25 (1998), 1–17. See also David Chandler, 'Great Power Responsibility

with the recently developed concept of responsibility to protect. Principle 1(B) of the 2001 Report of the International Commission on Intervention and State Sovereignty provides that:

> [W]here a population is suffering serious harm, as a result of internal war, insurgency, repression *or state failure*, and the state in question is unwilling or unable to halt or avert it, the principle of non-intervention yields to the international responsibility to protect.[521]

However, these accounts have proven to be dangerous from a policy perspective in so far as they potentially incentivise interventions in the domestic affairs of the State and potentially breach Article 2(7) of the UN Charter.[522] Therefore, intervention in the territory of an island State following partial submergence will be subjected to the prior valid consent of the government – whether still present in the territory or already evacuated. The purpose of the intervention may consist on assisting the State to prevent from collapsing and – in cases where collapse cannot be avoided – to provide the population with humanitarian relief.[523]

---

and Failed States: Strengthening Sovereignty?', in Julia Raue and Patrick Sutter, eds., *Facets and Practices of State-Building* (Leiden: Martinus Nijhoff, 2009), 15–30.

521 International Commission on Intervention and State Sovereignty, *Responsibility to Protect*, Report (Ottawa: International Development Research Center, 2001), XI. [Emphasis added].

522 On the perils of the responsibility to protect, see Jean Marie Crouzatier, 'Le principe de la responsabilité de protéger : avancée de la solidarité internationale ou ultime avatar de l'impérialisme?', in Jean-Marie Crouzatier et al., *La responsabilité de protéger* (Paris : Éd. des Archives Contemporaines, 2008), 13–32. The difficulties of integrating this principle into the operation of the collective security system is for instance explained by Barbara Delcourt, 'L'introduction de la notion de responsabilité de protéger dans les autorisations données par le Conseil de sécurité : enjeux politiques et paradoxes', in Karine Mollard-Bannelier et Cyrille Pison, eds., *Le recours à la force autorisé par le Conseil de sécurité : droit et responsabilité* (Paris: Pedone, 2014), 53–66.

523 On the application of the principle of responsibility to protect to situations of natural disasters, see Laurence Boisson de Chazournes, 'Responsabilité de protéger et catastrophes naturelles: l'émergence d'un régime?', in *Société française pour le Droit international*, 41st Colloque, Paris X- Nanterre (Paris: Pedone, 2007), 153–160.

3.1.2 *Total De-Territorialisation and De-Population: Pacific Island Governments in Exile*

A 'Traditional' Forms of Governments in Exile

Contrary to the controversy generated by the proposal to address the challenges of Pacific Islands national governmental capacity from the perspective of 'failing', 'failed' or eventually even 'collapsed' States, dealing with the evacuation of Pacific island governments beyond their national borders is a less questionable matter. Essentially, it calls upon interpreting how longstanding rules and State practice on governments in exile may be adapted and at least in some respects applicable to these situations. So far, the cross-border evacuation of Pacific Island governments is only a hypothetical scenario, based on projections of the acute consequences of climate change impacts.

Governments in exile are not, per se, a new concept in international law, nor a controversial one. In fact, as Stefan Talmon explains, the term 'governments in exile' as such does not correspond to a specific legal status but simply indicates that the (current) domicile of a government – that is, the authority through which the State displays and exercises its sovereign competences – is located outside the national territorial borders.[524] Talmon also criticises the hyphenated expression 'governments-in-exile', as it gives the impression that they constitute an independent subject of international law with a legal status distinct from that of governments. This consideration is indeed inconsistent with State practice, which shows that States have been called upon choosing between two options: either recognising an authority settled abroad as the government of a specific State; or not granting such recognition. What States have never done is to recognise an authority as a 'government in exile'. The fact that the government is outside the national territorial borders is thus a matter of fact which, albeit carrying a set of legal consequences pertaining to the exercise of sovereign State competences, does not transform the legal nature of the government, nor its legal status in international law.[525] Hence why, as Talmon recalls, there is no precedent of a State having denied a certain governmental competence to a government on the ground that its domicile was outside the national territorial borders of the State it represented.[526]

Traditional circumstances in which governments have been led to evacuate from their countries are related to the outbreak of violent conflict, leading to belligerent occupation or civil war and/or stemming from either decolonisation

524 Talmon, 16.

525 Ibid.

526 Ibid.

or even the defence of democratisation. Many of the prominent cases of governments in exile occurred during World War II, as the German invasion gradually spread over Europe. For instance, in 1939, the Polish Government escaped under the orders of General Sikorski from Paris to London; the same year the Czech president Eduard Benes settled in London and was recognised as the head of the Czech government in exile for two years. This also applied to many European royal families. Queen Wilhelmina of The Netherlands, the Grand Duchess of Luxemburg, King Peter II of Yugoslavia, King Haakon VII of Norway and King George II of Greece all moved to London with their cabinets between 1939 and 1941. More recent post-World War II examples of governments in exile include the aspirations of an exiled group known as the National Coalition Government of the Union of Burma to overthrow its home country's government; the case of the *Gouvernement Provisoire de la République Algérienne*, which resulted from post-colonial political instability; and even the claims of the Tibetan Government in Exile for a political status independent from China.[527] These circumstances imply that, when authorities of a State are forced to leave, one main political and legal issue arises, namely, the *competition* between two co-existing authorities claiming governmental status in respect of the State in question. This is particularly clear when it comes to situations of belligerent occupation, in which a de facto authority controlling the territory and population competes with the de iure government forced into exile, and seeks to obtain recognition from other States as the 'new' government of the occupied State. These cases imply that the de facto authority seeks to *shift* the recognition of other States from one authority to another, for only one government at a time can have the competence to bind its State.[528]

This situation of double authorities claiming international recognition is, however, highly unlikely to occur in the case of Pacific Island States. Rather, the situation they are most likely to find themselves in is one in which the same authorities that constituted the legitimate government, recognised as such by the international community, are forced to continue exercising their functions outside their national borders as a result of severe impacts of climate change, and to presumably do so along with the population, equally forced to leave the country. Therefore, as no competition of authority will take place, it seems more appropriate to refer to Pacific Islands' governments abroad as 'ex situ

527 See for instance, Helen L. Scanlon, *European Governments in Exile* (Washington, D.C.: Carnegie Endowment for International Peace, 1943), and Yossi Shain, ed., *Governments-in-Exile in Contemporary World Politics* (London: Routledge, 1991).

528 George Scelle, *Manuel de Droit international public* (Paris: Domat Montchrestien, 1948), 174–5.

governments', rather than as 'governments in exile'. This preference also takes into account the definition of the term 'exile' which, according to the Oxford Dictionary, is "the state of being barred from one's native country, typically for political or punitive reasons". The use of the term 'ex situ' instead of the word 'exile' allows to highlight that the cause of the evacuation is not political.[529]

Just as governments in exile do not constitute a category with a distinct legal status under international law, Pacific Islands' ex situ governments should not be different either. Issues pertaining to the recognition, by other States, of authorities abroad as the government of the Pacific Island State concerned are in principle simplified by the absence of such competing claims. Nonetheless, international recognition of the fact that the domicile of the government of a Pacific Island State has been established outside its national borders will arguably remain necessary. The next section therefore scrutinises the main legal consequences arising from the scenario of Pacific Islands' ex situ governments, ultimately relocating this projection within the overall challenge of climate change impacts to the continuity of Pacific Islands' statehood.

B 'New' Climate-Induced Governments in Exile

One of the fundamental characteristics of recognition of governments in exile is that the authority claiming governmental status does not have to exercise the effective control of the territory of the State it represents in order get recognition as government granted by another State.[530] Such a characteristic will be very important for Pacific Islands ex situ governments, which may ensure some level of protection of Pacific Islands' interests abroad, whether related to the defence of their maritime spaces from becoming part of the high seas, or associated with the needs and interests of Pacific Islanders abroad.[531] It is also worth noting that the possibility to fulfil such a function of protection of nationals abroad, described by Talmon as "one of the most noble tasks of States",[532] is not affected by the government's limited capacity or even complete incapacity to control its territory effectively.[533] Such protection may

529 The term 'ex situ' government is adapted from the previous proposal by Maxine Burkett on 'ex situ nations'. See Burkett, 345–346.

530 Marek, 99–101.

531 Talmon, 202.

532 Ibid.

533 As put by Marek, 99: "A such, they [the governments in exile] constituted the living application of the rule safeguarding the continuity of the occupied State. As such, they *were instrumental in carrying on, and giving expression to, that continuity. It is in this capacity that the exiled governments carried out acts of State sovereignty on foreign soil in every sphere -legislative, administrative and judicial*". [Emphasis added].

indeed be ensured by the commitment of other States, expressed through an inter-State agreement, to protect the nationals of the affected State – in the form of consular representation, the furnishing of passports and identity and, most importantly, through the conclusion of migration arrangements with other States in which Pacific Islanders may be relocated.

The need to conclude international arrangements, either with other States or with international organisations willing to help the evacuated populations, leads to question the legal effects of Pacific Islands' governments ex situ condition on the representation of their State before the international community. As only one government at a time is legally competent to bind the State, it follows that only if the authorities abroad have been recognised as governments may they validly sign, ratify, and accede to international agreements. Again, as no competition between de facto occupant authorities and de iure authorities in exile for recognition of governmental status is likely to take place in the case of Pacific Island States, all *formal* issues which recurrently arise when only some States parties to a multilateral agreement recognise the government of the State concerned that wishes to accede, are not an issue in our case. The same may be said of the problem, very frequent in some previous cases, of whether authorities abroad not recognised as governments may only enter into agreements in their own name or on behalf of the political movement they constitute, but not on behalf of the State they purport to represent.[534] In fact, considering that, in the case of Pacific Island States, the ex situ government would be the same government which previously resided at its own territory, un-replaced nor challenged by any other authority thereafter, one could even question whether a *de nuovo* recognition by other States would even be legally required, and necessary in practice. Given that these unprecedented cases do not entail an interruption of the constitutional continuity of Pacific Islands' governments, new forms expressing support of the ex situ government could potentially emerge from State practice, other than the well-established doctrine of 'recognition of governments'.

Notwithstanding this caveat, effects of recognition of authorities abroad as governments on the *substance* of the treaties that such authorities are competent to conclude, will be applicable to the case of Pacific Island States. To begin with, as the government of a State holds the exclusive and plenary capacity to bind the State it represents, only the authorities abroad recognised as governments will be able to conclude treaties that involve the exercise of State

534 Talmon, 203.

sovereignty, such as treaties which dispose of the State's territory.[535] Some scholars, such as Krystyna Marek, have nonetheless taken the view that the inherent limitations of an ex situ government make it incompetent to conclude treaties affecting the cession of a part of the State's territory.[536] Talmon, in contrast, holds that there is no halfway situation; either an authority is recognised as a government, becoming entitled to conclude *all* kinds of treaties, or it is not recognised as a government and cannot bind the State it seeks to represent by *any* treaty. What the absence of the government from the State territory may imply, Talmon specifies, is that the capacity of the State *to perform or comply with* the obligations of a treaty may be limited by the factual context in which it finds itself.[537]

Therefore, the ex situ condition is not, by itself, an impediment for Pacific Islands' governments evacuated abroad to continue concluding valid international agreements on behalf of their respective States. Rather, the main issue arising out of the specific circumstances of Pacific Islands' ex situ governments pertains to the fact that there will be no temporal limitation on such exceptional situations. As the general assumption that the absence of any government from the State territory is only temporary cannot be applied to Pacific Islands' ex situ governments, the limitation of the States' capacity to perform certain treaty obligations will eventually turn into a full incapacity *to implement* some treaties. Although this incapacity does not affect their competence to conclude them, it does limit the range of new treaties that Pacific Islands' governments will be able to conclude. In particular, they will not be able to conclude treaties that are supposed to be implemented, even partly, upon their return to the national territory. Alternatively, they may be called upon concluding treaties which are only in force for a certain number of years. This is actually common practice for 'traditional' governments in exile, which have concluded treaties governing military, defence, economic, trade, loan, aid, extradition, administration, and even political agreements that are enforceable for only twenty years.

Furthermore, what is being highlighted for future treaties is applicable to treaties that had been concluded *before* the evacuation of the government. Absence of the government from the State territory does not, per se, automatically suspend or terminate existing treaties. However, the conditions that cause a government to establish itself abroad may ground the termination

535 Ibid.

536 Marek, 458.

537 Talmon, 136.

or suspension of treaties for supervening impossibility of performance,[538] or fundamental change of circumstances.[539] In addition, even in cases when ex situ governments alone may not constitute an overriding impossibility or a fundamental change of circumstances sufficient to justify the termination or suspension of a treaty, they may yet constitute an *exonerating cause* of State liability for non-performance or non-compliance with treaty obligations.[540]

Closely connected to the importance of recognition – or support expressed through any new possible legal form – of Pacific Islands' authorities abroad as governments in the realm of treaty law – determinant in establishing whether such governments can exercise the rights of the State or perform State obligations and conclude new treaties, but also suspend, terminate or denounce existing treaties they may not be able to perform anymore – is the issue of privileges and immunities of such ex situ governments. In this sense, recognition by the State 'hosting' the ex-situ government may be required and necessary. This may not, however, be automatically extended to all other States.

538 VCLT, Article 61(1): "[A] party may invoke the impossibility of performing a treaty as a ground for terminating or withdrawing from it *if the impossibility results from the permanent disappearance or destruction of an object indispensable for the execution of the treaty*. If the impossibility is temporary, it may be invoked only as a ground for suspending the operation of the treaty". [Emphasis added].

539 VCLT, Article 62: "[A] fundamental change of circumstances which has occurred with regard to those existing at the time of the conclusion of a treaty, and which was not foreseen by the parties, may not be invoked as a ground for terminating or withdrawing from the treaty unless:

(*a*) The existence of those circumstances constituted an essential basis of the consent of the parties to be bound by the treaty; and.

(*b*) The effect of the change is radically to transform the extent of obligations still to be performed under the treaty.

2. A fundamental change of circumstances may not be invoked as a ground for terminating or withdrawing from a treaty:

(*a*) If the treaty establishes a boundary; or

(*b*) If the fundamental change is the result of a breach by the party invoking it either of an obligation under the treaty or of any other international obligation owed to any other party to the treaty.

3. If, under the foregoing paragraphs, a party may invoke a fundamental change of circumstances as a ground for terminating or withdrawing from a treaty it may also invoke the change as a ground for suspending the operation of the treaty.

540 Talmon, 115.

Similarly, authorities abroad recognised as governments benefit from the presumption that they possess *certain* competences, privileges and immunities, unless otherwise established by the recognising State.[541] This element has been very important in most precedents known so far, given that most situations of governments in exile have been the product of internal or international conflicts. 'Traditional' types of aid requested by governments in exile to regional and international organisations include the creation of new police forces or the modernisation of their existing armed forces; the freezing of the State foreign financial assets; the imposition, through the operation of the collective system of international peace and security, of economic sanctions and an arms embargo on the de facto, in situ authority; or the sending of civilian or diplomatic missions to their States. Yet, in the case of Pacific Island States, the importance of getting international recognition – or any new form of support or acknowledgement – of their evacuated authorities as governments stems from the fact that in traditional cases, recognised governments in exile are the only ones which may validly authorise other States to intervene within the territorial boundaries of the State for humanitarian purposes.[542] Here again a new outstanding question arises of whether Pacific Islands' governments will need to seek recognition or another form of international support *before* evacuating elsewhere, in order to validly consent to an international intervention and request humanitarian relief for the population remaining in the territory.

Besides, apart from authorisation to receive humanitarian or other sorts of assistance from the international community, Pacific Islands' ex situ governments will probably be called upon taking interim or provisional measures that will require, for instance, the development of national legislation. Therefore, the question of how the ex situ condition of a government may affect the State's jurisdiction requires an examination. A State's jurisdiction to prescribe exists in international law by virtue of having an authority with governmental status, irrespective of whether or not such government is absent from State territory. This point has two main consequences. First, governments in exile, Talmon explains, "do not exercise jurisdiction derived from the host State, but exercise *their own* jurisdiction in the host State" and therefore, secondly, "government recognition does not establish jurisdiction".[543] The State's jurisdiction to prescribe certain types of norms is thus not limited by the location of its government abroad. In contrast, pursuant to the principle of territorial

541 Ibid., 116.

542 Ibid., 126.

543 Ibid., 244. [Emphasis added].

sovereignty, the State's jurisdiction to enforce the laws it prescribes (whether new laws enacted ex situ or former laws still in force) is limited, and requires in any case the consent of the host State. Moreover, as Talmon reminds us, an ex situ government does not possess more powers than it does at home, but may nonetheless be allowed by the host State to exercise more of the existing powers than usual, in the light of special circumstances and by way of comity.[544] Finally, closely related to the limited jurisdiction of States with governments abroad to enforce is the jurisdiction to adjudicate. Traditional situations of governments in exile would not see their jurisdiction to adjudicate as such affected, since the judiciary is not evacuated with their governments. Therefore, the principle that applies in 'traditional' situations of governments in exile simply holds that host States may determine *to what extent* may the courts of the State exercise their jurisdiction – although they may not prescribe how such jurisdiction should be exercised. Moreover, the courts of the State hosting the ex-situ government are under no obligation to recognise and enforce the laws of another State within its territory. These courts will simply do so when the rule on conflict of laws of the host State so prescribes, which will generally depend on two factors: whether the law in question is valid in the State of origin according the rules of such State; and whether they are contrary to the public order of the host State.[545]

Last but not least, the evacuation of Pacific Islands' governments may have consequences on their access to their States' property. As a corollary of a government's exclusive and plenary right of representation of its own State in international relations, only authorities recognised as governments can validly make arrangements affecting the legal status of the property of their nationals abroad as well as of the State's public assets (in the former case, this competence corresponds to the exercise by the State of personal sovereignty over its nationals, be it in State territory or abroad). As most governments abroad have been at least partly financed by their respective host States and/or other States, they seek to secure access to ownership of property abroad as quickly as possible, so as to organise an effective administration in exile, send diplomatic missions to other States or multilateral fora to plead their cause and gain allies, and maintain a certain independence from their host State – all measures directed at regaining control over their national territory. Usually, one of the first actions taken is to secure control of their State's assets and sources of revenue abroad. As a matter of principle, a recognised government in exile is entitled to dispose of its State's property abroad – allowing the sale of national

544 Ibid.

545 Ibid.

property, drawing upon credits granted to their States; managing state-owned companies, occupying State property such like diplomatic and consular premises, and receiving sums due to its respective State.[546] Governments themselves are not the owners of their State's property, but only hold it in trust and administer it. Recognition of an authority abroad as a government implies that the right to administer property found at the territory of origin automatically passes to the newly recognised government. In the case of Pacific island ex situ governments, this right and capacity will presumably be capital for them to effectively organise the necessary measures to ensure the safety of evacuation or transfer of the population; the negotiation of agreements with other States concerning the right of entry of their nationals into the territory of another State; the eventual purchase of more land abroad to relocate their people; the payment of costs associated with the diplomatic missions to the regional and global international organisations seeking to gain support for their cause and international humanitarian assistance, etc.[547]

What can be extracted from this examination is that, in 'traditional' cases, governments in exile alone do not constitute a real challenge to the continuity of the States concerned. While the exercise of some of the State's powers – such as jurisdiction to enforce or the capacity to perform certain treaty obligations – is limited by the circumstances leading to the evacuation of State territory, an authority abroad, as long as it is recognised as the government in exile, can maintain the full and exclusive capacity to represent the State abroad, to exercise legislative jurisdiction over its national assets, and be endowed with all necessary privileges and immunities. It is thus still possible for the State to exercise its Weberian functions in favour of its population, in spite of having its representative organ located outside its national territory. However, the prospective situation of Pacific Island States offers a thoroughly different picture, for they do not comply with the general assumption that the ex situ condition is only temporary, nor does their situation stem from a competition between a de iure and a de facto authority from another occupying State, or from another government, with effective power in part or at all of the home territory. These distinctive characteristics result from the fact that the evacuation of Pacific island governments will presumably take place at a time when the two other dimensions of the State will be severely affected: that is, either partial or total de-territorialisation and/or de-population. The main characteristic of governments in exile in 'traditional' cases so far (driven by political reasons), namely their presumed temporality, would thus cease. In fact, even if the lands lost

546 Ibid., 192.

547 Rayfuse, 'International Law and Disappearing States', 12.

to sea-level rise, coastal erosion or other natural causes were to re-emerge in the long term, as Esteban and Yamamoto note,[548] the whole population would have been already evacuated anyway. Therefore, the problem of Pacific Islands' ex situ governments, when put in the context of the general deterioration of the other two dimensions of the State due to climate change impacts, devolves into a problem on the continuity of Pacific Island statehood altogether.

As James Crawford points out, international lawyers have generally paid little more attention to problems of State continuity or extinction than to those of the definition and operation of the concept of statehood itself.[549] The question of State continuity or extinction is rather underdeveloped theoretically and has been traditionally approached as an issue to be resolved through the law of State succession. According to Crawford, State continuity is thus assured when the same State can be said to continue to exist, despite changes of government, territory or population; whereas State succession implies that one State can be said to have replaced another – previously extinguished – State in the same territory.[550] Therefore, the concept of continuity implies that legal relations are preserved despite changes in the subject of such relations. Territorial changes – and the concomitant changes in population – such as loss of or acquisition of (some) territory – do not *per se* affect the continuity of the State; the presumption of continuity being particularly strong when the constitutional system of the State existing prior to the territorial change remains in force. The same can be applied to changes in the municipal constitution of the State, which do not affect by itself State continuity. Yet, problems arise where the constitutive elements or criteria of statehood undergo *multiple* and *substantial* changes. It is precisely in these cases that the "distinct element of legal fiction", as Crawford puts it, inherently present in the concept of continuity, is unveiled.[551] In such instances, the role of recognition or any other form of international political support, and the claims to continuity made by the concerned States have a decisive power. Thereafter, an incursion into the effects that climate change may exert on the Pacific Island State's capacity to

548 Lilian Yamamoto and Miguel Esteban 'Vanishing Island States and Sovereignty', *Ocean and Coastal Management*, vol. 53 (2010), 1–9. Extending the lines of this article, Yamamoto and Esteban, *Atoll Island States and International Law: Climate Displacement and Sovereignty*. See also Krystyna Marek, who rejects the possibility that extinguished States may 'miraculously' re-emerge afterwards, cited in Crawford, *The Creation of States in International Law*, 401 (chapter 16, note nº 7).

549 Crawford, *Creation of States*, 401.

550 Ibid., 400.

551 Ibid.

maintain or engage in international relations, context in which they are called to plead their case, becomes necessary.

3.2 The International Challenge: Impacts on Pacific Islands' Capacity to Maintain International Relations

3.2.1 *Pacific Island States in the International Community*

A General Active Participation in International Life

In contrast to the limited governmental capacity of Pacific Island States and their potentially increased struggle to fulfil the functions of the State, their participation in international life has proven to be not only steadily active, but also progressively more effective. This may be surprising if one considers that the governmental authorities exercising the sovereign power of Pacific Island States over their population and territory are the same ones equally tasked with establishing and maintaining international relations with other States and international organisations: arguably, the limited or total inability of Pacific Islands' governments to exercise their duties at a national level could be extended to the exercise of international rights and duties. Nonetheless, the history of Pacific Island States' conduct in the international arena is far from matching this assumption. Ever since their emergence as sovereign political entities, Pacific Island States have displayed many efforts to maintain and develop further their foreign affairs. It may be precisely because of their physical, political and economic inherent fragility, coupled with their geographical isolation, that they have increasingly sought the support of the international community.

The first and most visible example of the high level of dependence of Pacific Island States on the good state of their international relations can be found, as already analysed in chapter 1, in their request for full membership as United Nations Member *States*.[552] This stage was all the more important at the time of their independence that, for most of them, acceptance of their request for UN membership was equivalent to international recognition of their acquisition of statehood.[553] Besides, other than maintaining constant international relations with other States through multilateral diplomatic fora in relevant UN bodies and agencies, Pacific Island States have nurtured bilateral relations with individual States with which they have concluded international trade, commercial and investment agreements that are crucial for their respective

552 See chapter 1, section 1.1.2.B.

553 See Harris, 33–39.

economies. Sometimes, this dependency on foreign investments has been at the cost of controversial political decisions going against the general flow, such as the attempts by some Pacific Island States – Vanuatu, Kiribati, Nauru, the Solomon Islands and Tonga – to switch their allegiances, from China to the recognition of Taiwan as an independent State, a decision that could very much depend on the course of the 'dollar diplomacy'.[554] As Yongjin Zhang remarks, "the Pacific has been one of the most fiercely fought battle-grounds in the struggle between China and Taiwan for international recognition since the 1970s".[555] This example in turn raises the question of the political *roles* Pacific Island States have been and are recurrently called to play in the international community.

While the access to statehood was opened to them as a result of the political context of the time – which reflected the structural changes that the international community had undergone since the beginning of decolonisation – the role they played afterwards was still limited by their small leverage and high degree of dependence on external assistance to ensure their sustainability and development. The admission full UN membership did not provoke fundamental imbalances in the international order of the Cold War period, as some political leaders of the time feared. Yet, it cannot be denied that Pacific Island States have been and still are being used by other States (mostly the 'Great Powers') as a useful source of votes necessary in the voting procedure in the UN General Assembly. For instance, during the session at the UN General Assembly in which the resolution on 'Climate Change and its Possible Security Implications' was adopted, the Australian delegate made a straightforward acknowledgement of this recurrent pattern.[556] One of the latest and most paradigmatic examples of this political manipulation can be found in the session of the General Assembly where the admission of Palestine as a UN Observer *State* was voted. Of the nine votes against the proposal, which included the votes of Israel and the USA, four votes were from Pacific Island States – Nauru, the Marshall Islands, Federated States of Micronesia and Palau.

554 See Yongjin Zhang, 'A Regional Power by Default', in Wesley-Smith and Porter, eds., *China in Oceania: Reshaping the Pacific?* (Oxford and New York: Berghahn Books, 2010), 49–66, 51.

555 Ibid.

556 See the Verbatin Records of the 85th plenary meeting of the UN General Assembly, 3 June 2009, UN doc. A/63/PV.85, 9. Mr. Goledzinoski, representing Australia, referred to Pacific Island States as follows: "[T]hese are not countries that usually make a lot of noise in this Hall. They do not usually come to us asking for much. In fact, generally, our experience with these countries is that it is usually us asking them for things – usually to vote for this or that. But, on this occasion, they came to us asking for something fundamentally important, as eloquently expressed by the Permanent Representative of Nauru".

The later three island States were formerly administered by the USA under the Pacific Access Category and are now in Compact of Free Association with the same country. Interestingly enough, when Emanuel Mori, President of the Federated States of Micronesia, was interviewed on the reasons for his country's vote he replied in a rather compassionate tone:

> [I]srael is a minority in the Middle East and struggling to survive. [W]e are also out there. We have no enemies, only natural ones. Typhoons come, and we survive. Being surrounded by not-friendly neighbours, we kind of pity them.[557]

Moreover, other than lending their voting support to the interests of other States, Pacific Island States have a tendency not to be part of international agreements they feel they cannot implement. Their practice seems to indicate that they are highly self-conscious about the international obligations that statehood implies, and are therefore reluctant to enter into agreements they do not have the capacity to comply with. As already pointed out in chapter 2, this is one of the reasons why the Pacific has been identified by the Office of the Pacific of the High Commissioner on Human Rights as the region with the lowest number of ratifications of human rights instruments in the world.[558] However, this approach to international agreements had not precluded them from developing regional international co-operation nor from constructing, through their Pacific Islands Forum, an alternative and self-led platform on which, in spite of the numerous differences and cultural disparities among Pacific Island States, could serve to build up a solid and common 'regional identity'. Henceforth, in the last fifteen years, exponential growth and successful regional co-operation have developed in different areas such as security, exploitation of common marine resources, or sustainable development. As their common position is now increasingly being presented by the Secretary of the Pacific Islands Forum, instead of by each State individually, it has also helped building up a common voice for most Pacific Island States in international fora.

Furthermore, the reluctance of Pacific Island States to become parties of international agreements they may not be able to comply with has not affected their commitment to international environmental law. Indeed, *all* Pacific

557 The interview where the statement of Emanuel Mori is reproduced can be found at: <http://www.washingtonpost.com/blogs/worldviews/wp/2012/11/30/coalition-of-the-opposing-why-these-9-countries-voted-against-palestine-at-the-u-n/> (9 February 2015).

558 See chapter 2, section 2.2.2.A.

Island States are Parties to the UNFCCC, and some of them have also ratified other multilateral environmental agreements. For many, the UNFCCC is among the few treaties they have ratified. Their participation as State Parties to the international regime on climate change is, first of all, undoubtedly facilitated by the fact that the regime's structure is based on the principle of common but differentiated responsibilities. All Pacific Island States are non-Annex-I countries – that is, they do not have mitigation obligations – which are reserved to industrialised countries according to Article 4.2.a UNFCCC. Besides, pursuant to Articles 4.3, 4.4 and 4.5, they can benefit from climate change adaptation funds and transfer of technology offered by Annex II countries, both serving as complementary and indirect means of obtaining international investments necessary to cope with their high level of vulnerability to climate change impacts.

In fact, Pacific Island States – along with other highly active small island States, like the Maldives – can be said to be not only active 'players', but even 'powerful voices' in the context of the climate change negotiations, to the extent that they embody the conscience and underlying normative grounds and principles on which the international regime on climate change was built, and thus recall the ultimate purpose of much of the effort deployed to make international co-operation on this matter successful. Although this leverage cannot be equated to the economic and political strength of the USA, the EU and the five big emerging countries composing the 'BRICS' group – Brazil, Russia, India, China and South Africa – which hold the future of the regime, the role played by the Pacific Island States in the negotiations is nonetheless very significant. This role is all the more remarkable since it is accompanied by a highly efficient institutional structure through which it is displayed, namely the Alliance of Small Island States (AOSIS).[559] Through this political coalition, Pacific Island States and other small island States, such as from the Indian Ocean, have demonstrated that they are not simple instrumental votes in the hands of their closest and very important regional 'partners' – namely Australia and New Zealand, which are part of the same climate change

559 Created in 1990 during the Second World Climate Conference, the coalition of AOSIS gathers today forty four low-lying island States, some of which are also part of the group of Least Developed Countries. United by their high level of vulnerability and low resilience capacity to the common threat of climate change, this group is united by wide-embracing principles, such as the precautionary principle, the polluters pay principle, or the principle of common but differentiated responsibilities, and advocates for the maximum of 1,5°C increase of global temperature, as well as for the development of renewable energies.

political coalition as the USA, known as the UMBRELLA group.[560] But the most important action taken in the international political sphere to advance their plea in favour of their survival as States can be found, as detailed below, in the progressive construction of what will be referred to as the Climate Change and International Security Discourse.

B Launch of the Climate Change and International Security Discourse

When seeking to establish the appropriate global political framework to which the threat to extinction of Pacific Island States belongs, an account of the position taken by the international community is no less important than the political position expressed and the diplomatic actions taken by the affected States themselves. James Crawford for instance explains how the position of the former State concerning its own continuity or extinction in cases followed by succession has often been determinant for the international community to officially acknowledge the disappearance of the legal personality of the former State, as well as to recognise a new political entity as the successor of such State.[561] One of the most prominent examples of how the claims of the States involved bear a fundamental weight was when the Russian Federation assumed the seat in the Security Council that formerly belonged to the Soviet Union. Recalling such process, Michael Scharf noted:

> [T]he importance of Russia's *characterisation of itself* as the continuity of the USSR cannot be overstated. When, in the case of a substantial change in the State concerned, there are doubts as to the continued existence of the State, the position of the State itself on the issue can be the determining factor.[562]

560 The UMBRELLA group was launched in 1997 during COP. 3/MOP.3, as an evolution from the previous coalition JUZZCAN – gathering Japan, USA, Switzerland, Canada, Norway and New Zealand. Whilst Switzerland left the coalition, Russia, Iceland and Ukraine joined the newly baptised group. The UMBRELLA group is thus mainly composed of industrialised countries not part of the EU, and advocates for unrestricted use of market mechanisms in the regime, the development of carbon sinks, as well as the creation of mitigation obligations emerging countries not part of Annex I.

561 Crawford, *Creation of States*, 401. [Emphasis added]: "[I]t must be admitted that, in many of the marginal cases, as well as in the case of States re-established after illegal occupation, notions of continuation contain a distinct element of legal fiction: in such instances, *recognition, and the claims to continuation made by the parties*, may well be determinative".

562 Scharf, 'Musical Chairs: the Dissolution of States and Membership in the United Nations', footnote 114. [Emphasis added].

The fundamental weigh of the characterisation made by the affected States of themselves becomes all the more acute in extreme case-scenarios, or the category of cases that touch upon the undefined and blurred boundaries between international law and international life, in which the unprecedented case of Pacific Island States threatened to extinction by climate change impacts undoubtedly falls in. As Pacific Island States found themselves in a state of jeopardy due to climate change, their active participation in international life – consistently demonstrated over the years since they acquired statehood – particularly increased. Most importantly, it took a specific form – the Climate Change and International Security Discourse.

The term 'discourse' refers here to a shared way of apprehending the world.[563] Forged in the realm of political power and embedded in language, discourses construct meanings and relationships; they are forms of responding to the inherent complexity and inter-relatedness of problems of all kinds – including environmental ones.[564] Besides, to become recognisable and acquire self-standing, each new discourse needs to rely on a shared set of assumptions and contentions which help to define common sense and legitimate knowledge.[565] As the reconstruction and analysis of the discourse undertaken below reveals, this political move developed in parallel to the mainstream climate change negotiations and involving a great majority of State actors, has been channelled through a wide range of regional and global international organisations.

The Climate Change and International Security Discourse can be conceived as a 'revival' of a political process started in the 1980s, whereby the introduction of environmental considerations into the notion of security was effected. As a product of the Cold War, both the United States of America and the Soviet Union actively participated in this contextualisation of the notion of security, although they each initiated it for different reasons and purposes, within different arenas and at different times. The first calls for redefining and 'greening' security sounded in American doctrinal circles closely related to the governmental foreign affairs and defence agencies, in reaction to a rising concern over 'resource wars' since 1973.[566] The coining of the notion of 'international environmental security' yet emerged from the opposing Soviet bloc – driven

563 John Dryzec, *The Politics of the Earth: Environmental Discourses* (New York: Oxford University Press, 2nd ed. 2005), 9.

564 Ibid.

565 Ibid.

566 See for instance, Joseph Nye, 'Collective economic security', *International Affairs*, vol. 50 (1974): 4, 584–598. Jessica Tuchman Mathews, 'Redefining security', *Foreign Affairs*, vol. 68

by the Soviet Union and supported by some closely allied communist States, such as Czechoslovakia. Such a concept is indeed the combined product of two manifestations of the Soviet 'New Thinking' promoted by Gorbachev in the mid-1980s: a newly discovered Soviet national environmentalism, on the one hand;[567] and the reification of international law, on the other hand.[568] The term 'international environmental security' emerged as one of the conceptual vehicles of the fundamental change of strategy advanced by the Soviet Union's President of the time, Mikhail Gorbachev, before the United Nations.[569] There, the notion was presented under an ambivalent blueprint, as much as an engine of international environmental co-operation (discussion in the Second Committee of the General Assembly, on Economic and Financial Affairs)[570] as of international disarmament co-operation (dealt with by the First Committee of the General Assembly, on Disarmament and International

(spring 1989): 2, 162–177; and Michael Myers, 'Environment and Security', *Foreign Policy*, vol. 74 (spring 1989), 23–41.

567 On the state of environmental deterioration of the USSR, see for instance: Murray Feshback and Alfred Friendly Jr., *Ecocide in the USSR: Health and Nature under Siege* (New York: Basic Books, 1992). Marshall I. Goldman, *The Spoils of Progress: Environmental Pollution in the Soviet Union* (Cambridge, Mass.: MIT Press, 1972). On the link between Perestroika and the environmentalist wave of the Soviet Union in the early 90s, see Nicholas A. Robinson, 'Perestroika and Priroda: Environmental Protection in the USSR', *Pace Environmental Law Review*, vol. 5 (spring 1988): 2, 351–423; and Oleg Kolbasov, 'Modern Ecological Policy and the Utilisation of a Global Environmental Protection Strategy', *Pace Environmental Law Review*, vol. 5 (spring 1988): 2445–454.

568 See Mikhail Gorbachev, *Perestroika: A New Thinking for Our Country and the World* (New York: Harper and Row, 1987). Specifically dealing with how Perestroika enacted a new age in the approach of the Soviet Union to international law, see Rein Müllerson, 'Sources of International Law: New Tendencies in Soviet Thinking', *American Journal of International Law*, vol. 83 (1989), 494–512.

569 See Statement of President Mikhail Gorbachev, 'Reality and Safeguards for a Secure World', 18 September 1987, and attachment of the Letter of the Deputy Head of the Delegation of the Union of Soviet Socialist Republics to the 42nd session addressed to the Secretary-General, UN doc. A/42/574. Conceptual explanations from Soviet scholars on the meaning and legal content of this newly coined concept emerged thereafter. See for instance: Alexandre S. Timoshenko, 'The Global Environmental Security: International Legal Aspects', *Soviet State and Law*, vol. 1 (1989), [page numbers not available]; from the same author, 'Ecological Security: the International Aspect', *Pace Environmental Law Review*, vol. 7 (autumn 1989): 1, 151–160. Oleg Kolbasov, 'The Concept of Environmental Security: Legal Aspects', *Soviet State and Law*, vol.12 (1988) [page numbers not available].

570 See the Verbatim Record of the 21st meeting of the Second Committee of the General Assembly, 23 October 1987, UN doc A/C.2/42/SR.21; followed by the Draft Resolution presented to the Second Committee by Czechoslovakia and Ukrainian Soviet Socialist

Security).[571] Unsurprisingly, in spite of the national doctrinal move that had previously shown an incipient openness towards expanding the scope of security to non-military issues (including environmental ones), the United States of America fiercely opposed the policy-based and conceptual Soviet innovation put before UN organs. This rejection was nonetheless more a preventive and defensive reaction against the wider Soviet attempt to launch a new debate on the system of international peace and security, than against the notion of international environmental security as such.[572] Then, the disintegration of the Soviet Union brought about a double-edged effect on the notion of 'international environmental security'.[573] Whilst direct references to the concept before UN organs dissolved, the positive dimension – which called for the development of international environmental co-operation was pursued – for instance, with the holding, in 1992, of the Rio Conference on Environment and Development.[574] In contrast, proving to be a useful compensation tool for the lack of an 'enemy syndrome', the notion of environmental security was (ironically) embraced by American governmental agencies after the end of the Cold War. Important conceptual innovations of the notion of security and its relation to the environment mushroomed amongst the schools of international security studies,[575] and important funding reorientation gave rise to the first

Republic, entitled 'International Ecological Security', 30 October 1987, UN doc. A/C.2/42/L.34.

571 Verbatim Record of the 47th to 51st meetings of the First Committee of the General Assembly, 23–30 November 1988, UN docs.: A/C.1/43/PV.47–PV.54.

572 On the wider Soviet proposal where the notion of international environmental security belonged, see UN General Assembly Resolution 41/92, 'Establishment of a Comprehensive System of International Peace and Security', adopted on 4 December 1986, 96th plenary meeting, UN doc. A/RES/41/92. More detailed explanations can be found, for instance, in Pietr Vladimirsky, 'Comprehensive Security Equal for All', *USSR Report: International Affairs*, (October 1989), 9–19, and Vladimir Fyodorov, 'The UN and a Comprehensive International Security System', *USSR Report: International Affairs* (September 1987), 90–95.

573 The end of the debate on this concept can be found in the Speech of Mikhail Gorbachev, Verbatim Record, 8 December 1988, UN doc. A/43/PV.72; followed by that of Eduard Shevardnadze, Verbatim Record, 1989, UN doc. A/44/PV.6, pp. 26–50.

574 UN General Assembly Resolution 42/184, 'International Co-operation in the Field of Environment', adopted during the 96th plenary meeting, on 11 December 1987, UN doc A/RES/42/184.

575 See, for instance, Lothar Brock, Peace through Parks: The Environment on the Peace Research Agenda', *Journal of Peace Research*, vol.28 (November 1991): 4, 407–423; Simon Dalby, 'Geopolitics and Ecology: Rethinking the Contexts of Environmental Security', in Miriam R. Lowi, ed. *Environment and Security: Discourses and Practice* (New York: Palgrave

empirical attempts to link environmental degradation or stress and violent conflict.[576] This reinforced the link between environment and security and provided a basis for the Climate Change and International Security Discourse which was yet to be launched.

The 'second wave' of the environment and security link unfolded within the specific context of climate change, and corresponds to the birth, circulation and limits of the Climate Change and International Security Discourse within and among regional organisations. Whilst the core contentions, innovative language (for instance referring to climate change as a "threat multiplier") and region-based methodology were born and settled in Germany,[577] the Climate Change and International Security Discourse can be truly identified as such since the European Union institutionalised it.[578] It is from such a political cradle that legitimate knowledge on climate change and international security was reinforced through the multiplication and mushrooming of EU

McMillan, 2000), 84–100. Critical voices include: Niels P. Gleiditsch, 'Armed Conflict and the Environment: a Critique on the Literature', *Journal of Peace Research*, vol. 35 (May 1998): 3, 381–400; Daniel Deudney, 'The Case against Linking Environmental Degradation and National Security', *Millennium*, vol. 19 (winter 1991): 3, 461–476.

576 On American post-Cold War funding reorientation, see Michael B. Wallerstein, 'Whither the Role of Private Foundations in Support of International Security Policy?', *The Nonproliferation Review*, vol. 9 (spring 2002): 1, 83–92. Empirical studies on the link between environmental stress and violent conflict include, for instance, Stephan Libiszewski, 'What is an Environmental Conflict?', Occasional Paper nº 1, *Center for Security Studies and Conflict Research* (Bern: Swiss Peace Foundation and Zurich, July 1992), and Thomas Homer-Dixon, 'Environmental Scarcities and Violent Conflict: Evidence from Cases', *International Security*, vol. 19 (winter 1994): 1, 5–40.

577 Hans Günter Brauch, 'Climate Change, Environmental Stress and Conflict', AFES-PRESS Report for the German Federal Ministry for the Environment, Nature Conservation and Nuclear Safety, reproduced in 'Climate Change and Conflict: Can Climate Change Impacts Increase Conflict Potentials? What is the Relevance of This Issue for the International Process on Climate Change?', *Environmental Policy*, Part II (November 2002), 9–113; and German Advisory Council on Climate Change (WBGU), *World in Transition: Climate Change as a Security Risk', Summary for Policy-Makers* (Berlin: WBGU, 29 May 2007), 1–13.

578 Council of the European Union, 'Presidency Conclusions by M. Hans Pöttering', 21–22 June 2007, document reference: 11177/1/07 – Rev.1- Conc. 2;'Climate Change and International Security: Paper from the High Representative and the European Commission to the European Council', 14 March 2008, document reference: S/113/08; Javier Solana (High Representative of the EU), 'Joint Process Report and Follow-up Recommendations on Climate Change and International Security', 18 December 2008, document reference: CL08–241EN. Council of the European Union, 'Council Conclusions on Climate Change and Security', 2985th Foreign Affairs Council Meeting, 8 December 2009.

contract reports, elaborated by the same community which had previously been tasked by the German Government with the early formulations of the links between climate change and security.[579] Simultaneously, as individual EU Member States such as the UK, Denmark and The Netherlands mainstreamed the Climate Change and International Security Discourse in the realm of their national security strategies, more research institutions, such as the International Institute for Sustainable Development or the Royal United Services Institute, joined the pioneering one – Adelphi Consult.[580]

These reports (along with the political will promoting their elaboration) originally tailored the Climate Change and International Security Discourse as a widely embracing framework under which the manifestations of the adverse impacts of climate change – from food security to water stress, sea-level rise or the advance of desertification – were mapped in different regions of the world. Yet, the specificity of the Climate Change and International Security Discourse which allowed to differentiate it from related discourses on climate change adaptation and resilience was its underlying purposes, namely: 1) anticipating how those changes may exacerbate local or regional conflict; and 2) correlatively measure whether and to what extent the geo-political status of the regional organisation and the Member States promoting the Discourse may be subject to transformations.

579 The following reports on the regional security consequence of climate change were contracted by the Directorate-General of the European Commission: Alexander Carius, Achim Maas and Janina Barkemeyer, *Climate Change and Security: Two Scenarios for the Indian-Pacific Ocean Island States* (Berlin: Adelphi Research, 2009); Alexander Carius and Achim Maas, *Climate Change and Security: Three Scenarios for Middle America* (Berlin: Adelphi Research, 2009); Alexander Carius, Achim Maas and Kerstin Fritzsche *Climate Change and Security: Three Scenarios for South West Asia* (Berlin, Adelphi Research, 2009); from the same authors, *Climate Change and Security: Three Scenarios for South West Asia* (Berlin: Adelphi Research, 2009); and *Climate Change and Security: Scenarios for South East Asia* (Berlin: Adelphi Research, 2009).

580 The reports prepared by these two organisations include, on the one hand, Shiloh Fetzek, *Climate-Related Impacts on National Security in Mexico and Central America* (London: Royal United Service Institute, 2009); On the other hand, Oli Brown and Alec Crawford, *Assessing the Security Implications of Climate Change for West Africa: a Country Case Studies of Ghana and Burkina Faso* (Winnipeg: IISD, 2008); from the same authors, *Rising Temperatures, Rising Tensions: Climate Change and The Risk of Violent Conflict in the Middle East* (Winnipeg: IISD, 2009), and *Climate Change and Security in Africa: A Study for the Nordic-African Foreign Ministers Meeting* (Winnipeg: IISD, March 2009).

Once settled, the EU sought to consolidate the Climate Change and International Security Discourse by extending it to the OSCE and NATO areas (two regional security organisations of which the EU is a member),[581] and by enabling the formalisation of a transatlantic dialogue on climate change and international security with the United States.[582] Mainstreamed in EU foreign policy and consolidated as a discursive product of hegemonic regions, the Climate Change and International Security Discourse was then circulated to areas beyond the scope of the European Union which are extremely vulnerable to the adverse impacts of climate change. Presumably, such circulation permits downgrading of the strong geo-political footprint of the Climate Change and International Security Discourse as originally conceived, and thereby legitimises it. The contrast between the incorporation of the Discourse into the agenda of the Pacific Islands Forum as well as into that of the African Union with its restrained acceptance in Asian and Latin American regional organisations makes it clear that a set of conditions is required for a successful circulation to take place. First of all, the political will of the promoter of the Discourse (the EU) to circulate it to a specific organisation needs to crystallise in two ways: the creation or revival of a political partnership, coupled with the association of such an inter-regional co-operation framework with a financial flow (for instance in the form of financial assistance for adaptation to climate change or investments in renewable energy projects). Secondly, the forum for the reception of the Discourse needs to fulfil some requirements.

581 See for instance, Goran Svilanovic, 'The Security Implications of Climate Change in the OSCE region: Opening Remarks', OSCE Chairmanship Conference, Bucharest (Romania), 5 October 2009, document reference: SEC.GAL/168/09; Mats Aberg, Swedish EU Presidency, 'Climate Change and Security: Challenges and Opportunities in the OSCE region', 19 October 2009, document reference: PC.DEL/805/09; followed by the contracted report of Achim Maas et al., *Shifting Bases, Shifting Perils: A Scoping Study on Security Implications of Climate Change in the OSCE Region and Beyond* (Berlin: Adelphi Consult, 2010). On the introduction of the Discourse within NATO's agenda, see 'NATO Secretary General Debates Climate Change Security Threats in Copenhagen', NATO Press Release, 15 December 2009; and Speech of Secretary General A. Rasmussen on 'Emerging Security Risks: Climate Change, Piracy and Cyber Defence', Addressed by NATO's Secretary General at a Lloyd's Conference in London, 1 October 2009.

582 International Institute for Strategic Studies, *The IISS Transantlantic Dialogue on Climate Change and Security', Report to the European Commission Directorate for External Relations* (Winnipeg: IISD, January 2011). The two most prominent American studies of the security implications of climate change are Peter Schwartz and Doug Randall, *An Abrupt Climate Change Scenario and Its Implications for United States Security* (Washington, D.C.: The Pentagon, October 2003); followed by the report of the Center for Naval Analysis, *National Security and the Threat of Climate Change* (Washington, D.C.: CNA, 2007).

It seems important that the members of the regional organisation concerned have a rather homogeneous level of vulnerability to climate change impacts and share a similar or even common position in climate change negotiations. Also, the existence of a common interest with the promoters of the Discourse or at least the absence of competing economic interests in key areas embraced by the Climate Change and International Security Discourse, such as food or energy security, is fundamental.

Mapping the birth and inter-regional circulation of the Climate Change and International Security Discourse serves, above all, to understand which was the *original* underlying rationale of this narrative. However, the later appropriation by Pacific Island States before international organisations of a global scope truly re-signified the Discourse. Thereafter, it became the vehicle of self-characterisation of their situation, and embodied the position and claims of Pacific Island States regarding their threat to extinction.

When moving on from the regional to the global level of analysis, one is called upon reconstructing the introduction of the Climate Change and International Security Discourse into the policy agenda, and the operation, of the two core organs of the United Nations: the Security Council and the General Assembly. The process of operation of the Climate Change and International Security Discourse before universal organisations can be divided into three main stages. First, the Discourse was introduced before the Security Council as a result of an initiative of the United Kingdom, one of the main 'drivers' of the Discourse which, as explained above, originated in Germany and was then institutionalised at the EU.[583] This was the first contact of the Discourse with the UN organ primary responsible for the maintenance of international peace and security, and took place at a time when the process of institutionalisation within the EU was still at a very early stage of development. Its presentation before the Security Council, confronted with other UN Member States that would not be necessarily reached by the inter-regional circulation of the Discourse, was thus closely attached to a geo-political view on how the consequences of the phenomenon could affect several key factors of international security – with special emphasis on its effects on the energy sector.[584] The introduction of the Discourse at a time when the 2007 Bali Summit was under preparation, coupled with its original hegemonic bias, conveyed the idea that

583 UN Security Council, Letter from the Permanent Representative of the UK of Great Britain and Northern Ireland to the United Nations addressed to the President of the Security Council, and its Annex 'Energy, Security and Climate', dated 5 April 2007, UN doc. A/2007/186.

584 UN Security Council, UN doc. S/PV.5663, 17 April 2007.

it essentially served to exert influence and pressure on the mainstream climate change negotiations through an alternative (and arguably illegitimate) way. Despite the fact that States threatened by the most extreme impacts of climate change supported the hegemonic initiative for its potential capacity to raise awareness and increase the sense of urgency of the phenomenon, this first attempt to 'securitise' climate change unsurprisingly encountered fierce opposition from States seeking to preserve the originally grounded developmental rationale of international climate change co-operation.[585] The turning point in the history of this narrative yet comes when Pacific Island States took the command of Discourse and, after intense lobbying in UN corridors, managed to introduce the topic of 'Climate Change and International Security' in the agenda of the General Assembly.[586] An important shift in orientation took place during the negotiations that lead to the endorsement in 2009 of General Assembly Resolution 63/281 on 'Climate Change and its Possible Security Implications'.[587] The existential plea of these low-lying island States, let by the Pacific Islands Forum, tainted the core understanding of this narrative and truly re-signified the understanding of the association of climate change with international security. Besides, the definition of the respective roles that UN organs and agencies were to play in this connection was also established. Thereafter,

585 The G77+China group, as well as the Non-Aligned Movement, rejected to hold the debate before the Security Council. The general arguments of these groups can be found in the letter of Pakistan laying down the objections to the UK concept paper on behalf of the G77+China group, as well as in the letter of Cuba submitted on behalf of the Non-Aligned Movement. Both groups, as well as States acting in their individual capacities such as Egypt, Bangladesh or the Philippines, upheld the view that the debate on climate change within its 'original context', namely, under the concept of sustainable development. See UN Security Council, Letter dated 16 April 2007 from the Permanent Representative of Pakistan to the United Nations addressed to the President of the Security Council, UN doc.: S/2007/211; and UN Security Council, Letter dated 12 April 2007 from the Chargé d'Affaires of the Permanent Mission of Cuba to the United Nations addressed to the President of the Security Council, UN doc.: S/2007/203.

586 See UN General Assembly, Draft Resolution on 'Security and Climate Change', 27 October 2008, UN doc.: A/63/L.8; followed by: UN General Assembly, Amendments to Draft Resolution A/63/L.8, 17 November 2008, UN doc.: A/63/L.30.

587 UN General Assembly, Resolution 63/281, 'Climate Change and its Possible Security Implications', 3 June 2009, 85th plenary Meeting, UN doc.: A/RES/63/281. For a detailed account of the discussions see UN General Assembly, UN doc.: A/63/PV.85, 3 June 2009. The resolution tasked the Secretary General to produce a Report on the matter, which was released three months later. See General Assembly, 'Climate Change and its Possible Security Implications: Report of the Secretary-General', 11 September 2009, UN doc.: A/64/150.

the security implications of climate change would be associated, in particular, with the territorial loss and population migration resulting from sea-level rise in small island States. This newly created understanding of the meaning of the Discourse, as well as the division of competences over this matter among the relevant organs and organisations, was finally consolidated when climate change entered the agenda of the Security Council for the second time in 2011, upon initiative of Germany.[588] The last stage – so far discernible – of the operation of the Climate Change and International Security Discourse before universal organisations only implied that climate change would from now on be considered by the UN Secretary-General in his description of the contextual information of Security Council operations and did not give rise to concrete measures directed towards UN Member States.[589]

Yet, although so far the Discourse has not turned into actions that may significantly modify the behaviour of either the UN or individual Member States towards climate change, the introduction of the Discourse before universal organisations had a double-barrelled and reciprocal impact. On the one hand, the purpose-based incorporation of the Climate Change and International Security Discourse into the realm of work of universal organisations reveals the process through which an agenda-setting strategy is turned into concrete political action, either by transforming old patterns of international co-operation or by generating innovative trends. On the other hand, when confronted with a genuinely unlimitedly multilateral setting as the United Nations, the

588 UN Security Council, Letter of the German Presidency of the Security Council to the UN Secretary-General, 'Maintenance of International Peace and Security: Impacts of Climate Change', Concept Note, 1 July 2011, UN doc.: S/2011/408, para. 6: "[I]n his 2009 report, the Secretary-General identified sea-level rise as the "ultimate security threat" for some small island States, with some possibly set to "disappear over the next 30 years". While complete inundation may take years and the rise in the sea level may differ in different regions of the world, this is not only a future risk but a current reality: on some islands, the situation is already dire enough to require the evacuation of the resident population now. Furthermore, even before sea-level rise actually submerges an island, its impacts may render it uninhabitable, requiring permanent resettlement".

589 UN Security Council, Verbatim Record S/PV.6587, 6587th meeting, 20 July 2011, 10 a.m. The session concluded with the adoption by unanimity of the first Presidential Statement on the matter. See UN Security Council, Presidential Statement on Maintenance of International Peace and Security and Impacts of Climate Change, 20 July 2011, document reference: S/PRST/2011/15, para. 8: "[T]he Security Council expresses its concern that possible security implications of loss of territory of some States caused by sea-level-rise may arise, in particular in small low-lying island States".

Climate Change and International Security Discourse itself underwent a process of transformation resulting from political exchange, negotiation and compromise. The 'shared understanding of the world', finally embodied by the Discourse after going through the prism of universal organisations, may thus not fully correspond to its original conception; thereafter, the concrete actions into which the Discourse leads may not match the original purpose of its proponents either. But it is indeed at this level of analysis that the Climate Change and International Security Discourse has finally come to be understood by a majority of States, as embodying the threat of climate change to the survival of low-lying island States.

Notwithstanding that security is an inherently contested notion, and that some scholars have warned against the perils of the 'securitisation' of climate change,[590] the existence of the Climate Change and International Security Discourse – a political narrative developed and endorsed by the most affected actors – cannot be denied; neither its relevance for the resolution of the issue of State extinction for climate change impacts. The successful dissemination of the Discourse among relevant global international organisations and the progressive acceptance by a majority of States of such a threat in a perspective, impacting on the survival of small island States had, first and foremost, the political effect of introducing this matter onto the international agenda and of mainstreaming some of the related issues in the mandates of the relevant UN bodies and agencies. Specific effects of the Discourse in the form of policy action can be expected to develop from now on in different areas, including, inter alia, those of human rights and refugee law and policy, or through further development of international scientific survey assistance enabling low-lying island States to conclude maritime delimitation agreements. In addition, beyond this concrete short-term effect, the Climate Change and International Security Discourse will arguably be fundamental when the worst-case scenarios of de-territorialisation or de-population lead to the serious questioning of the future of Pacific Islands' statehood.

---

590 See for instance, McAdam, *Climate Change, Forced Migration and International Law*, 4: "[T]his has fed populist notions of 'sinking nations' and mass influxes of 'climate refugees', which in turn have fuelled the idea that climate change-related displacement is a threat to international security. This is both unfortunate and ironic. For while an underlying objective of those who take an alarmist perspective is typically to highlight the deleterious effects of climate change on human society – in particular, on those who are displaced – their focus on mass movement also leverages an anti-immigration-security agenda".

The Discourse plays in fact a double-edged role: not only does it help in pointing out the fundamental legal question underlying the Discourse – namely, the continuity of small island States threatened by climate change impacts – it can also become a component of the solution to this legal question. The Climate Change and International Security Discourse could arguably serve as an official proof, registered in UN deliberations and known in advance by the international community, of Pacific Island States' struggle against the looming and jeopardising effects of climate change on their continuity as States; as evidence of their resistance to their eventual extinction. And yet, if the position of the States concerned is key to the resolution of their future as States, how the international community behaves and reacts to the present and future prospects of such a situation is equally important – both from a political and from a legal perspective.

3.2.2 *The International Community and Pacific Island States*

A Creation of Pacific Island States: Past Reflection of the De-Colonisation Movement

From a formal point of view, Pacific Island States could be recognised as autonomous political entities different from one another, as well as from their formal colonial or administering States, especially given that their acquisition of statehood was in most cases preceded by a referendum, which secured the appropriate exercise of Pacific Islanders' right to self-determination. And yet, as pointed out on several occasions above, their constitution as truly autonomous, self-sustained political entities was never pretended to be fully achieved. Such a high level of dependence on other countries – which led some of them, such as the Marshall Islands, Palau and the Federated States of Micronesia, to opt for "free association", instead of independence, as their preferred form of governance – almost prevented their acceptance as full UN Member States.

However, as the core characteristic of international society had changed at the time, giving an unprecedented weight and importance to new normative developments, adjustments to the criteria for UN membership ensued. The entry of Pacific Island States into the UN as full Member States therefore mirrored the structural changes that the international community was undergoing. That development also reflected the underlying forces that were changing the normative grounds of the international system, one in which the self-determination of peoples would seemingly crystallise human liberation from colonial oppression. Moreover, a new sense of universality – encompassing all new independent States – simultaneously arose.

B Extinction of Pacific Island States: Future Reflection of the 'Limits to Growth'?

If the access of Pacific Island States to UN membership as a consequence of the attribution of statehood was a reflection of the structural changes that the international community was undergoing at the time, it can be argued that the international community's treatment of the issue of the extinction of these same States may follow the same pattern. When dealing with the future of statehood in general terms, Marti Koskenniemi calls upon considering that the *substratum* of statehood is evolutive:

> [T]he status of statehood can be associated with various sets of rights and duties. It carries no given, determinate, normative implications. [W]e cannot deduce the extent of a State's freedom of action from the mere fact of its statehood. [T]he construction of rights and duties – giving substance to statehood – remains a matters of *political adjustment* and of finding a *contextual equilibrium in particular circumstances*.[591]

As much as the contextual framework of international life was fundamental for the Pacific Island States *to acquire* statehood and be admitted as full Member States of the UN, the same contextual variable may arguably become just as fundamental for these States *to maintain* their sovereignty and statehood, irrespective of the actual level of deterioration of their geographical, human and political dimensions.

Today's threat of Pacific Island States' physical disappearance, followed by the extinction of their international legal personality, can first of all be seen as a tangible reflection, if not a proof, of the core argument of any political discourse that has ever embraced the idea of the 'limits to growth' – and which John Dryzec referred to in general as 'the Survivalists'.[592] Since the publication in 1972 of the report 'The Limits to Growth', produced by the Massachusetts Institute of Technology and commissioned by the Club of Rome, the idea that the human demands exceed the carrying capacity of the Earth was introduced in the international political sphere, and served to promote inter-state cooperation on transnational and global environmental matters.[593] Yet, the

591 Martii Koskenniemi, 'The Future of Statehood', *Harvard Journal of International Law*, vol. 32 (1991), 397–410, 409. [Emphasis added].

592 John S. Dryzek, *The Politics of the Earth: Environmental Discourses* (Oxford and New York: Oxford University Press, 2nd ed. 2005), 27–46.

593 Donnella H. Meadows et al., *The Limits to Growth: A Report for the Club of Rome's Project on the Predicament of Mankind* (Potomac Associates, 1972).

strength of the earlier deeply rooted and widely spread 'promethean discourse', which denies the existence of environmental limits;[594] assumes that 'growth is good', that 'growth in progress'; and conceives of environmental goals and subordinated to economic goals,[595] has resulted in disastrous macro-consequences.

The metaphorical power of the critical environmental condition of Pacific Island States results from the fact that it dramatically exposes the failure of humanity to respond adequately to the challenge of a changing Earth system. This may in turn lead to the failure to preserve the fundamental role of the State as a value-system whose primary focus was "human loyalty and communal allegiance".[596] Therefore, as the State faces the possibility of an unprecedented form of extinction due to climate change impacts, the dominance that this form of political organisation held in the "ontological landscape of politics", as Andrew Hurrell puts it, may be subject to re-evaluation. Ironically, the value of decentralisation through empowerment of individuals and communities, which can often been found in ecological thinking for sustainability purposes, may be revived.[597]

Considering that the lack of reaction of other States to the extinction of Pacific Island States would be tantamount to ignoring the beginning of the disintegration of their *own* nature, what can be done to preserve the continuity of Pacific Islands' statehood? The next stage to be addressed is whether the *justification* of the continuity of Pacific Islands' statehood may be based on arguments other than mere utilitarian and practical purposes. A more detailed incursion into the realms of both international responsibility, the institution of State recognition, and the other new possible prescriptive avenues becomes necessary.

3.3 Exploring Normative Grounds for the Continuity of Pacific Island States

3.3.1 *Remedies for Climate Change Loss and Damage*

A Principal Remedy: Economic Compensation

In cases of State extinction as known so far – that is, when succession ensues – all States have an obligation *not to recognise* the extinction of the

594 Dryzek, 51–73.

595 Dryzec, 58.

596 Andrew Hurrell, 'The State', in Andrew Dobson and Robyn Eckersley, eds., *Political Theory and the Ecological Challenge* (New York: Cambridge University Press, 2006), 165–182, 165.

597 Dryzek, 72.

former State if substantial international illegality, such as an illegal annexation, is involved.[598] Therefore, evaluating the *circumstances* in which State continuity may be at stake is just as fundamental as examining the *effects* of the challenge in question in each of the dimensions of statehood. An incursion into the field of State responsibility may thus unveil the existence of particular rights of Pacific Island States that could possibly help protecting, ensuring or even defend their continuity as States.[599] To be sure, the first fundamental difference between cases of illegal annexation and the threats due to climate change impacts on Pacific Island States is that, while the former refers to the breach of the peremptory norm – the general prohibition of the use of force – by means such as military invasion by a foreign State in the context of *debellatio*, the impact on Pacific Islands' territorial integrity results from the adverse effects of acts that do not violate a *ius cogens* rule: the emission of greenhouse gases by State(s), also affecting a global common good, namely, the atmosphere. In fact, greenhouses gas emissions are not, *per se*, an act prohibited by international law.[600] On the one hand, the quest for the establishment of an international liability regime for the protection of the global commons – explored almost two decades ago by Michael Reisman and Manhoush Arsanjani – remains today a normative project largely belonging to the realm of utopia.[601] On the other hand, as developed below, the possibility for a *particular* State to claim

598 Ibid., 416. [Emphasis added].

599 See for instance, Peter Cashman and Ross Abbs, 'Liability in Tort for Damage Arising from Human-Induced Climate Change', in Rosemary Lyster, ed., *In the Wilds of Climate Law* (Brisbane: Australian Academic Press, 2010); one recent account of the difficulties of applying the law of State responsibility for unlawful acts to the context of 'sinking islands' can be found in McAdam, *Climate Change, Forced Displacement and International Law*, 92–96.

600 See, for instance, Katak Malla, 'Climate Change Loss and Damage Compensation', *Indian Journal of International Law*, vol. 53 (2013), 183–205.

601 Although the evolution of international law since the end of World War II has come to recognise the existence of 'global commons' – a concept which began in the law of the sea with the idea, argued by Malta at the UN General Assembly in 1967, that ocean seabed beyond national jurisdiction and its resources were a 'common heritage of mankind', eventually formulated in Part XI of the 1982 UNCLOS, and was applied afterwards to the global atmosphere in the 1985 Vienna Convention on the Protection of the Ozone Layer – it still falls short of recognising the existence of 'global responsibilities' for damage caused by all States but affecting, in a particularly acute way, a specific group of States. This would be a development of the law on co-operation which has not yet crystallised. See W. Michael Reisman and Manhoush H. Arsanjani, 'The Quest for an International Liability Regime for the Protection of the Global Commons', in Karel C. Wellens, ed., *International Law: Theory and Practice; Essays in Honour of Eric Suy* (The Hague: Martinus Nijhoff Publishers, 1998), 469–492.

compensation for damages on its territory derived from climate change impacts against *another particular State*, remains substantially challenged by a number of predominant political dynamics, as well as by important conceptual and legal uncertainties.

First, there is the difficulty of establishing a causal link between the activity and the damage. This includes not just the need to establish general causation, that is, a causal link between the activity (greenhouse gas emissions) and the general outcome (change in the radiative forcing in the atmosphere); but also the need to establish specific causation, that is, a causal link between the activity attributable to a concrete State, and the damage suffered by another concrete State in its territory.[602] At the moment, general causation is not subject to controversy, as there is almost universal consensus on the fact that climate change results from a disproportionate release of anthropogenic greenhouse gas emissions and is not just the result of natural climate variability.[603] Specific causation is yet a big challenge, primarily because the multiplicity of both the wrong-doers and the victims makes it impossible to attribute specific emissions to a specific country to specific impacts on another specific country. No rule of international law yet exists that can solve this attribution issue, which will ultimately be resolved by considerations of fairness arising from the specific case.[604]

Second, the difficulty of establishing what kind of damages would be compensated (whether non-market impacts on health and ecosystems are included or not),[605] and how are the multidimensional types of harm evaluated and quantified presumably by judicial institutions. As noted by Tol and Verheyen, the determination of these categories "makes billions of dollars of difference in the actual compensation paid".[606]

Third, the difficulty of clearly identifying a primary obligation, the violation of which would constitute a harmful activity attributable to a State. Some scholars such as Richard Tol and Roda Verheyen consider it clear that State responsibility can be triggered – at least – by the "failure to stop, reduce or

602 This differentiation is drawn from Richard S.J. Tol and Roda Verheyen, 'State responsibility and Compensation for Climate Change Damages: a Legal and Economic Assessment', *Energy Policy*, vol. 32 (2004), 1109–1130, 1112.

603 Ibid.

604 Christina Voigt, 'State Responsibility for Climate Change Damages', *Nordic Journal of International Law*, vol. 77 (2008): 1–2, 1–22, 21.

605 Tol and Vehreyen, 1128.

606 Ibid.

regulate emitting activities with due care".[607] In contrast, Voigt notes that from a treaty-based perspective, ambiguous and broad terminology creates uncertainty; whilst from a customary international law perspective, the definition of an objective standard of care necessary to apply the principle of prohibition of transboundary harm is almost impossible.[608] However, despite this set of difficulties, litigation in the realm of State responsibility for climate change impacts is likely to develop in the future, as the impacts of climate change on the most vulnerable States push them to the edge of extinction.

So far, the possibility of obtaining remedies for climate change damage has mainly been dealt with in the context of State responsibility for injurious acts not prohibited by international law – rather than within the context of State liability for unlawful acts.[609] Ever since the beginning of the 1980s, the International Law Commission (ILC) has worked on the codification and development of both the law on prevention of transboundary environmental damage (primary rule), and the law on compensation for such damage (secondary rule).[610] As extensively developed by Roda Vereyen, the origins of the law on the prevention of transboundary environmental damage can be found in the 'no-harm rule', whose foundations in turn derive from the principle of good neighbourliness between States formally equal under international law.[611] A few years later, the 1949 UN Survey of International Law concluded that, on the basis of the principle of territorial integrity, States shall

607 Ibid., 1112.

608 Voigt, 20.

609 For a general explanation of the international liability regime, see for instance Karl Zemanek, 'Causes and Forms of International Liability', in Bing Cheng and Eduard D. Brown, eds., *Contemporary Problems in International Law: Essays in Honour of Georg Schwartzenberger on His Eightieth Birthday* (London: Stevens and Sons, 1988); and Mohammed Bedjaoui, 'Responsibility of States: Fault and Strict Liability', *Encyclopedia of Public International Law*, vol. 4 (1987), 358–362.

610 See Jean Combacaud and Denis Allan, 'Primary and Secondary Rules in the Law of State Responsibility: Categorising Legal Obligations', *Netherlands Yearbook of International Law*, vol. 16 (1985), 81–109.

611 Roda Verheyen, *Climate Change Damage and International Law: Prevention Duties and State Responsibility* (The Hague and Leiden: Martinus Nijhoff, 2005). Pierre-Marie Dupuy, 'International Liability for Transfrontier Pollution', in Michael Bothe, ed., *Trends in International Environmental Policy and Law* (Gland: International Union for the Conservation of Natural Resources, 1980), 369–379. Both the no harm rule and the principle of good neighbourliness are today part of customary international law. They can be traced back to the 1941 *Trail Smelter* arbitration case, which significantly influenced the formulation and content of the no-harm rule and set the basis of the duty to prevent environmental damage, as well as of the law on State responsibility for such damage – not

not permit the use of their territory for purposes injurious to the interest of other States in a manner contrary to international law.[612] The customary no-harm rule, enshrined and expressly recognised in the Stockholm Declaration, served as the benchmark for the ILC to work of the possibility to develop treaty law on this matter. Therefore, until recently, the work of the Commission on the theme of Responsibility and Liability for Injurious Consequences of Acts not Prohibited by International Law had been primarily centred on the codification and development of the primary rule of prevention.[613]

The appointment in 1997 of Pemmaraju Sreenivasa Rao as the third Special Rapporteur resulted in the report containing seventeen articles exclusively dedicated to the issue of prevention of transboundary harm.[614] The ILC then adopted the final text of the project, consisting of nineteen articles on 'Prevention of Transboundary Harm Resulting from Dangerous Activities', which was endorsed in 2001 by the UN General Assembly, along with the recommendation of the ILC to elaborate a Convention based on the draft Articles adopted by the ILC.[615] This topic remains still today under the agenda

only with regard to States in dispute, but also with regard to the territories of other States, common spaces of mankind and the environment as a whole.

612 Following these 'founding' stage, the no-harm rule was then enshrined in Principle 21 of the 1972 Stockholm Declaration on the Human Environment. It was completed by Principle 22, which calls upon States to develop international law on State liability for transboundary harm. The next stage that finally reaffirmed the existence and content of the no-harm rule, and in which the significant impact of the *Trail Smelter* case became evident when, in 1996, the ICJ endorsed the principle in the *Nuclear Weapons* case. The Court found that the general obligation of States to ensure that activities within their jurisdiction and control respect the environment of other States or of areas beyond national control "is now part of the corpus of international law relating to the environment". See *Legality of the Threat or Use of Nuclear Weapons*, para. 29.

613 See International Law Commission, 'First Report on Prevention of Transboundary Damage from Hazardous Activities', Special Rapporteur: Pemmaraju Sreenivasa Rao, 18 March 1998, UN docs.: A/CN.4/487 and Add.1.

614 International Law Commission, 'Third Report on International Liability for Injurious Consequences Arising out of Acts not Prohibited by International Law (Prevention of Transboundary Damage for Hazardous Activities)', Special Rapporteur: Pemmaraju Sreenivasa Rao, 9 June 2000, UN doc.: A/CN.4/510.; General Assembly Resolution 56/82, 'Report of the International Law Commission on the work of its fifty-third session', 12 December 2001, UN doc.: A/RES/56/82.

615 General Assembly Resolution 56/82, 'Report of the International Law Commission on the work of its fifty-third session', 12 December 2001, UN doc.: A/RES/56/82. However, the Assembly limited itself to expressing its gratefulness for the great task undertaken by the Commission, inviting it to re-launch its examination of the issues pertaining to State liability and reparation, and drawing the attention of governments to it in 2007 – without

of the General Assembly which, after inviting governments to submit further comments on both the articles on prevention of transboundary harm and the principles on the allocation of loss for transboundary harm arising out of hazardous activities, requested to the Secretary-General a compilation of the decisions of international courts and tribunals referring to them and included the item in the agenda of its 2013 session. By 2006, the work of the Commission on both prevention and liability was concluded, and a set of eight draft principles on the allocation of loss for transboundary harm with commentaries was submitted to the General Assembly.[616]

It is against the background of the development of the law of State responsibility for transboundary harm that one is called upon examining the more specific question of State responsibility for climate change damages. In addition to being a customary of prohibition of transboundary harm rule, Articles 2 and 4.2 of the UNFCCC could arguably establish a specific primary obligation to prevent climate change – although, as Voigt notes, this point leaves a wide margin of interpretation.[617] Indeed, as the pressure of industrialised countries prevented any provision dealing with – or even hinting – at the question of responsibility for climate change damages from being included in the UNFCCC, the Association for Small Island States tried to include in the text a reservation establishing that:

> [S]ignature of the Convention shall, in no way, constitute a renunciation of any rights under international law concerning state responsibility for the adverse effects of climate change.[618]

Although this reservation was not included in the final text, it is noteworthy that four Pacific Island States – Fiji, Kiribati, Nauru and Papua New Guinea – adopted it as a declaration of interpretation upon their ratification of the UNFCCC.[619]

---

yet calling for further action. See General Assembly Resolution 62/68, 'Consideration of Prevention of Transboundary Harm from Hazardous Activities and Allocation of Loss in the Case of such Harm', 6 December 2007, UN doc.: A/RES/62/68.

616 General Assembly Resolution 61/36, 'Allocation of loss in the case of transboundary harm arising out of hazardous activities', 4 December 2006, UN doc.: A/RES/61/36.

617 Voigt, 20.

618 Submission of Vanuatu on behalf of the Association of Small Island States, 'Elements for a Framework Convention on Climate Change', 18 June 1991, UN Doc. A/AC.237/Misc.1/Add.3, 22–32, 24.

619 All declarations of interpretation to the UNFCCC can be found at: <http://unfccc.int/essential_background/convention/items/5410.php> (3 February 2015).

More recently, in the light of the progressive level of gravity that climate change impacts constitute for the most vulnerable States, former President of the Republic of Palau, Johnson Toribiong, embarked on a new legal and political initiative undoubtedly inspired, if not triggered, by the prior development of the Climate Change and International Security Discourse. In September 2011, Toribiong defined the impacts of climate change as "an existential threat that exemplifies the issue of transboundary harm", and recalled how the international community had just addressed and acknowledged the security threats of climate change in the Security Council, where it was acknowledged by Presidential Statement that the survival of low-lying island States is at stake because of climate change impacts.[620] Importantly, he also recalled that in Resolution 63/281 on 'Climate Change and its Possible Security Implications' of June 2009, the General Assembly had invited "*the relevant organs of the UN* to intensify their efforts in considering and addressing climate change, including its possible security implications".[621] Considering the International Court of Justice as the principal judicial organ of the UN, Palau and the Marshall Islands thus proposed:

> [T]o seek, on an urgent basis, pursuant to Article 96 [of the UN Charter] an Advisory Opinion from the International Court of Justice on the obligations and responsibilities of States under international law to ensure that activities carried out under their jurisdiction or control that emit greenhouse gases do not damage other Sates.[622]

Although the diplomatic moves pressing for this initiative went very far, the request seems to have been dormant since Toribiong left office in early 2013. Whilst current President Remengesau has not departed from this initiative, it seems difficult for Palau to continue pressing for it, particularly given the

620 See UN Security Council, Presidential Statement on Maintenance of International Peace and Security and Impacts of Climate Change, 20 July 2011, UN doc.: S/PRST/2011/15; and Security Verbatim Record S/PV.6587, of the 6587th meeting of the Security Council, 20 July 2011, 10 a.m.

621 UN General Assembly Resolution 63/281, para. 1. [Emphasis added].

622 See Johnson Toribiong, President of the Republic of Palau, Statement to the 66th Regular Session of the United Nations General Assembly (22 September 2011), available at the official website of Palau's Permanent Mission to the United Nations: <http://gadebate.un.org/66/palau> (3 February 2015). For a recent explanation and defence of the position of the Republic of Palau, see Aaron Korman and Giselle Barcia, 'Rethinking Climate Change: Towards an International Court of Justice Advisory Opinion', *Yale Journal of International Law Online*, vol. 37 (2012), 35–42.

opposition of the USA – which administered Palau between 1940 and 1994 and is now in Compact of Free Association with it – to any developments on the law of State responsibility for climate change impacts outside of the UNFCCC framework.[623]

Nonetheless, as McAdam rightly noted, in addition to asserting legal rights, litigation on the law of State responsibility – or any attempts in this sense – may still be seen as serving a political function.[624] In this case, the mere initiative of seeking an ICJ Advisory Opinion on State responsibility for climate change impacts could potentially promote international efforts to come to an agreement within the UNFCCC negotiation process.[625] And indeed, the concerns underlying Palau's initiative have been partly addressed within the UNFCCC, in particular through the negotiations that run from the official introduction of loss and damage in the UNFCCC during COP.16, celebrated in Cancun in 2010 as part of the Cancun Adaptation Framework, to its final institutionalisation in 2013 through the creation of the International Mechanism for Loss and Damage at COP.19, held in Warsaw.[626] As previously noted, pressure of industrialised countries to make reference to the two other frameworks dealing in parallel with loss and damage – namely, the Hyogo Framework for Action and International Strategy for Disaster Reduction – served neutralising the discussions on State responsibility for transboundary harm. Yet, the very inclusion of the topic in the UNFCCC framework was an unprecedented achievement of AOSIS diplomacy.[627]Ahead of Paris COP 21 at the end of 2015, efforts of vulnerable countries focus on including loss and damage in the new climate agreement for the second commitment period.[628] Although under its current design the Mechanism falls short of constituting the cradle of a future climate change compensation mechanism, the review of the Mechanism's structure,

623 See Sir Ronald Sanders, 'A Case for Compensation for Climate Change Destruction?', Press release, 2 January 2014, available at: http://www.sirronaldsanders.com/viewarticle.aspx?ID=416 (24 February 2015).

624 McAdam, *Climate Change, Forced Displacement and International Law*, 95.

625 See Jaspreet Kindra, 'Island Nation Takes on the World's Polluter', Press release, available at: http://www.irinnews.org/report/99578/island-nation-takes-on-the-world-s-polluters (24 February 2015).

626 See chapter 2, section 2.1.1.B above.

627 M.J. Mace and Michiel Schaeffer, 'Loss and Damage under the UNFCCC: What Relationship with the Hyogo Framework', (Bonn: Germanwatch/ The Loss and Damage in Vulnerable Countries Initiative, October 2013), 1–16, 5.

628 See the negotiating text for the Paris Climate Change Summit, prepared by the Ad-Hoc Working Group on the Durban Platform for Enhanced Action, 15 February 2015, doc. FCCC/ADP/2015/1.

mandate and effectiveness in 2016 leaves a door open for hope of bringing about developments in this sector.[629]

Despite the underdevelopment of the international law on State responsibility for climate change impacts, two interesting initiatives were initiated by Micronesia and the Marshall Islands. First, in the early 2010s, around the same time when Palau's former President proposed for an ICJ Advisory Opinion on State responsibility for transboundary harm, Micronesia challenged the extension of the Czech coal-fired power station Prunerov II, the largest single source of carbon dioxide of the country.[630] Micronesia requested the Czech Republic to commission a transboundary environmental impact assessment, as the retrofit of one of the largest sources of CO2 emissions in the world could severely impact the entire environment of the low-lying island State.[631] In doing so, Micronesia virtually introduced a new understanding of the concept of neighbourliness, for the term 'transboundary' had until then been seen as referring to neighbouring States sharing a common border only. Yet, following the decision of Czech Prime Minister Jan Fisher, and considering that 70% of the controlling shares of the plant are owned by the Czech State, the plant was allowed to keep operating until 2035 with aged technology, instead of closing down in 2020.[632] This decision was taken after the independent evaluation report commissioned by the Czech State and prepared by the Det Norske Veritas (DNV) found that by using better technology,[633] the plant's energy efficiency could reach 42%, instead of 39% – saving 200,000 tonnes of CO2 emissions per year. Although Micronesia's complain did not reach the Czech courts or any other judicial institution, it had an important political

629 Decision 2/CP.19, para. 15.

630 For a historical account of the Prunerov II power station, see http://www.cez.cz/en/power-plants-and-environment/coal-fired-power-plants/cr/prunerov.html (25 July 2015).

631 See, for instance, Geofrrey Lean, 'Micronesia Lawsuit Highlights Climate Change Litigation', The Telegraph, Press Release, 24 May 2011, available at: http://www.telegraph.co.uk/news/earth/environment/climatechange/8533367/Micronesia-lawsuit-highlights-climate change-litigation.html (24 February 2015).

632 An explanation of the process can be found in the online article by the non-profit organisation Bellona, at: http://bellona.org/news/climate change/2010–03-comment-czech-environmental-minister-resigns-in-dust-up-over-coal-plant-retrofits (24 February 2015).

633 See the press release by the DNV on this matter, explaining the scope and methodology of the report: http://www.dnv.com/press_area/press_releases/2010/dnvtoassesstherefurbishmentofczechpowerplant.asp (24 February 2015); as well as the press release by the Polish Environment Ministry http://www.mzp.cz/en/news_100208_prunerov (24 February 2015).

impact, as Jan Dusik – the Green Party appointed Minister of Environment – resigned from his office a few months before the holding of general elections.

It is also worth mentioning the recent move by the Republic of the Marshall Islands, which in July 2014 initiated three separate proceedings before the International Court of Justice against the United Kingdom, Pakistan, and India, on the obligation to negotiate the cessation of the nuclear arms race and the obligation of nuclear disarmament.[634] Whilst the applications were submitted on the basis of customary international law that binds all (nine) nuclear states (USA, UK, Russia, China, France, India, Pakistan, Israel and North Korea), five of which are in addition Parties to the Non-proliferation Treaty (USA, UK, Russia, France and China), for jurisdictional reasons regarding the acceptance of compulsory jurisdiction, the institution of proceedings had to be limited to only three States (United Kingdom, Pakistan and India). As a an non-nuclear weapon State Party to the Non-proliferation Treaty (NPT), the Marshall Islands filed the three applications for the institution of proceedings highlighting how the country had particularly suffered the consequences of nuclear radiation, as it was a historical site for nuclear testing during the time when it was placed under the trusteeship of the United States of America, that is, between 1946 and 1958. Yet, most interestingly, the three applications contain a virtually identical paragraph in which the Marshall Islands associate the decision to institute proceedings on the violation of the obligation to conclude negotiations leading to nuclear disarmament with the recent efforts to combat the impacts that climate change have for its 'survival' as a State:

> [w]ith regard to the RMI's [Republic of the Marshall Islands] interest in bringing this application to the Court, the following should be added. It is well known that over recent years the RMI has been preoccupied with combating the extremely harmful consequences that the effects of climate change have for its very survival. *While focusing on the problem of climate change, the RMI has come to realise that it cannot ignore the other major threat to its survival*: the ongoing threat posed by the existence of large arsenals of nuclear weapons the use of which, according to the Court, "seems scarcely reconcilable with respect for [...] requirements

634 Republic of the Marshall Islands v. Republic of India, Application for the Institution of Proceedings to the ICJ, *Obligation to Pursue in Good Faith and Conclude Negotiations Leading to Nuclear Disarmament*, submitted on 24 April 2014. The applications for the institution of proceedings Republic of the Marshall Islands v. Pakistan and Republic of the Marshall Islands v. United Kingdom were submitted to the ICJ on the same day and under the same title.

> [of the principles and rules of law applicable in armed conflict]". It is obvious that the RMI's participation in the common struggle against climate change needs to lead to firm commitments by all States, which must include not only moral, but also legal obligations aimed at realising concrete, clear-cut goals in order to remove the threat of devastation caused by continued reliance on the use of fossil fuel energy sources. It is from this perspective of striving to reach agreement on such commitments in the struggle against climate change that the RMI has concluded that it is no longer acceptation simply to be a Party to the NPT while total nuclear disarmament pursuant to Article VI and customary law remains at best a distant prospect.[635]

These three applications are thus to be considered as having been promoted by the general action lately taken by Pacific Island States to defend their continuity as States, a move which was initially purported by the threat of climate change. By connecting climate change impacts to the obligation to negotiate nuclear disarmament, the Marshall Islands has conceptualised climate change as akin to the nuclear threat, in the sense that it has with the potential to result in the State's absolute destruction and ultimate disappearance.

B Subsidiary Remedy: Recognition of State Continuity or New Forms of International Political Support

The underdevelopment of international law on State responsibility and liability for transboundary harm makes it difficult, so far, to ascertain whether Pacific Island States may pursue claims of economic *compensation* for the loss and damage of the material elements composing them, as opposed to the already existing channels of economic *aid* for adaptation to climate change impacts. Yet, as the reality of Pacific Island States further deteriorates and their future becomes more uncertain, the central and defining role of the institution of recognition under international law will be all the more reinforced. A new era in the understanding of this institution and its use is thus likely to be opened by the unprecedented challenges to statehood experienced by Pacific Island States. As defined by Martha J. Peterson, the term 'recognition' refers here to the "*acknowledgement of the existence of an entity or a situation* indicating that the full legal consequences of that existence will be respected".[636]

---

635 *Marshall Islands v. Pakistan*, Application, para. 9; *Marshall Islands v. India*, Application, para. 9; *Marshall Islands v. United Kingdom*, Application, para. 10. [Emphasis added].

636 Martha J. Peterson, *Recognition of Governments: Legal Doctrine and State Practice, 1815–1995* (New York: Macmillan, 1997), 1. [Emphasis added].

While the recognition of governments and the recognition of States may be seen as two closely related forms of the wider institution of recognition,[637] they have *so far* been regarded as two distinct legal acts. On the one hand, recognition of States touches upon the conditions for the very creation and existence of the State in a system where the westphalian approach to membership to the international community remains predominant. Impinging on the "foundation stone of the edifice of international law",[638] recognition of States belonged for long to the sole realm of power politics, until efforts by legal scholars – such as Sir Hersch Lauterpacht and Ti-Chiang Chen – to frame recognition of States within legal parameters gradually developed.[639] In doing so, much focus was put on the question of the *nature* of recognition, which consequently also led to question of nature of the State itself. As it is widely known, adherents to the constitutive school argued that recognition *creates* the State; as opposed to proponents of the declaratory theory, who argued that recognition by other States simply consists of an *acknowledgment* of an already existing factual reality, for the State 'is' since the requirements of statehood are met. 'In between' or eclectic approaches such as that of Lauterpacht considered that, notwithstanding the constitutive nature of recognition, if a political entity meets the criteria of statehood as laid down in article 1 of the Montevideo Convention, other States are under the *duty* to grant recognition to that entity.[640] Nowadays, it is generally accepted that the nature of recognition is merely declaratory of a 'primary fact' – namely, the existence of a State,[641] so that recognition essentially serves as an instrument for the *validation* in the political sphere of effectiveness to statehood.[642]

637 Ibid., 2.

638 John Dugard, *Recognition and the United Nations* (Cambridge, UK: Grotius, 1987), 1.

639 Sir Hersh Lauterpacht, *Recognition in International Law* (Cambridge: Cambridge University Press, 1947); Ti-Chiang Chen, *The International Law of Recognition – with Special Reference to Practice in Great Britain and the United States* (London: Stevens, 1951).

640 See Lauterpach, 7: "[A]lthough recognition is thus declaratory of an existing fact, such declaration, *made in the impartial fulfillment of a legal duty*, is constitutive, as between the recognising State and the community so recognised, of international rights and duties associated with full statehood". [Emphasis added].

641 See Théodore Christakis, 'The State as a "primary fact": Some Thought on the Principle of Effectiveness', in Marcelo G. Kohen, ed., *Secession: International Law Perspectives* (Cambridge: Cambridge University Press, 2006), 138–170, 170.

642 John Dugard and David Raic, 'The Role of Recognition in the Law and Practice of Secession', in Kohen, ed., 94–137, 94.

On the other hand, as previously mentioned, the second form of the institution of recognition – that is, recognition of governments, refers to the acknowledgment by other States that a specific entity is regarded as the agent of (an already existing) State. In the absence of a central international authority defining the eligibility criteria for an entity to become an agent, the governments of other States have thus been called upon assuming such function of control.[643] Thus, in principle, recognition of governments operates *after* recognition of States, once statehood has been established (and only when necessitated by the specific conditions, such as caused by the lack of constitutional continuity). However, while these two forms of recognition are distinct, as they operate at different stages, and produce different legal consequences, they have yet been subject to similar questions. The question of the constitutive or declaratory nature of recognition has also been raised in the context of recognition of governments;[644] so much as whether States have full political discretion of what governments of other States they may recognise, or whether there are bound by legal rules in this context.[645] Besides, the answers relating to both the nature of recognition and the duties attached to it are also similar. First, the declaratory theory also applies nowadays to recognition of governments, as differentiations between recognised and unrecognised governments erode.[646] Second, while States are not under a *duty* to grant recognition to either a State or a government, they are bound – in both areas – by the prohibition not to prematurely (nor belatedly) recognise a new entity as a State or a government.[647]

As previously developed in section 3.1 above, while acute de-territorialisation and de-population accelerate, Pacific Islands' governments will also be called upon to evacuate their respective territories. At an early stage – that is, whilst some territorial basis and partial population of the State remains in place – traditional rules on recognition of governments in exile would be – to some extent – applicable to cases of the recognition of Pacific Islands' authorities as ex situ governments.[648] Yet, doubtfulness on the status of such authorities is likely to be raised when the total disintegration of these elements ensues. As Crawford notes, "the rules for determining identity and continuity are variants upon the basic criteria of statehood, with the addition

643 Peterson, 2.

644 Ibid., 12.

645 Ibid., 28–50.

646 Ibid., 12.

647 Ibid.

648 See section 3.1.2.B above.

of certain specific qualifying or particularising rules", with a decisive power of recognition and the claims of the States concerned.[649] Such extreme and marginal cases are thus likely to call for an unprecedented form of recognition, where both recognition of government and of State would merge into one single legal process. Simply put, the *non-termination* of other States' recognition of Pacific Islands' ex situ governments would, in scenarios of total de-territorialisation and/or de-population, arguably amount to recognition of the continuity Pacific Islands States. The recognition of Pacific Islands' ex situ governments initially granted would thus have the potential to *derive into* recognition of State continuity. Yet, in order to substantiate the legality and practical feasibility of this potential new form of recognition, the respective weight of two opposing factors – effectiveness v. legitimacy – ought to be balanced.

First of all, the main obstacle to the possibility of not suspending recognition to Pacific Islands' ex situ governments in scenarios of total de-territorialisation and/or de-population lies in the criterion of effectiveness. Borrowing the explanation given by Hans Kelsen at The Hague Academy of International Law, Théodore Christakis recalls that the "birth and death" of the State are meta-juridical facts not explained by legal rules.[650] The only relevant factor in both contexts is, according to Christakis, the principle of effectiveness, which sets the standard from which international law elevates certain real facts by attributing to them set of legal consequences – thereby operating a transformation of simple facts into 'legal facts' (*faits juridiques*).[651] Such understanding of the birth and death of States as belonging to the realm of facts goes along with the evolution towards a general acceptance that the nature of recognition is merely declaratory, and that States have an objective existence. In the context of birth of States, the weight of the principle of effectiveness can be particularly determinant in controversial cases of State secession outside of the context of decolonisation. As Christakis explains, the presumption against secession – which protects the sacrosanct principle of territorial integrity of States – may in some cases be overturned by the capacity of a secessionist movement to establish a clear effectiveness over its territory *and* to impose it over an apparently irreversible situation (the so-called 'ultimate success' theory).[652] Similarly, in the realm of recognition of governments, the principle of effectiveness has come to set limits to the margin of political discretion that persists in this area. First, the prohibition of premature recognition (which

649 Crawford, *Creation of States in International Law*, 411.

650 Christakis, 139.

651 Ibid., 139 and 141.

652 Ibid., 147.

also operates for recognition of States) implies that recognition shall not be granted to a government before it exercises sufficient control over its territory.[653] Secondly, and most importantly, States are also under the *duty not to prolong* recognition of a government once it has become defunct – either by national elections or by loss of sufficient control over its territory.[654] This rule would potentially operate against the proposal of non-termination of recognition of Pacific Islands' ex situ governments, as the material disappearance of territory and/or migration of population a fortiori implies the loss of control over these criteria of statehood. In other words, effective control cannot be exercised over a territory that has ceased to exist from a factual perspective, or over a population that has migrated and is dispersed elsewhere. Thus, an incursion in the principle of effectiveness would in principle lead to conclusion that States would be under a duty to terminate their recognition of Pacific Islands' ex situ governments, which, in scenarios of total territorialisation and depopulation, would be tantamount to the end of their recognition as States – that is, a confirmation of the extinction of their international legal personalities following the loss of their territory and migration of their population. At best, defenders of Pacific Islands' State continuity in the extreme scenario of total de-materialisation could argue that the termination of recognition ought to be *postponed* as much as possible – ideally, until the consequences of the loss of statehood on the population – in particular, the need for relocation and of avoiding statelessness – are duly addressed by the ex situ governments in co-operation with other States and the relevant international organisations. Yet, such a conclusion shall be balanced against the weight of other criteria equally relevant in the realm of State creation and State extinction. In particular, the role of legitimacy and the claims of the States affected on their own situation may counter-balance the impact of the principle of effectiveness.

The prohibitions of premature and prolonged recognition both help avoiding the creation of inadequacies between the realm of facts and the realm of law; to some extent they prevent the creation of legal fictions which cannot be sustained by tangible facts. However, on occasions the institution of recognition has served precisely the opposite function. In the context of creation of States, political entities which manifestly fell short of meeting the criteria of effectiveness were yet recognised as States. And in fact, as previously developed in chapter 1, Pacific Island States are the paramount example of how normative and political considerations may legitimise the overlooking of the principle of effectiveness – or even the viability requirement.

653 Peterson, 95.

654 Ibid.

In this case, the flexible approach to the conditions for the emergence of a State laid down in Article 1 of the Montevideo Convention resulted from the weight of two equally important factors: the right to self-determination and political independence from colonial domination, coupled with the aspiration of the United Nations to become a truly universal organisation.[655] As noted by John Dugard, in these cases statehood was more evidenced and validated by admission to the United Nations and acceptance or certification by the colonial "parents", than by strict compliance with the traditional requirements of statehood.[656] Although recognition and non-recognition of States and governments remains a unilateral endeavour,[657] the admission of Pacific Island States to the United Nation amounting to collective recognition was effected through the agency of the Organisation. Borrowing Dugard's words, Pacific Island States "in political fact, if not legal abstraction, are *products* of the United Nations".[658] Pacific Island States therefore *embody* both the maximum political aspiration of the UN as an international organisation, and the ethical aspiration to justice of the legal system. As a corollary, the continuity or eventual extinction of Pacific Island States, as the total de-materialisation scenario progresses, will essentially depend on the treatment given by the United Nations to the fruits of its own value-system it once 'produced'.

If effectiveness can be overruled by non-legal criteria in the realm of State creation, what can be the development in the cases possibly leading to State extinction? Mirroring the history of their creation as States, the continuity of Pacific Island States may thus be preserved as long as their membership in the United Nations is maintained. Although the extinction of States belongs to the universe of facts, their expulsion from the organisation, an irrefutable evidence of the loss of their statehood, would tantamount to the collective termination of their recognition as States by the international community.[659] In contrast, the continuity of membership to the organisation, even after de-materialisation, would corroborate the inherent connection between the fate of Pacific Islands States, and the evolutions of both the underlying value-system of the international community, and the challenges it is called to face. In this context, the natural bias of the international legal system towards

655 See chapter 1, section 1.1.2.

656 Dugard, *Recognition and the United Nations*, 3.

657 Ibid., 4. Contrary to what was expected by Lauterpacht and Chen, who both considered that the move towards collective recognition of States was necessary for the increase of the political integration of the international community.

658 Ibid. [Emphasis added].

659 UN Charter, Article 18(2).

stability will likely help in maintaining Pacific Islands' full UN membership for a while, along with their recognition of statehood. Also, the acknowledgment by the two main organs of the United Nations of the security implications of climate change may well serve to the continuity of membership, as the loss of statehood is being considered the political responsibility of the international community as a whole.

Considering that even in such extreme circumstances maintenance of the continuity of statehood ought to be granted implies that the possibility of a new form of statehood – the 'metaphysical State' – would indirectly be suggested. In order to substantiate such an extreme proposal, a more detailed account of its sources of legitimacy is necessary.[660]

3.3.2 *Sources of Legitimacy for the Continuity of Pacific Islands States*

A The Wider Context: A Global, Southern, and Post-Colonial Issue

As the shortcomings of a legally based argument in favour of the continuity of Pacific Island States have been reviewed, this closing section presents the basis for a succinct but important prescriptive note. This indeed implies entering "the realm of the utopia"[661] and reflecting upon the interaction between the "social is" and the "social ought",[662] where the international lawyer, standing in the present, is called upon thinking where and how international law *should* evolve. The first and foremost difficulty when embarking on such a path is being confronted with the task of making epistemological choices, in the midst of a complex system in which the accepted normative values that underline international legal discourse have been dramatically challenged by newly developing circumstances, as much as by intellectual arguments. This "time of perplexity"[663] carries with it the heavy burden of constant uncertainty, which is also among the key features of the Anthropocene. And, in fact, it may not seem very coherent to close this work with a prescriptive note whilst taking the view that knowledge is culturally determined.

660 Legitimacy may indeed explain not only why States obey powerless rules, but even why States can adopt a specific political behavior which is not even prescribed by a rule. See Thomas Franck, *The Power of Legitimacy among Nations* (New York and Oxford: Oxford University Press, 1990).

661 Koskenniemi, 397.

662 Franck, *The Power of Legitimacy among Nations*, 7.

663 Thomas Frank, 'Epistemology at a Time of Perplexity', *European Journal of International Law*, vol. 13 (2002): 4, 1025–1030, 1026.

Yet, as this pressing topic will remain, in the years to come, a mutable field of enquiry, a reference to the ethical parameters in which it may be subject to further evolution seems necessary. While a full-fledged and comprehensive conceptualisation of this issue in ethical or philosophical terms falls outside the scope of this work, this section attempts nonetheless to pinpoint the main *structural* characteristics of the ethical parameters in which the issue of State extinction for climate change impacts could and should be understood. In fact, this endeavour seems all the more required that the question of whether – and, in the affirmative, when – a Pacific Island State will cease to exist will presumably be resolved in the political realm before legal arguments are called into play and accommodate the political resolution of the issue at stake.

As noted by James Nickel and Daniel Magraw "the philosophy of environmental law overlaps considerably with philosophical work in environmental ethics".[664] Similarly, the issue of State extinction due to or resulting from climate change impacts overlaps and crosscuts more than one philosophical demand. One way to approach the ethical parameters of this topic is therefore to conceive it as a structure comprising three different layers. The first layer, corresponding to the global scale of analysis, allows to address this issue as pertaining to the broader debate on inter-generational fairness. The second layer, in turn, can be grounded in the context of the North/South divide, colouring the global scale of analysis with additional considerations from the third-world approaches to international law on global environmental justice. Finally, the third layer adds complexity by reminding us of how the Climate Change and International Security Discourse pointed out at the limits of the North/South divide in this case-scenario, for the developmental rationale of emerging countries clashed with the existential claims of Pacific Island States.

The image of 'drowning' or 'sinking' States has certainly come to be equated with one of the most paradigmatic images of today's post-modern global environmental crisis; an image which, Jane McAdam highlights, "has become the canary in the coalmine – the litmus test for the dramatic impacts of climate change on human society".[665] Before considering the consequences of the loss of international legal personality from a geopolitical, State-centred perspective, it is worth noting that this issue represents the most stringent manifestation of the need to take the interests of future generations into account

664 James Nickel and Daniel Magraw, 'Philosophical Issues in International Environmental Law', in Samantha Besson and John Tasioulas, eds., *The Philosophy of International Law* (Oxford: Oxford University Press, 2010), 453–471, 453.

665 McAdam, *Climate Change, Forced Migration and International Law*, 119. [Emphasis added].

when States participate in decision-making processes. The principle of inter-generational justice, which traditionally arose in the context of decision-making involving the exploitation of natural resources or the prevention and management of pollution,[666] is also referred to in the Preamble as well as in Article 3(1) of the UNFCCC.[667] One of the main problems of this principle is that it requires governments to adopt long-term perspectives about future peoples and future needs about which they lack factual knowledge. Although, as Nickel and Magraw state, we can nonetheless "be confident that they will need a liveable earthly home,"[668] this inherent looseness or "non-identity problem"[669] surely impinges on the capacity of the principle to effectively influence States and relevant actors during decision-making processes. Yet, in many respects, the case of Pacific Island States highly reduces the need to make uncertain projections; somehow, it makes 'real', and close to the present time, the needs already of the present and, indeed, future generations of Pacific Islanders.[670] It is therefore not surprising that most political statements of Pacific Island States' representatives continuously make direct references to the next two generation of islanders (that is, to the grandsons and grand-daughters of present politicians, some of whom may already be born), and the fear that, as early as within these next two generations, their descendants will be forcibly set apart from their traditional lands and loose a fundamental source of their cultural heritage. Therefore, the sinking State paradigm is not only an issue which fits within the moral demands laid down by the principle of inter-generational justice, now quite solidly settled in international environ-

666 See, for instance, the early landmark work of Edith Brown Weiss, *In Fairness to Future Generations: International Law, Common Patrimony and Intergenerational Equity* (Tokyo: United Nations University, 1989).

667 While the last paragraph of the Preamble of the UNFCCC reads "[The Parties to this Convention] Determined to protect the climate system for present and future generations"; Article 3(1) stipulates: "[T]he Parties should protect the climate system for the benefit of present and future generations of humankind, on the basis of equity and in accordance with their common but differentiated responsibilities and respective capabilities. Accordingly, the developed country Parties should take the lead in combating climate change and the adverse effects thereof".

668 Nickel and Magraw, 457.

669 Robert Crisp, 'Ethics and International Environmental Law', in Besson and Tasioulas, eds., 473–490, 477.

670 See Marc D. Davidson, 'Wrongful Harm to Future Generations: the Case of Climate Change', *Environmental Values*, vol. 17 (2008), 471–488. On the connection between present and future generations, Davidson recalls that future generations' entitlements to property originate in our present entitlements.

mental law; it also embodies such principle in a way that is closest to a present reality than any other mid- or long-term scenario to which the principle has been applied.

The principle of inter-generational justice goes hand in hand with the view – particularly widespread among Northern countries – that the defining and ultimate purpose of international environmental law is to improve the state of the global environment. As much as the sinking State paradigm can be approached from this ethical global standpoint, it is necessary to accept, as Adil Najam explains, that environmental concerns "need to be contextualised within the broader politics of sustainable development, which is itself contextualised within the even broader context of North-South politics".[671] In addition, considering that in the formulation of environmental concerns, the core of the issue is the environmental *condition*,[672] one is called upon treating today's extreme vulnerability of Pacific Island States and the threat to the continuity of their statehood as part of a broader context that considers their Southern and postcolonial identity as inherent components of such environmental condition. This is a means to effect the objective, highlighted by Mark Stallworthy, of finding *global policy* and legal frameworks of environmental justice in order to address *localised impacts*.[673] For this reason, it is a matter which naturally lends itself to inspection in terms of Third World approaches to international law (TWAIL), a movement which, Karin Mickelson explains, engages with the predominant discourse of international law and its failure to take into account the perspectives and concerns of the Third World by incorporating the post-colonial critique.[674] Indeed, the sinking island paradigm is embodied by underdeveloped States that have acquired independent legal personality following decolonisation, and whose extreme vulnerability to the impacts of climate change results not only from their geographical characteristics, but is also bound to the colonial exploitation by their former colonial rulers. As Ulrich Beck puts it:

---

671 Adil Najam, 'Why Environmental Politics Looks Different from the South', in Peter Dauvergne ed., *Handbook of Global Environmental Politics* (Northampton: Edward Elgar Publishers, 2005), 111–126, 123.

672 Ibid. [Emphasis added].

673 Mark Stallworthy, 'Environmental Justice Imperatives for an Era of Climate Change', *Journal of Law and Society*, vol. 36 (March 2009): 1, 55–74, 73.

674 See Karin Mickelson, 'Critical Approaches', in Dan Dobansky, Jutta Brunée and Ellen Hey, eds., *The Oxford Handbook of International Environmental Law* (Oxford: Oxford University Press, 2007), 262–290, 273.

> [T]he exposure to risk is replacing the class as the main inequality of modern society, for risk is defined in a reflexive way by the actors. The exposure to risk is basically a power game.[675]

Therefore, in this sense, the threat to the continuity Pacific Island States, due to climate change impacts, cannot be understood as disconnected from the earlier colonial encounter. Rather, their current situation could even be seen as the latest manifestation of such encounter. This approach implies embracing Antony Anghie's view on colonialism – who expressed the unwillingness to relegate colonialism to the past and conceived it, instead, as a phenomenon that "*reproduces itself* through its victims and continuously creates and represses new subjects".[676]

In addition to the consequences of Pacific Island States' post-colonial identity for the understanding of their present environmental condition – and largely connected to these – one should also bear in mind that in many respects the sinking State paradigm must also be understood within the context of sustainable development and the North–South divide, generally pertaining to the debate on how global environmental justice should be materialised in an unequal world.[677] For indeed, the image of Pacific Island States, which has penetrated today's collective imagination as the paramount example of vulnerability, strikingly contrasts with their almost non-existent responsibility for the on-going climate change. As southern-hemisphere countries, their image also accounts for the unjust and dark faces of an industrial revolution that largely took place within northern-hemisphere industrial countries. This point, as already mentioned above, is also relevant when considering the

675 Ulrich Beck, 'Vivir en la Sociedad del Riesgo Mundial', *Documentos CIBOB-Dinámicas Interculturales*, vol. 8 (2007), 5–32, 11 [My own translation from Spanish]: "[L]a exposición al riesgo está reemplazando a la clase como principal desigualdad de la sociedad moderna, por ser el riesgo definido reflexivamente por los actores. La definición del riesgo es básicamente un juego de poder".

676 See Anthony Anguie, 'The "Heart of my Home": Colonialism, Environmental Damage and the Nauru Case', (1993) *Harvard International Law Journal*, vol. 34 (1993): 2, 445–473, 505, (cited in Mickelson, 'Critical Approaches', 273).

677 Literature on this issue is of course very extensive and essentially deals with the distributional issues relating to the fair allocation of burdens and benefits of environmental (or, more precisely, climate change) policies. See, for instance, Henry Shue, Global Environment and International Inequality', *International Affairs*, vol. 75 (1999): 3, 531–545; and Michael Grubb, 'Seeking Fair Weather: Ethics and the International Debate on Climate Change', *International Affairs*, vol. 71 (1995): 3, 463–496.

obligations – ethical and/or political – of the international community of not withdrawing the recognition of statehood of Pacific Island States.

And yet, the introduction of the Climate Change and International Security Discourse before universal international organisations revealed that the big emerging countries, leaders of the southern world, initially stood against the 'securitisation' of climate change and the development of such an approach to the phenomenon in UN organs. It can thus be inferred that the sinking State paradigm partly cross-cuts and defies the traditional North–South divide. This, however, is not a circumstance wiping away the fact that the case of Pacific Island States falls within this context for, as Najam explains:

> [I]t would be all-too-easy to find examples within specific developing country experiences in global environmental politics that defy the general characterisations above [the North-South tension]. As soon as one shifts one's focus from the Southern collective to individual developing countries [...] one could reasonably argue that a fascination with a broad term such as "sustainable development" is inappropriate.[678]

However, as the author responds to this reality of international politics, equally acknowledging that "there are many important distinctions within the South"[679] she highlights the fact that developing countries still project a collective image.

All in all, the sinking State paradigm should not be approached or located within one single ethical stream that seems to require a sui generis ethical approximation, in which elements of global ethical parameters, such as the principle of inter-generational justice, must be contextualised within the North–South divide and post-colonial critique. Two main considerations can be extracted from this triple-edged ethical ground to facilitate the understanding of future policy and legal developments. On the one hand, the stream on inter-generational justice invites exploring the situation of Pacific Island States outside the Statist framework, and turning towards the peoples of the region. On the other hand, the relevance of some of the critiques laid down by the TWAIL movement implies that such a turn towards the peoples should be done by taking into account their southern and post-colonial conditions as elements of their identity. And thus, the sum of these two points leads us to explore, in the last section of this work, the relevance of Pacific Islanders' right

678 Najam, 123.

679 Ibid.

to self-determination in the future resolution of the situation of Pacific Island States threatened by extinction as a result of climate change impacts.

B The Turn to the Peoples: Self-Determination and the Future of Pacific Islands' Statehood

Irrespective of the legal, political and ethical grounds in favour of the continuity of Pacific Island States, three scenarios may ultimately crystallise. In the first scenario – the one best preserving the status quo – international recognition of Pacific Island States and independent political entities is simply maintained. For the interim period in which de-territorialisation and de-population remain partial, this is the scenario that will presumably take place for all twelve Pacific Island States. However, when total (or at least *en masse*) de-population finally crystallises – particularly in the three Pacific Island States exclusively consisting of low-lying coral atolls (Tuvalu, Kiribati and the Marshall Islands), recognition of such entities as a State may be gradually withdrawn by other States.[680] As already pointed out, as a political and discretionary act, granting or withdrawing recognition cannot be imposed on any State other than in exceptional cases where the creation of a State results from a previous unlawful use of force or breach of a *ius cogens* rule, in which cases – which are not relevant for the situation of Pacific Island States – other States are bound not to recognise the new entity as a State.

As already developed in chapter 1, the Holy See and the Order of Malta exemplify how international law's flexibility may allow for alternative forms of sovereignty to be recognised, even outside the statist framework. In addition, if the international legal personality of Pacific Island States were to become extinct and alternative forms of non-State sovereign statehood were discarded, the future of the political organisation of Pacific Islanders could take various forms, including federation, merger or association with neighbouring countries. Although the decision of whether – and, in the affirmative, which – alternative and exceptional path may be taken to accommodate these unprecedented cases will surely depend, to a great extent, on the reactions of the international community, the will of the actors concerned is at least equally fundamental. This work has consistently taken this view, considering that the actors to be taken into account were the Pacific Island States themselves, whose view on their own situation can be, so far, primarily found in the reconstruction of the Climate Change and International Security Discourse.

680 McAdam, *Climate Change, Forced Migration and International Law*, 158.

This principled approach is not surprising, given that, as James Crawford recalls, the general rule still holds that the State, as a community of persons, has rights in international law which are moderated through a government that is "not necessarily representative, but legally the representative, of the people of the State".[681] Nonetheless, the position taken by the government could eventually diverge from the interests or the will of the people of the State that a government represents. Also, as in the case of Pacific Island States, when the continuity of the State is questioned the interests of the people of the State may emerge independently of the official position of the State, as expressed through its government. An exploration of Pacific peoples' rights in the context of the possible extinction of their States seems, therefore, both suitable and justified. Two consecutive sorts of rights of Pacific island peoples are likely to become relevant in this context: first, the right to self-determination, and the right to physical and cultural existence as peoples (likely in the territory of one or several other States).

By virtue of the right to self-determination, defined in Article 1 of the International Covenant on Civil and Political Rights, Pacific island peoples were called upon to determine their political status following decolonisation.[682] Now, in the midst of ever more pressing forces exerted over the territory and the population of the State – arguably constituting, as pointed out in the previous section, the latest manifestation of the colonial encounter – it could be argued that such right may be revived – outside of the original timely framework (the decolonisation movement), yet still conceptually connected to it. Most importantly, the great political controversy and consequent legal constraints arising out of the possibility to exercise the right to self-determination outside of the context of decolonisation may not be upheld in the case of Pacific Island States. In fact, as McAdam highlights, the idea of the State is in fact not particularly stringent in the minds of Pacific island peoples:

> [F]or many Tuvaluans and I-Kiribati, the issues of key importance to them is the retention of 'home' – land, community and identity – rather than preserving the political entity of the State itself.[683]

---

681 James Crawford, 'The Rights of Peoples: "Peoples" or "Governments"?', in James Crawford, ed., *The Rights of Peoples* (Oxford, UK: Clarendon Press, 1988), 55–67, 56.

682 International Covenant on Civil and Political Rights, Article 1(1), which reads: "All peoples have the right of self-determination. By virtue of that right they freely determine their political status and freely pursue their economic, social and cultural development."

683 McAdam, *Climate Change, Forced Migration and International Law*, 154.

Two main divergent conclusions can be extracted from this fact. On the one hand, one should in principle expect that if the preservation of the State itself is not of major importance for Pacific island peoples, the exercise of their right to self-determination would most likely *not* involve a claim to statehood and secession impinging upon the territorial integrity of another State.[684] Rather, it would more likely result in the choice of various models of *self-government or autonomy* listed by Ian Brownlie, such as federalism, association with another State (as Niue, Tokelau and the Cook Islands have done with New Zealand), or the creation of a sui generis form of management similar to the traditional forms of UN trusteeships.[685] On the other hand, the fundamental importance of the land for the cultural identity of Pacific island peoples indicates that, although the State as a political construct is not an inherent part of the identity of Pacific island peoples, the *land* in which they live – which corresponds to the *territory* of the State – is a key component of it. Following Jeremy Waldron's dual conceptual approach to self-determination, it seems that Pacific island peoples' claim to self-determination would thus be closer to the identity-based – rather than the territorially based – approach, in so far as they regard themselves as culturally – and sometimes also ethnically – distinct.[686] Possibly forced to continue their existence as peoples as part of one or several other political communities, the maintenance of group identity – which, according to Brownlie, is the second form in which groups' rights are guaranteed after self-determination[687] – will thus be essential for Pacific island peoples to avoid their final disappearance, not only as a State, but also as a culturally united people.

3.4 Conclusions

As the third dimension of the State, besides territory and population, the government embodies the State's distinctive sovereign powers and provides

684 Ian Brownlie, 'The Rights of Peoples in Modern International Law', in Crawford, ed., *The Rights of Peoples*, 1–16, 6.

685 Brownlie, 6.

686 Jeremy Waldron, 'Two Conceptions of Self-Determination', in Besson and Tasioulas, eds., 397–413, who explains that, according to the territorially-based conception of self-determination, 407: "[T]he phrase people of a territory simply refers to all the individual men and women who happen to inhabit the territory in question. It is not assumed that they have any relation to the land much stronger than habitation".

687 Brownlie, 6.

it with the necessary organs to act and to display such powers, both in the national realm and with regard to other States in the international arena. Challenges to the governmental capacity of the State are therefore challenges to the State's capacity to fulfil the range of duties expressing the legitimacy of its existence. The potentially adverse impacts that climate change may have on a State's governmental capacity have therefore been explored from two standpoints. The first was an analysis of how climate change impacts affect the capacity of the State to fulfil the Weberian functions of the State vis-à-vis its population, offering them above all security, as well as meeting other basic survival needs. While the controversial concept of 'failed' States is not used in this thesis to define what Pacific Island States may become in the future, it is acknowledged that Pacific Islands' capacity to fulfil those functions will probably be undermined as the pressure on the other two pillars of the State continues to increase. Such a progressive decrease in Pacific Islands' governmental response capacities may in turn be particularly worsened as the impacts of climate change on the States' territory and population become so acute that the government itself is forced to evacuate the State territory and settle abroad. Recognition of Pacific Islands' prospective authorities abroad as ex situ governments is then fundamental for the affected States to be able to conclude the international bilateral or multilateral treaties necessary to secure the protection of the population (for instance, through migration agreements); provide valid authorisation allowing international assistance to enter their territory; enact legislation necessary to adapt to the new exceptional circumstances and needs; establish all sorts of disaster-response measures (including, for instance, the adoption of national and/or transnational relocation plans); control State assets in the territory and abroad, as an important source of revenues needed to finance States' responses to acute levels of de-territorialisation and de-population.

Although, prima facie, the means by which recognition of Pacific Islands' authorities abroad as ex situ governments does not seem to raise particular controversy when compared to traditional cases of governments in exile, the special circumstances in which Pacific Islands' governments will presumably be forced to evacuate State territory will in the long run generate unprecedented problems. This alludes to the instances in which the severe levels of de-territorialisation (leading to possible complete submergence, in the case of the low-lying Pacific Island States of Tuvalu, Kiribati and the Marshall Islands) and de-population (likely to affect most Pacific Island States as progressive de-territorialisation leads to the inhabitability of the State territory) take place at a time when Pacific Islands' governments are located outside of the territorial boundaries of the State. In such a scenario, the general assumption of the temporality of ex situ governments would no longer apply to Pacific Island States. This would generate a myriad of specific legal consequences regarding, for instance, the suspension or termination of existing treaties for reasons of impossibility of compliance or of fundamental change of circumstances. However, the paramount consequence is that, through the conjunction of the threat to the three dimensions of the State, the continuity of Pacific Islands' statehood will be seriously at stake. Both the response of the international community and the claims of the concerned States regarding their own situations become all the more determinant in defining the future of these States in such extreme and marginal situations.

This last chapter has shown how, in contrast to the limitations that the adverse impacts of climate change may generate in Pacific Islands' governmental capacity with regard to their national sphere, their participation in the international arena, always characterised as highly active and engaged, is likely to be pursued. Indeed, as the history of Pacific Island States shows, their creation as States was closely connected with the underlying normative evolutions that the international legal order underwent as a result of de-colonisation and the correlative quest to secure the United Nations as an organisation with a genuinely universal scope. For this reason, the admission of Pacific Island States to the United Nations membership in effect implied their recognition as States. Ever since then, their engagement with the international community has remained highly active, for it is precisely by such a bond to the international community that the inherent governmental limitations are counterbalanced. The development of a regional co-operation network through the establishment of the Pacific Islands Forum, their active participation in multilateral environmental agreements – in particular the UNFCCC, through AOSIS – are only a few examples of how Pacific Island States remain far from being the mere puppets of the Great Powers. Now that it is their continuity as States which is at stake, it can be presumed that their active participation in international affairs would, if anything, be all the more increased. As already pointed out, the final elements that will prove to be determinant in the resolution of their status in the future are continuing recognition by other States and their own claims to statehood. It is thus, at this point, that the launch of the Climate Change and International Security Discourse deserves close consideration and acquires a special relevance.

Nonetheless, this chapter has sought to explore whether the continuity of Pacific Island States could be based on other arguments than those of temporal practical utility or necessity. The first natural incursion is thus focused on the law on State responsibility for climate change loss and damage. There, the present underdevelopment of such legal avenue seems to indicate that the possibility of Pacific Island States to obtain economic compensation for climate change impacts, as recently put forward by Palau's attempt to gather a majority before the UN General Assembly to request from the International Court of Justice an Advisory Opinion on the matter, remains questionable. Yet, it has been suggested that, as an alternative remedy for climate change damage in the case of Pacific Island States, international recognition of their continuity as States could be raised. Last but not least, as the present state of positive law on the matter remains limited, some ethically based arguments in favour of an evolution in this sense have been raised. By conceptualising the future of Pacific Islands' statehood as a matter of global environmental justice within the post-colonial North–South context, the path of future evolution in international law in this context is traced from a normative perspective. And thus, after coming to terms with the blurred and slippery borders between international law and international politics in the realm of State extinction, one is ultimately – and perhaps also paradoxically – called upon to pierce the veil of the State and consider that the voices which should decide on the future of Pacific Island States, particularly of the three among those States most acutely at risk, may be, also, if not primarily, those of the Pacific Island peoples.

Conclusions

Physical Reality and Legal Construct: A Correlation Revisited

The well-grounded, traditional – for some, outdated – idea of the State's 'privileged' position in the international arena, as well as its historically rooted ontological monopoly, lay at the heart of the introduction to this book. The display of such background essentially sought to strengthen, if not even justify, the importance of conducting an in-depth study on how statehood may conceptually disintegrate for unprecedented reasons related to the physical disintegration of its subject – the State.

The reflection on the fate of statehood, benchmark of this scholarly journey, has been conducted in the context of the Anthropocene and with the understanding that ongoing climate change and sea-level rise, in which human activities have played a major role, constitute some of the manifestations of this new geological epoch. One of the outstanding attributes of the Anthropocene is that it may challenge the foundations on which the present international legal order is built, since the onset of the Anthropocene alters the generally stable conditions of the Holocene, upon which the stability of the international legal order relies.[688] The set of questions that one faces when dealing with contemporary issues through the lenses of the Anthropocene has marked a new stage in our scholarly efforts to grasp contemporary complexities associated with international subjectivity. Through its manifestations of climate change and, in particular, of the resultant sea-level rise – the Anthropocene is already prompting us to rethink not only the State or the international law on statehood but, more precisely, the correlation between *physical reality* and *legal construct* – two equally constitutive elements of the notion of statehood. In this sense, the present study offers a reminder that the legal world not only *relies on* but also *reflects* a material existence – one to which it then ascribes legal attributes.

688 See Davor Vidas et al., 'International Law for the Anthropocene?', 4, who explain that: "[T]he international legal order will always be in search of stability and, ultimately, solutions to facilitate peace and prevent conflict. However, with a fundamental change of the context in which international law operates – and with the challenges increasingly recognised as the consequences of natural, not only political, change – new legal axioms will have to evolve. That transformation will have to embrace the fundamental principles of the international law architecture – with geography firmly and deeply embedded in its core".

 | DOI 10.1163/9789004321618_006

The look into such a correlation between physical reality and legal construct has been carried out by focusing on a particular type of State actors – post-colonial, low-lying island States. The choice of these States as 'samples' for our study is explained, first of all, by a rather simple and pragmatic fact that they are among the very first international subjects to be dramatically affected by climate change impacts – to such an extent that their continuity is already jeopardised. However, a second reason must be highlighted as well, namely, that the post-colonial condition of those States stressed the need to put their *creation* as States into the appropriate normative and political context. The study of their *continuity* as States should, therefore, correlatively be approached with the same identification and consideration of the meta-legal elements that may determine the need for the law on statehood to show its most flexible face. In addition to accentuating the importance of context when dealing with contemporary issues in the law on statehood, low-lying island States' main characteristics clash with the classical view of the State as a monolithic structure with a vocation to persist in time. By their very existence, now jeopardised by climate change impacts, low-lying island States contribute to a more complex but also more nuanced understanding of the challenges to be faced by the law on statehood in the Anthropocene.

The Looming Transformation of the Three Dimensions of the State

Several main conclusions can be drawn from the scrutiny carried out in this book of the impacts of climate change and, in particular, of sea-level rise, on each of the three dimensions of the State. First, it was noted that the three dimensions – territory, population and government – are not equally affected by these phenomena of Earth system change. The physical degradation of the State does not have a homogenous level of repercussion in each of its dimensions – to the extent that the degree of interaction between physical reality and a legal construct is variable from one dimension of statehood to another.

The territorial dimension of the State undoubtedly constitutes the locus at which the correlation between physical reality and legal construct is more salient. Several scholarly writings have considered that the population is, in contrast, the most determinant dimension of the State, and the first to be studied in the context of the impacts of climate change and sea-level rise. This approach is mainly substantiated by the idea that the extinction of a State's legal personality is likely to result from the transnational displacement of all or a great part of its population, a situation which would chronologically precede the total submergence of the territory. Without seeking to challenge the idea of the population's central role (which is at the heart of the argument developed in chapter 2), the present study has nonetheless departed from this mainstream approach. It has instead pointed out that, as transnational displacement *results from* the degradation of the population's environment or 'habitat', this challenge should be considered as a *subsequent* consequence of the impacts of climate change and sea-level rise on the territorial dimension of the State. In other words, the challenge of de-population, as central to the continuity of the State as it may be, remains essentially a socio-economic manifestation of the previously triggered process of de-territorialisation. Operating as the first clinging domino, it is the process of de-territorialisation that constitutes the original source of the State's physical and legal disintegration, challenging its durability.

The third main conclusion of this study regards the role of the population in the quest of low-lying island States to sustain their continuity. Whilst the challenge of de-territorialisation is the *source* of the overall threat to Pacific Island States, the challenge of de-population lies at the heart of the future of these States. Indeed, the second challenge raised by climate change and sea-level rise has revealed that the protection of the population subjected to climate-induced displacement, whether individual or collective, is an area with extensive legal lacunae. Ultimately, the protection of the population of the affected State will undoubtedly be an important variable in the recognition of either the continuity of Pacific Islands' statehood, even in the most extreme circumstances, or the justification of a new form of international subjectivity. The multi-layered scheme proposed in this book thus sought to reflect and respond to the variety of the legal gaps currently existing in this area.

Considering the overwhelming and unprecedented impacts of climate change on the territorial and the human dimensions of the State, the challenge of government failure may perhaps be seen, at first glance, as comparatively less striking and less complex to respond to. Given that precedents of government failure or of governments in exile already exist, the challenge of climate-induced ex situ government essentially lies in the need to adapt to the fact that, in the case of low-lying island States threatened by sea-level rise, the 'exceptional' and temporary character of the ex situ location of the government will at some point turn into a permanent situation. By its very nature, the political dimension of the State is perhaps the one less predisposed to be apprehended through legal tools. Although, in this sense, it is the more uncertain dimension of the State, at the same time it has the advantage of being the most flexible one. Most importantly, it is in the governmental dimension of the State that one is compelled to consider the challenges to the three dimensions of the State *jointly*.

As much as the joint consideration of the three challenges (de-territoralisation, de-population, and government failure) exposes the limits of international law on statehood as it currently stands, it also marks the need to turn to the 'meta-legal realm'. Indeed, as recalled by the International Court of Justice in the 1949 *Reparations* case, "throughout its history, the development of international law has been influenced by the requirements of international life".[689] The ultimate aim of the present study has thus been to unveil the new direction that the law on statehood might take or should be taking. Such a task, related to the phenomenon of sea-level rise and approached within the context of the Anthropocene, invites the suggestion that new possible ways to represent the correlation between the physical world and the legal construct, other than by statehood, may have to be taken into consideration.

From Statehood to International Legal Personality? The Adaptation of the Law on Statehood to the Anthropocene

It is clear that 'meta-legal' and/or 'para-legal' factors will be determinant elements of the response of the international community to this extreme ontological challenge to statehood as to be posed, in a perspective, by Earth system change phenomena such as sea-level rise. Indeed, it is in the ethical and political realms – where the international law on statehood currently operates – that this area of law can find the flexibility needed to adapt to unprecedented situations. However, in view of a fundamental change of the context, some serious obstacles emerge. To begin with, maintaining the idea of a 'de-territorialised' and/or 'depopulated' State will be at odds not only with predominant scholarly positions, but also with the competing interests of third States. From a scholarly perspective, the pressure on and stretching of such legal principles as 'the land dominates the sea' can only serve on a temporary basis. Moreover, from a geopolitical perspective, Pacific Island States' jurisdiction over vast maritime spaces rich in fishery and other natural resources, has already given rise to several submissions for an extension of their respective continental shelves beyond 200 nautical miles. Hence, paradoxically, it is at the dawn of their physical existence that the geopolitical interest of Pacific

689 *Reparation for Injuries Suffered in the Service of the United Nations*, Advisory Opinion, ICJ Reports 1949, p. 174, at 8.

Island States emerges, and is 'discovered', both by themselves as well as by third States.

Whilst acknowledging the obstacles to the defence of their cause, the question of whether the continuity of Pacific Island States will be upheld by the international community even in scenarios of total de-materialisation has, purposely, been left open. Two main reasons sustain the view that giving a categorical answer would, at this stage, be premature. First, by not 'closing the box', time remains left for the normative power of 'the sinking island paradigm' to reach its full potential in the coming decades. On a similar line, the Anthropocene could become the cradle in which such an exponential normative development may progress further. The forces, impacts and consequences of both the allegory of Pacific Island States, and the macro-context in which it can be expected to be increasingly understood by the international community, are all yet to be seen.

Bearing this precaution in mind, it is nonetheless possible to suggest that the continuity of Pacific Islands' statehood would be sustainable for as long as Pacific Island governments remain actively involved in international diplomatic arenas, and while the de-territorialisation or total de-population remains *on-going*. Once in the 'post-submergence' or 'post-displacement' scenario, the continuity of Pacific Islands' statehood will certainly be difficult to maintain. Yet, the political and ethical case made by these States, coupled with their power of leverage as 'universal symbols', may allow for a new form of international subjectivity to see the light of day. As a particularly flexible concept which has the vocation to reflect the developments of international law, the preservation of their 'international legal personality'[690] (not necessarily limited to statehood) could thus accommodate the interests of third States with the fundamental quest for survival of Pacific Island States, and the fundamental interests and needs of their population.

---

690 As described, for instance, by Janne Elisabeth Nijman, *The Concept of International Legal Personality: an Inquiry into the History and Theory of International Law* (The Hague: T.M.C. Asser Press, 2004), at 13 : "The focus will primarily be on the concept of international legal personality as the legal concept that functions at the crossroads of the political, the moral and the legal: the concept as the hinge pin of the meta-legal and the legal realm."

Bibliography

Introduction: The Law on Statehood in the Anthropocene

Articles and Book Chapters

Aznar, Mariano, 'The Extinction of States', in Rieter, Eva and De Waele, Henri, eds., *Evolving Principles of International Law: Studies in Honour of Karel C. Wellens* (Leiden: Nijhoff, 2012), pp. 25–51.

Bethelhem, Daniel, 'The End of Geography: the Changing Nature of the International System and the Challenge to International Law', *European Journal of International Law*, vol. 25 (2014), pp. 9–24.

Burkett, Maxine, 'The Nation Ex-Situ: On Climate Change, Deterritorialised Nationhood and the Post-Climate Era', *Climate Law*, vol. 2 (2011): 1, pp. 345–374.

Bodansky, Dan, 'The Copenhagen Climate Change Conference: a Post-Mortem', *American Journal of International Law*, vol. 104 (2010): 2, pp. 230–240.

Craven, Matthew, 'The Problem of State Succession and the Identity of States under International Law', *European Journal of International Law*, vol. 9 (1998), pp. 142–162.

Crutzen, Paul J. and Stoermer, Eugene F., *The Anthropocene, IGBP Global Change News*, vol. 41 (2000), pp. 17–18.

Crutzen, Paul, 'Geology of Mankind', *Nature*, vol. 415 (2002), p. 23.

Dalby, Simon, 'Anthropocene Geopolitics: Globalisation, Empire, Environment and Critique', *Geography Compass*, vol. 1 (2007): 1, pp. 103–118.

Davis, Heather and Turpin, Etienne, eds., *Art in the Anthropocene: Encounters Among Aesthetics, Politics, Environments and Epistemologies* (London: Open Humanities Press, 2015).

D'Aspremont, Jean, 'The International Law on Statehood: Craftsmanship for the Elucidation and Regulation of Births and Deaths in the International Society', *Connecticut Journal of International Law*, vol. 29 (2014): 2, pp. 201–224.

Fasternrath, Ulrich, 'States, Extinction', *Encyclopedia of Public International Law*, vol. 10 (1987), pp. 465–467.

Finney, Stanley C., 'The "Anthropocene" as a Ratified Unit in the ICS International Chronostratigraphic Chart: Fundamental Issues that Must be Addressed by the Task Group', *Geological Society of London (Special Publications)*, vol. 395 (2014), pp. 23–28.

Knop, Karen, 'Statehood: Territory, People, Government', in Crawford, James and Koskenniemi, Martti, eds., *The Cambridge Companion to International Law*, (Cambridge: Cambridge University Press, 2012), pp. 95–117.

Krabbe, Huggo, 'L'Idée Moderne de l'État', *Recueil des Cours de l'Académie de Droit International*, vol. 13 (1926): III, pp. 509–583.

Lewis, Simon L. and Maslin, Mark A., 'Defining the Anthropocene', *Nature*, vol. 519 (2015), pp. 171–180.

Palumbi, Stephen R., 'Humans as the World's Greatest Evolutionary Force', *Science*, vol. 293 (2001), pp. 1786–1790.

Rayfuse Rosemary, 'International Law and Disappearing States: Maritime Zones and Criteria for Statehood', *Environmental Policy and Law*, vol. 41 (2011), pp. 281–287.

Rudiman, William F., 'The Anthropogenic Greenhouse Gas Era Began Thousands of Years Ago', *Climate Change*, vol. 61 (2003), pp. 261–293.

Rudiman, William F., 'The Anthropocene', *Annual Review of Earth and Planetary Sciences*, vol. 41 (2013), pp. 45–68.

Running, Steven W., 'A Measurable Planetary Boundary for the Biosphere', *Science*, vol. 337 (2012), pp. 1458–1459.

Scott, Karen N., 'International Law in the Anthropocene: Responding to the Geoengineering Challenge', *Michigan Journal of International Law*, vol. 34 (winter 2013), pp. 309–358.

Steffen, Will and Lambin, Eric, 'Earth System Functioning in the Anthropocene: Human Impacts on the Global Environment', in Interactions between Global Change and Human Health, (Vatican City: Pontificiae Academiae Scientiarum Scripta Varia, 2006), pp. 112–144.

Steffen, Will, Crutzen, Paul and McNeill, John R., 'The Anthropocene: Are Humans Now Overwhelming the Great Forces of Nature?', *Ambio*, vol. 36 (2007), pp. 614–621.

Steffen, Will, Grinevald, Jacques, Crutzen, Paul and McNeill, John, 'The Anthropocene: Conceptual and Historical Perspectives', *Philosophical Transactions of the Royal Society A*, vol. 369 (2011), pp. 842–867.

Tickell, Crispin, 'Societal Responses to the Anthropocene', *Philosophical Transactions of the Royal Society-A*, vol. 369 (2011), pp. 926–932.

Vidas, Davor, 'The Anthropocene and the International Law of the Sea', *Philosophical Transactions of the Royal Society-A*, vol. 369 (2011), pp. 909–925.

Vidas, Davor, 'Sea-Level Rise and International Law: at the Convergence of Two Epochs', *Climate Law*, vol. 4 (2014), pp. 70–84.

Vidas, Davor, Fauchald, Ole K., Jensen, Øystein and Tvedt Morten, W., 'International Law for the Anthropocene? Shifting perspectives in Regulation of the Oceans, Environment and Genetic Resources', *Anthropocene*, online resource, 4 July 2015, available at: http://www.sciencedirect.com/science/article/pii/S2213305415300084 (9 August 2015).

Vidas, Davor, Zalasiewicz, Jan and Williams, Mark, 'What is the Anthropocene and Why is it Relevant for International Law?', *Yearbook of International Environmental Law*, forthcoming (2015).

Waters, Colin N., Zalasiewicz, Jan A., Williams, Mark, Ellis, Michael A. and Snelling, Andrea M., 'A Statigraphical Basis for the Anthropocene?', *Geological Society of London* (*Special Publications Online First*), 24 March 2014, available at: http://sp.lyellcollection.org/content/early/2014/03/21/SP395.18 (5 May 2015).

Wong, Derek, 'Sovereignty Sunk? The Position of "Sinking States" at International Law', *Melbourne Journal of International Law*, vol. 14 (2013), pp. 346–391.

Zalasiewicz, Jan, 'Our Brave New World', *New Scientist*, 8 November 2014, pp. 26–27.

Zalasiewicz, Jan et al., 'Are We Now Living in the Anthropocene?', *GSA Today*, vol. 18 (2008), pp. 4–8.

Zalaziewicz, Jan, Williams, Mark, Haywood, Alan and Ellis, Michael, 'The Anthropocene: a New Epoch of Geological Time?' *Phil. Trans. R. Soc. Lond. A*, vol. 369 (2011), pp. 835–841.

Zalaziewicz, Jan, Williams, Mark, Smith, Alan et al., 'The Anthropocene', in Gradstein, Felix M. et al., eds., *The Geologic Time Scale 2012* (Leiden: Elsevier, 2012), pp. 1033–1040.

Zalasiewicz, Jan et al., 'When Did the Anthropocene Begin? A Mid-Twentieth Century Boundary Level is Optimal', *Quaternary International*, (2014), 1–8, online publication available at: http://ib.berkeley.edu/labs/barnosky/Zalasiewicz%20et%20al%20QI2015.pdf (5 May 2015).

Zimele, Ineta, 'States, Extinction of', in Rüdiger Wolfrum, ed., *Max Planck Encyclopedia of Public International Law* (Oxford University Press, 2008–), online edition; entry updated as of May 2007, available at: http://0-opil.ouplaw.com.catalogue.ulrls.lon.ac.uk/view/10.1093/law:epil/9780199231690/law-9780199231690-e1111?rskey=gQDrxB&result=1&prd=EPIL (5 May 2015).

Books

Chalesworth, Hillary and Chinkin, Christine, *The Boundaries of International Law* (Manchester: Manchester University Press, 2000).

Crawford, Emily and Rayfuse, Rosemary, *International Law in the Era of Climate Change* (Cheltenham, Northampton: Edward Elgar, 2012).

Davis, Heather and Turpin, Etienne, eds., *Art in the Anthropocene: Encounters Among Aesthetics, Politics, Environments and Epistemologies* (Ann Arbor, Mich.: Open Humanities Press, 2015).

Gerrard, Michael B. and Wannier, Gregory E., *Threatened Island Nations: Legal Implications of Rising Seas in a Changing Climate* (Cambridge: Cambridge University Press, 2013).

Marek, Krystina, *Identity and Continuity of States in International Law* (Geneva: Librairie Doz, 2nd ed. 1968).

McAdam, Jane, ed., *Climate Change and Displacement: Multidisciplinary Perspectives* (Oxford and Portland: Hart Publishing, 2012).

McAdam, Jane, *Climate Change, Forced Migration and International Law* (Oxford: Oxford University Press, 2012).

Munavvar, Mohamed, *Ocean States: Archipelagic Regimes in the Law of the Sea* (Dordrecht: Martinus Nijhoff, 1995).

Nordquist, Myron H., ed., *United Nations Convention on the Law of the Sea 1982: A Commentary* (Dordrecht, Boston, and Lancaster: Martinus Nijhoff, 1985).

Plato, *Timeos and Critias* (Oxford: Oxford University Press, 2009).

Rousseau, Charles, *Droit international public* (Paris: Sirey, 1970).

Yamamoto, Lilian and Esteban, Miguel, *Atoll Island States and International Law: Climate Change Displacement and Sovereignty* (Berlin: Springer, 2013).

Official Documents

International Law Association, Final Report of the ILA International Committee on Baselines under the International Law of the Sea, adopted at the 75th General Conference of the ILA, held in Sofia (Bulgaria), 26–30 August 2012.

International Law Association, Resolution nº 1/2012, 'Baselines under the International Law of the Sea', adopted in Sofia (Bulgaria), 26–30 August 2012.

United Nations, Division for Ocean Affairs and the Law of the Sea (UNDOALOS), *The Law of the Sea: Practice of Archipelagic States* (New York: United Nations, 1992).

Chapter 1: The Challenge of De-Territorialisation

Articles and Book Chapters

Ago, Roberto, 'Occupazione Bellica dell' Italia e Trattato Lateranense', *Comunicazioni e Studi*, 1946, pp. 129ff.

Ancel, Jacques, 'Les frontières: étude de géographie politique', *Recueil des Cours de l'Académie de Droit international de La Haye*, vol. 55 (1956): 1, pp. 203–298.

Barberis, Julio, 'Les liens juridiques entre l'État et son territoire: perspectives théoriques et évolution du Droit international', *Annuaire français de Droit international*, vol. XLV (1999), pp. 132–147.

Bardonnet, Daniel, 'Les frontières terrestres et la relativité de leur tracé (problèmes juridiques choisis)', *Recueil des Cours de l'Académie de Droit international de La Haye*, vol. 153 (1976), pp. 9–166.

Bassin, Mark, 'Race Contra Space: The Conflict Between German "Geopolitik" and National Socialism', *Political Geography Quarterly*, vol. 6 (1987): 2, pp. 115–134.

Burkett, Maxine, 'The Nation Ex-Situ: On Climate Change, Deterritorialised Nationhood and the Post-Climate Era', *Climate Law*, vol. 2 (2011): 1, pp. 345–374.

Caron, David, 'Climate Change, Sea-Level Rise and the Coming Uncertainty in Oceanic Boundaries: A Proposal to Avoid Conflict', in Hong, Seoung-Yong, and Van Dyke,

Jan M., *Maritime Boundary Disputes, Settlement Processes and the Law of the Sea* (Leiden and Boston: Martinus Nijhoff, 2009), pp. 1–17.

Charney, Jonathan, 'Rocks that Cannot Sustain Human Habitation', *American Journal of International Law*, vol. 93 (1999): 4, pp. 863–878.

Conforti, Bennedetto, 'The Theory of Competence in Verdross', *European Journal of International Law*, vol. 5 (1994), pp. 70–77.

Degan, Vladimir-Duro, 'Consolidation of Legal Principles on Maritime Delimitation: Implications for the Dispute between Slovenia and Croatia in the North Adriatic', *Chinese Journal of International Law*, vol. 6 (2007), pp. 601–634.

De La Pradelle, Paul, 'Notions de territoire et d'espace dans l'aménagement des rapports internationaux contemporains', *Recueil des Cours de l'Académie de Droit international de La Haye*, vol. 157 (1977), pp. 415–444.

Di Leva, Charles, and Morita, Shachiko, 'Maritime Rights of Coastal States and Climate Change: Should States Adapt to Submerged Boundaries?', *World Bank, Law and Development Working Paper Series*, vol. 5 (2008), pp. 1–52.

Elferink, Alex G. Oude, 'Clarifying Article 121(3) of the Law of the Sea Convention: the Limits Set by the Nature of International Legal Processes', *Boundary Security Bulletin*, vol. 6 (1998): 2, pp. 58–69.

Farran, Olivier, 'The Sovereign Order of Malta in International Law, *International and Comparative Law Quarterly*, vol. 2 (1954), pp. 217–234.

Foisneau, Luc, 'Omnipotence, Necessity and Sovereignty: Hobbes and the Absolute and Ordinary Powers of God', in Spingborg, Patricia, ed., *The Cambridge Companion to Hobbes* (Cambridge: Cambridge University Press, 2007), pp. 271–290.

Freestone, David and Pethick, John, 'Sea Level Rise and Maritime Boundaries: International Implications of Impacts and Responses', in Blake, Gerald H., ed., *Maritime Boundaries* (London: Routledge, 1994), vol. 5, pp. 73–90.

Grant, Thomas, 'Defining Statehood: The Montevideo Convention and its Discontents', *Columbia Journal of Transnational Law*, vol. 37 (1999), pp. 403–457.

Grant, Thomas, 'Micro States', *Max Planck Encyclopedia of Public International Law* (Oxford University Press, 2008-), online edition; entry updated as of February 2013, available at: http://o-opil.ouplaw.com.catalogue.ulrls.lon.ac.uk/view/10.1093/law:epil/9780199231690/law-9780199231690-e1068?rskey=UbAZu5&result=1&prd=EPIL (5 May 2015).

Gross, David, 'The Temporality of the Modern State', *Theory and Society*, vol. 14 (1985): 2, pp. 53–82.

Grote Stoutenburg, Jenny, 'Implementing a New Regime of Stable Maritime Zones to Ensure the (Economic) Survival of Small Island States Threatened by Sea-Level Rise', *International Journal of Maritime and Coastal Law*, vol. 26 (2011), pp. 263–311.

Gunter, Michael M., 'The Problem of Mini-state Membership in the UN System: Recent Attempts Toward a Solution', *Columbia Journal of Transnational Law*, vol. 12 (1973), pp. 470–472.

Harris, William L., 'Microstates in the United Nations: a Broader Purpose', *Columbia Journal of Transnational Law*, vol. 9 (1970): 23, pp. 23–53.

Hayashi, Moritaka, 'Sea-Level Rise and the Law of the Sea: Legal and Policy Options', Proceedings of the International Symposium on Islands and Oceans (Tokyo: Ocean Policy and Research Foundation, 2009), pp. 78–90.

Karski, Karol, 'The International Legal Status of the Sovereign Military Hospitaller Order of St John of Jerusalem, of Rhodes and of Malta, *International Community Law Review*, vol. 14 (2012): 1, pp. 19–32.

Kasenetz, Eric D., 'Desperate Times Call for Desperate Measures: the Aftermath of Argentina's State of Necessity and the Current Fight in the ICSID', *The George Washington International Law Review*, vol. 41 (2010): 3, pp. 709–747.

Kolb, Robert, 'L'interprétation de l'article 121, paragraphe 3, de la Convention de Montego Bay sur le droit de la mer: les rochers qui ne se prêtent pas à l'habitation humaine ou à une vie économique propre', *Annuaire français de Droit international*, vol. XL (1994), pp. 876–909.

Kwiatkowska, Barbara and Soons, Alfred H.A., 'Entitlement to Maritime Areas of Rocks Which Cannot Sustain Human Habitation or Economic Life of Their Own', *Netherlands Yearbook of International Law*, vol. 21 (1990), pp. 139–181.

Kunz, Joseph L., 'The Status of the Holy See in International Law', *American Journal of International Law*, vol. 46 (1952), pp. 308–314.

Lusthaus, Jonathan, 'Shifting Sands: Sea-Level Rise, Maritime Boundaries and Inter-State Conflict', *Politics*, vol. 30 (2010): 2, pp. 113–118.

Orakhelashvili, Alexander, 'International Law and Geopolitics: One Object, Conflicting Legitimacies?', *Netherlands Yearbook of International Law*, vol. 39 (2008), pp. 155–204.

Orlow, Daniel, 'Of Nations Small: the Small State in International Law', *Temple International and Comparative Law Journal*, vol. 9 (1995), pp. 115–140.

Osterud, Oyvind, 'The Uses and Abuses of Geopolitics', *Journal of Peace Research*, vol. 2 (1988), p. 191–199.

Radnitzky, Ernst, 'Die rechtliche Natur des Staatsgebiets' [Juridical Nature of the State], 1906, *Archiv für öffentliches Recht* XX, pp. 313–356.

Rapoport, Jacques, 'The Participation of Ministates in International Affairs', *Proceedings of American Society of International Law*, vol. 62 (1968), pp. 155–163.

Rayfuse, Rosemary, 'International Law and Disappearing States: Utilising Maritime Entitlements to Overcome the Statehood Dilemma', *University of New South Wales Faculty of Law Research Series*, paper nº 52 (2010), pp. 1–15.

Ruiz Fabri, Hélène, 'Génèse et disparition de l'État à l'époque contemporaine', *Persée* (1992), pp. 153–178.

Scelle, George, 'L'Obsession du territoire', in Van Asbeck, Frederik et al., eds., *Symbolae Verzijl* (The Hague: Nijhoff, 1958), pp. 347–361.

Schoenborn, Walther, 'Nature juridique du territoire', *Recueil des Cours de l'Académie de Droit international de La Haye*, vol. 30 (1929): 5, pp. 81–190.

Schofield, Clive, Arsana, Andi and Van De Poll, Robert, 'The Outer Continental Shelf in the Asia-Pacific Region: Progress and Prospects', in Vidas, Davor, ed., *Law, Technology and Science for Oceans in Globalisation* (Leiden and Boston: Martinus Nijhoff, 2010), pp. 539–576.

Scovazzi, Tullio, 'Maritime Delimitation Cases before International Courts and Tribunals', in Rudiger Wolfrum, ed., *Max Planck Encyclopedia of Public International Law* (Oxford: Oxford University Press, 2006).

Silverstein, A., 'Okinotorishima: Artificial Preservation of a Speck of Sovereignty', *Brooklyn Journal of International Law*, vol. 16, issue 2 (1990), pp. 409–432.

Soons, Alfred H.A., 'The Effects of a Rising Sea Level on Maritime Limits and Boundaries', *Netherlands International Law Review*, vol. 37 (1990), pp. 207–232.

Suy, Eric, 'Réflexions sur la distinction entre la souveraineté et la compétence territoriale', in Marcic, René et al., eds., *Internationale Festschrift für Aldred Verdross zum 80. Geburststag* (Munich: Flink, 1971), pp. 493–508.

Tsalsas, Grigoris, Bourtzis, Tilemachos and Rhodotheatos, Gerasimos, 'Artificial Islands and Structures as a Means of Safeguarding State Sovereignty against Sea-Level Rise: a Law of the Sea Perspective', in *Proceedings of the 6th ABLOS Conference "Contentious Issues in UNCLOS – Surely Not?"*, celebrated at the International Hydrographic Bureau, Monaco, January 2010.

Van Der Vyver, Johan D., 'Statehood in International Law', *Emory International Law Review*, vol. 5 (1991): 1, pp. 9–102.

Van Dyke, John M. and Brooks, Robert A., 'Uninhabited Islands: Their Impact on the Ownership of the Oceans' Resources', *Ocean Development and International Law Journal*, vol. 12 (1983): 3–4, pp. 265–300.

Vidas, Davor, 'Consolidation or Deviation? On Trends and Challenges in the Settlement of Maritime Delimitation Disputes', in Boschiero, Nelina et al., eds., *International Courts and the Development of International Law: Essays in Honour of Tullio Treves* (The Hague: T.M.C. Asser Press, 2013), pp. 325–334.

Weber, Max, 'Politics as Vocation', in Gerth, Hans H. and Wright Mills, Charles, eds., *Essays in Sociology* (New York: Oxford University Press, 1946), chapter 4.

Weil, Prosper, 'Geographic Considerations in Maritime Delimitation', in Charney, Jonathan J. and Alexander, Leslie M., eds., *International Maritime Boundaries* (Dordrecht: Nijhoff, 1993), pp. 115–130.

Xue, Guifang, 'How Much Can a Rock Get? A Reflection on the Okinotorishima Rock', in Norquist, Myron, ed., *Law of the Sea Convention: US Accession and Globalization* (Charlottesville: Center for Oceans Law and Policy, 2012), pp. 341–388.

Books

Bartelson, Jens, *The Critique of the State* (Cambridge: Cambridge University Press, 2001).

Baty, Thomas, *International Law* (New Jersey: The Law Book Exchange Ltd., rev. ed. 2005).

Becker Lorca, Arnulf, *Mestizo International Law* (Cambridge: Cambridge University Press, 2015).

Blake, Gerald H., ed., *Maritime Boundaries* (London: Routledge, 1994).

Bobbio, Norberto, Matteuci, Nicola and Pasquino, Gianfranco, eds., *Diccionario de Política* (Méjico DF: Siglo XXI, 2008).

Bowett, Derek W., *The Legal Regime of Islands in International Law* (New York: Oceana, 1979).

Brierly, James L., *The Law of Nations: an Introduction to the International Law of Peace* (Oxford: Clarendon Press, 5th ed., 1955).

Charney, Jonathan J. and Alexander, Lewis M., eds., *International Maritime Boundaries* (Dordrecht: Martinus Nijhoff, 1993).

Churchill, Robin R. and Lowe, Alan V., *The Law of the Sea* (Manchester: Manchester University Press, 3rd ed. 1999).

Christopher, Anthony J., *Atlas of States: Global Change 1900–2000* (London: John Wiley and Sons, 1999).

Crawford, James, *The Creation of States in International Law* (Oxford: Clarendon Press, 2nd rev. ed. 2006).

Crawford, James and Rothwell, Donald R., eds., *The Law of the Sea in the Asian Pacific Region: Developments and Prospects*, Publications on Ocean Development, vol. nº 21 (Dordrecht, Boston, and London: Martinus Nijhoff, 1995).

De Visscher, Charles, *Theory and Reality in Public International Law*, revisited edition translated from the French by P.E. Corbett (Princeton, NJ: Princeton University Press, 1968).

Donati, Donato, *Stato e Territorio* (Rome: Athenaeum, 1924).

Duguit, Léon, *Traité de Droit constitutionnel* (Paris: Fontemoign and Cie., 1921–25).

Duursma, Jorri, *Fragmentation and the International Relations of Micro-States: Self-determination and Statehood* (Cambridge: Cambridge University Press, 1996).

Faucille, Paul, *Traité de droit international public* (Paris: Rousseau & Cie éditeurs, 8th ed. 1922).

Grewe, Wilhelm G., *The Epochs of International Law* (Berlin and New York: Walter de Gruyter, 2000).

Harden, Sheila, ed., *Small is Dangerous: Micro-States in a Macro-World* (London: Frances Pinter Publishers, 1985).

Hong, Seoung-Yong and Van Dyke, John M., eds., *Maritime Boundary Disputes, Settlement Processes, and the Law of the Sea*, Publications on Ocean Development, vol. 65 (Leiden and Boston: Martinus Nijhoff, 2009).

Jayerwardene, Hiran W., *The Regime of Islands in International Law* (Dordrecht: Martinus Nijhoff, 1990).

Jennings, Sir Robert, *The Acquisition of Territory in International Law* (Manchester: Manchester University Press, 1963).

Kearns, Gerry, *Geopolitics and Empire* (Oxford: Oxford University Press, 2009).

Kelsen, Hans, *Pure Theory of Law*, 2nd ed. (Berkeley: University of California Press, 1970).

Lauterpacht, Hersch, *Private Law Sources and Analogies of International Law* (London: Longman, Greens and Co., 1927).

Maroto Camino, Mercedes, *Producing the Pacific: Maps and Narratives of Spanish Exploration (1567–1606)* (Amsterdam and New York: Rodopi, 2005).

McIntyre, David W., *British Decolonization*: 1946–1997 (New York: St Martin's, 1998).

Nussbaum, Arthur, *A Concise History of the Law of Nations* (Michigan: McMillan, rev. ed. 1954).

Oppenheim, Lassa, ed., *John Westlake: Collected Papers on Public International Law* (Cambridge: Cambridge University Press, 1914).

Papadakis, Nikos, *The International Legal Regime of Artificial Islands* (Leiden: Nijhoff, 1977).

Prescott, Victor and Schoefield, Clive, *The Maritime Boundaries of the World* (Leiden and Boston: Martinus Nijhoff Publishers, 2nd ed. 2005).

Ratzel, Friedrich, *Der Lebensraum: Eine biogeographische Studie* (Darmstad: Wissenschaftliche Buchgesellschaft, 1901).

Sorensen, Georg, *Changes in Statehood: The Transformation of International Relations* (New York: Palgrave McMillan, 2001).

Tanaka, Joshifumi, *The International Law of the Sea* (Cambridge: Cambridge University Press, 2012).

Tang, Chenxi, *The Geographical Imagination of Modernity* (Stanford: Stanford University Press, 2008).

Taylor, Hannis, *International Public Law: A Treatise* (London: Sweet & Maxwell, Limited, 1902).

Von Listz, Franz, *Das Völkerrecht: Systematisch Dargestellt* (Berlin: Häring, 1904).

Walker, Erik A., *The Cambridge History of the British Empire* (Cambridge: Cambridge University Press, 1963).

Weber, Max, *Economy and Society* [*Wirtschaft und Gesellschaft*] (New York: Free Press, 1922).

Weil, Prosper, *The Law of Maritime Delimitation: Reflections* (Cambridge: Grotius Publications, 1989).

Wesley-Smith, Terence and Porter, Edgar A., eds., *China in Oceania: Reshaping the Pacific?* (Oxford and New York: Bergham Books, 2010).

Official Documents and Reports

Climate Vulnerability Monitor, *State of the Climate Crisis, 2010 Report of the Climate Vulnerability Initiative* (Madrid: DARA, 2012).

Commission on the Limits of the Continental Shelf, 'A Partial Submission of Data and Information of the Outer Limits of the Continental Shelf of the Kingdom of Tonga Pursuant to Part VI and Annex II to the United Nations Convention on the Law of the Sea', May 2009.

Commission on the Limits of the Continental Shelf, 'Outer limits of the continental shelf beyond 200 nautical miles from the baselines: Submissions to the Commission: Submission by the Cook Islands', 16 April 2009, available at: http://www.un.org/depts/los/clcs_new/submissions_files/submission_cok_23_2009.htm (5 May 2014).

Commission on the Limits of the Continental Shelf, 'Preliminary information submitted by the Republic of Vanuatu, Solomon Islands and the Republic of Fiji Islands on the North Fiji Basin', 19 April 2009, available at: http://www.un.org/depts/los/clcs_new/submissions_files/preliminary/fji_slb_vut_2009_preliminaryinfo.pdf (5 May 2014).

Commission on the Limits of the Continental Shelf, 'A Partial Submission by the Republic of the Fiji Islands for the Establishment of the Outer Limits of the Continental Shelf of Fiji Pursuant to Article 76, paragraph 8 of the United Nations Convention on the Law of the Sea', 20 April 2009, available at: http://www.un.org/depts/los/clcs_new/submissions_files/fji24_09/fji_2009exsummary.pdf (5 May 2015).

Commission on the Limits of the Continental Shelf, 'Preliminary information Indicative of the Outer Limits of the Continental Shelf Beyond 200 Nautical Miles for the Eauripik and Mussau Ridge Areas, submitted by the Federated States of Micronesia', May 2009, available at: http://www.un.org/depts/los/clcs_new/commission_preliminary.htm (15 September 2015).

Commission on the Limits of the Continental Shelf, 'Preliminary information submitted by the Republic of Vanuatu to the Commission on the Limits of the Continental Shelf', 10 August 2009, available at: http://www.un.org/depts/los/clcs_new/submissions_files/preliminary/vut_2009_preliminaryinfo.pdf (5 May 2014).

Commission on the Limits of the Continental Shelf, 'Preliminary information Indicative of the Outer Limits of the Continental Shelf Beyond 200 Nautical Miles for the Mussau Ridge and Eauripik Rise Areas submitted by Papua New Guinea', 5 May 2009, available at: http://www.un.org/depts/los/clcs_new/submissions_files/preliminary/fsm_preliminaryinfo.pdf (5 May 2015).

Commission on the Limits of the Continental Shelf, 'Joint Submission by Federated States of Micronesia, Papua New Guinea and the Solomon Islands', 5 May 2009: http://www.un.org/depts/los/clcs_new/submissions_files/submission_fmpgsb_32_2009.htm (5 May 2015).

Commission on the Limits of the Continental Shelf, 'Republic of Palau, Submission on the Limits of the Continental Shelf Pursuant to Article 76 of the United Nations Convention on the Law of the Sea', 2009.

Commission on the Limits of the Continental Shelf, 'Joint Submission by Tuvalu, France and Tokelau (New Zealand)', 18 December 2012, available at: http://www.un.org/depts/los/clcs_new/submissions_files/submission_tft_62_2012.htm (5 May 2015).

Commission on the Limits of the Continental Shelf, Submission by Kiribati, 24 December 2012, available at: http://www.un.org/depts/los/clcs_new/submissions_files/submission_kir_64_2012.htm (5 May 2015).

Commission on the Limits of the Continental Shelf, Submission by the Federated States of Micronesia, 30 August 2013, available at: http://www.un.org/depts/los/clcs_new/submissions_files/submission_fsm_67_2013.htm (5 May 2015).

Commission on the Limits of the Continental Shelf, Submission by Tonga, 25 April 2014, available at: http://www.un.org/depts/los/clcs_new/submissions_files/submission_ton_73_2014.htm (5 May 2014).

Intergovernmental Panel on Climate Change, J. Dronkers et al., *Strategies for Adaptation to Sea Level Rise*, Report of the IPCC Coastal Zone Management Subgroup: Intergovernmental Panel on Climate Change (Geneva: Intergovernmental Panel on Climate Change, 1990).

Intergovernmental Panel on Climate Change, Watson, R.T. et al. (eds.), *Regional Impacts of Climate Change: an Assessment of Their Vulnerability* (Cambridge: Cambridge University Press, 1997).

Intergovernmental Panel on Climate Change, *Regional Impacts Report, supra*. See also, UNEP–WCMC, *Annual Sea-Surface Temperature 2003–2007 (derived from the NASA 'OceanColor' database)* (Cambridge: Monitoring Centre, 2008).

Intergovernmental Panel on Climate Change, Field, Christopher B. et al., eds., *Managing the Risks of Extreme Events and Disasters to Advanced Climate Change Adaptation* (Cambridge, New York: Cambridge University Press, 2012).

Solomon Islands Mission to the United Nations, Note 13/09, 4 May 2009: http://www.un.org/depts/los/clcs_new/submissions_files/preliminary/slb_preliminaryinfo.pdf (5 May 2015).

United Nations, Secretary General, *Annual Report on the Work of the Organization*, 1960, 15 UN GAOR 332.

United Nations, Secretary General, *Annual Report on the Work of the Organization*, 1968, 22 UN GAOR, Supp. 1 A, UN doc.: A/6701/Add.1.

United Nations, Rapoport, Jacques et al., *Small States and Territories: Status and Problems*, 1969, UN Institute for Training and Research (UNITAR), Series n° 3.

United Nations, General Assembly Resolution 1514/XV, 'Declaration on the Granting of Independence to Colonial Countries and Peoples', adopted on 14 December 1960, 947th plenary meeting, UN doc.: A/RES/1514 (XV).

United Nations, United Nation Environment Programme–World Conservation Monitoring Centre (WCMC), *Global Distribution of Coral Reefs – Extracted from Version 7.0 of the Global 1km Raster Dataset Compiled by the UNEP – World Conservation Monitoring Centre* (Cambridge: ReefBase, 2002).

World Resources Institute, Burke, Lauretta et al., *Reefs at Risk in Southeast Asia* (*RRSEA*) (Washington DC: World Resources Institute, 2002).

World Resources Institute, Burke, Lauretta and Maidens, Jonathan, *Reefs at Risk in the Caribbean* (Washington D.C.: World Resources Institute, 2004).

Chapter 2: The Challenge of De-Population

Articles and Book Chapters

Alland, Denis, 'Le Dispositif international du droit de l'asile', in Société Française de Droit International, Colloque de Caen, *Droit d'Asile et des Réfugiés* (Paris: Pedone, 1997), pp. 13–83.

Barnett, John and Webber, Michael, 'Migration as Adaptation: Opportunities and Limits', in McAdam, Jane, ed., *Climate Change and Displacement: Multidisciplinary Perspectives* (Oxford, Portland: Hart Publishing, 2012), pp. 57–81.

Bates, Diane C., 'Environmental Refugees: Classifying Human Migration Caused by Environmental Change', *Population and Environment*, vol. 23 (2002): 5, pp. 465–477.

Bedford, Richard, 'Population Movement in the Pacific: A Perspective on Future Prospects' (Wellington: Labour and Immigration Research Center, February 2012), pp. 1–110.

Bedford, Richard and Hugo, Graeme, 'International Migration in a Sea of Islands: Challenges and Opportunities for Insular Pacific Spaces', *Population Studies Centre*, vol. 69 (July 2008), pp. 1–40.

Bedford, Richard, 'Population Movement in the Pacific: Future Prospects', Labor and Immigration Research Centre, Department of Labor of the New Zealand Government (with collaboration of the Australian Government), February 2012, full report available at: http://www.dol.govt.nz/publications/research/population-movement-pacific-perspective-future-prospects/04.asp (5 May 2015).

Betts, Alexander, 'Survival Migration: A New Protection Framework', *Global Governance*, vol. 16 (2010), pp. 361–382.

Biermann, Frank and Boas, Ingrid, 'Protecting Climate Refugees: The Case for a Global Protocol', *Environment Magazine*, vol. 50 (2008): 6, pp. 10–16.

Biermann, Frank and Boas, Ingrid, 'Global Adaptation Governance: The Case of Protecting Climate Refugees', in Biermann, Frank, Pattberg, Philipp and Zelli, Fariborz, eds., *Global Climate Governance Beyond 2012: Architecture, Agency and Adaptation* (Cambridge: Cambridge University Press, 2010), pp. 255–269.

Boege, Volker, 'Challenges and Pitfalls of Resettlement Measures: Experiences in the Pacific Region', *COMCAD Arbeitspapiere Working Papers*, nº 102 (Series on Environmental Degradation and Migration, 2011), pp. 1–40.

Cooper, Jessica B., 'Environmental Refugees: Meeting the Requirements of the Refugee Definition', *New York University Environmental Law Journal*, vol. 6 (1998), pp. 480–529.

Cournil, Christel, 'Les réfugiés écologiques: Quelle(s) protection(s), quel(s) statut(s)?', *Revue de Droit Public*, vol. 4 (2006), pp. 1035–1066.

Cournil, Christel and Mazzega, Pierre, 'Catastrophes écologiques et flux migratoires: Comment protéger les "réfugiés écologiques"?', *Revue Européenne de Droit de l'Environnement*, vol. 6 (December 2006), pp. 417–427.

Docherty, Bonnie and Giannini, Tyler, 'Confronting a Rising Tide: A Proposal for a Convention on Climate Change Refugees', *Harvard Environmental Law Review*, vol. 53 (2009), pp. 49–403.

Ellie, Jérôme, 'The Historical Roots of Co-operation between the UN High Commissioner for Refugees and the International Organization for Migration', *Global Governance*, vol. 12 (2010), pp. 345–360.

Gemenne, François and Shen, Shawn, 'Environmental Change and Forced Migration Scenarios (EACH-FOR): Tuvalu and New Zealand Case Study', 2008, pp. 1–32, (paper resulting from a research project financed by the European Commission), available at: www.each-for.eu. (5 May 2015).

Gocke, Katja, 'Stateless Persons', *Max Planck Encyclopedia of Public International Law* (New York: Oxford University Press, 2008-), internet resource, entry last updated August 2013, available at: http://oopil.ouplaw.com.catalogue.ulrls.lon.ac.uk/view/10.1093/law:epil/9780199231690/law-9780199231690-e878?rskey=LEouou&result=1&prd=EPIL (5 May 2015).

Goodwin-Gill, Guy S. 'The Right to Leave, the Right to Return, and the Question of a Right to Remain', in Gowlland-Debbas, Vera, ed., *The Problem of Refugees in the Light of Contemporary International Law Issues* (The Hague, Boston, London: Martinus Nijhoff Publishers, 1995), pp. 82–93.

Gowlland-Debbas, Vera, 'La responsabilité de l'État d'origine pour des flux de réfugiés', in Société Française de Droit International, Colloque de Caen, *Droit d'Asile et des Réfugiés* (Paris: Pedone, 1997), pp. 93–131.

Haddad, Emma, 'The Refugee: The Individual between Sovereigns', *Global Society*, vol. 17 (2003), pp. 297–232.

Haines-Sutherland, Kelly, 'Balancing Human Rights and Customs in the Pacific Region: A Pacific Charter of Human Rights', *ANU Undergraduate Research Journal*, vol. 2 (2010), pp. 125–141.

Hall, Margaux J. and Weiss, David C., 'Avoiding Adaptation Apartheid: Climate Change Adaptation and Human Rights Law', *Yale Journal of International Law*, vol. 37 (2012), pp. 309–364.

Hampson, Françoise, 'Human Rights Implications of Indigenous Peoples in States and Territories Threatened with Extinction for Environmental Reasons', Commission on

Human Rights, Working Group on Indigenous Peoples, 13 July 2004, document reference: E/CN.4/Sub.2/AC.4/2004/CRP.1., pp. 1–8.

Haslan, Emily, 'Population, Expulsion and Transfer', *Max Planck Encyclopaedia of Public International Law*, (New York: Oxford University Press, 2008-), last entry updated April 2011, internet resource, available at: http://o-opil.ouplaw.com.catalogue.ulrls.lon.ac.uk/view/10.1093/law:epil/9780199231690/law-9780199231690-e861?rskey=em3IKU&result=1&prd=EPIL (5 May 2015).

Herr, Richard and Bergin, Anthony, *Our Near Abroad: Australia and Pacific Islands Regionalism* (Barton: Australian Strategic Policy Institute, November 2011), pp. 1–84.

Hodkinson, David et al., 'Towards a Convention for People Displaced by Climate Change', *New Critic*, vol. 8 (2008), pp. 1–6, available at: http://www.ias.uwa.edu.au/new-critic

Hodkinson, David et al., 'Copenhagen, Climate Change "Refugees" and the Need for a Global Agreement', *Public Policy*, vol. 4 (2009), pp. 155–174.

Hodkinson, David et al., 'The Hour When the Ship Comes In: A Convention for Persons Displaced by Climate Change', *Monash University Law Review*, vol. 36 (2010), pp. 69ff.

Hodkinson, David and Young, Lucy, 'In the Face of a Looming Catastrophe: A Convention for Climate Change Displaced Persons', 2012, available at: www.ccdpconvention.com (5 May 2015).

Hulme, Mike, 'Climate Refugees, Cause for a New Agreement?', *Environment*, vol. 50 (November/December 2008): 6, pp. 50–52.

Kälin, Walter, 'Guiding Principles on Internally Displaced People: Annotations, Studies in Transnational Legal Policy', *American Journal of International Law*, vol. 38 (2008), pp. 1–78.

Keane, David, 'The Environmental Causes and Consequences of Migration: A Search for the Meaning of "Environmental Refugees"', *Georgetown International Environmental Law Review*, vol. 16 (2004), pp. 209–223.

Kolmannskog, Vikram and Myrstad, Finn, 'Environmental Displacement in European Asylum Law', *European Journal of Migration and Law*, vol. 11 (2009): 4, pp. 313–326.

Limon, Marc, 'Human Rights and Climate Change: Constructing a Case for Political Action', *Harvard Environmental Law Review*, vol. 33 (2009), pp. 439–476.

Loughry, Maryanne and McAdam, Jane, 'Kiribati: Relocation and Adaptation', *Forced Migration Review*, vol. 31 (October 2008), pp. 51–52.

Mace, M.J. and Schaeffer, Michiel, *Loss and Damage under the UNFCCC: What Relationship with the Hyogo Framework?* (Bonn: Germanwatch/The Loss and Damage in Vulnerable Countries Initiative, October 2013), pp. 1–16.

Marchi, Sergio, 'Global Governance: Migration's Next Frontier', *Global Governance*, vol. 16 (2010), pp. 323–329.

McAdam, Jane, 'Environmental Displacement Governance', *University of New South Wales Faculty of Law Legal Research Paper Series*, Paper n° 1 (2008), pp. 1–33.

McInerney-Lankford, Siobhan, 'Climate Change and Human Rights: an Introduction to Legal Issues', *Harvard Environmental Law Journal*, vol. 33 (2009), pp. 431–437.

McCue, Gregory S., 'Environmental Refugees: Applying International Environmental Law to Involuntary Migration', *Georgetown International Environmental Law Review*, vol. 6 (1993–1994), pp. 151–190.

McInerney-Lankford, Siobhan, Darrow, Mac and Rajamani, Lavanya, *Human Rights and Climate Change: A Review of the International Legal Dimensions* (Washington: World Bank, May 2011), pp. 1–162.

McNamara, Karen E., 'Conceptualising Discourses on Environmental Refugees at the United Nations', *Population and Environment*, vol. 29 (2007), pp. 12–24.

Marauhn, Thilo, 'The Changing Role of the State', in Bodanski, Dan, Brunée, Jutta and Hey, Ellen, eds., *The Oxford Handbook of International Environmental Law* (Oxford, New York: Oxford University Press, 2007), pp. 727–747.

Martin, Susan, 'Climate Change, Migration and Governance', *Global Governance*, vol. 16 (2010), pp. 397–414.

Montini, Massimiliano, 'Reshaping Climate Governance for Post-2012', *European Journal of Legal Studies*, vol. 4 (2011): 1, pp. 8–24.

Myers, Norman, 'Environmental Refugees', *Population and Environment*, vol. 19 (1997): 2, pp. 167–182.

Page, Edward, 'Theorising the Link between Environmental Change and Conflict', *Review of European Community and International Environmental Law*, vol. 9 (2000): 1, pp. 33–43.

Redgwell, Catherine, 'Life, the Universe and Everything: A Critique of Anthropocentric Rights', in Boyle, Alan and Anderson, Michael R., eds., *Human Rights Approaches to Environmental Protection* (Oxford: Clarendon Paperbacks, 1996), pp. 71–108.

Renaud, Fabrice, Bogardi, Janos, Dun, Olivia and Warner, Koko, 'Control, Adapt or Flee: How to Face Environmental Migration?', *InterSecTions Publication Series of United Nations University Institute for Human Security*, vol. 5 (2007), pp. 1–48.

Schäfer, Laura and Kreft, Sönke, 'Loss and Damage: Roadmap to Relevance for the Warsaw International Mechanism', Briefing Paper (first version), (Bonn: Germanwatch, March 2014), pp. 1–24.

Trebilcock, Michael J., 'Customary Land Reform in Papua New Guinea: Law, Economics and Property Rights in a Traditional Community', *The Adelaide Law Review*, vol. 9 (1983): 1, pp. 191–228.

Weisbrodt, David and Collins, Clay, 'The Human Rights of Stateless Persons', *Human Rights Quarterly*, vol. 28 (2006), pp. 245–276.

Williams, Angela, 'Turning the Tide: Recognising Climate Change Refugees in International Law', *Law and Policy*, vol. 30 (October 2008): 4, pp. 502–529.

Books

Almeida, Vera Martins, *Overview on South–South Migration and Development in the Pacific: Trends and Research Needs* (Geneva: International Organization for Migration/ACP Observatory on Migration, 2011).

Gammeltoft-Hansen, Thomas, *Access to Asylum: International Refugee Law and the Globalization of State Control* (Cambridge: Cambridge University Press, 2012).

Gare, Arran, *Postmodernism and the Environmental Crisis* (London: Routledge, 1995).

Glover, Leigh, *Postmodern Climate Change* (London: Routledge, 2006).

Goodwill-Gill, Guy S. and McAdam, Jane, *The Refugee in International Law* (Oxford: Oxford University Press, 3rd ed., 2007).

Gowlland-Debbas, Vera, ed., *The Problem of Refugees in the Light of Contemporary International Law Issues* (The Hague, Boston, and London: Martinus Nijhoff Publishers, 1995).

Hathaway, James C., *The Rights of Refugees under International Law* (Cambridge: Cambridge University Press, 2005).

Humphreys, Stephen, ed., *Human Rights and Climate Change* (Cambridge: Cambridge University Press, 2010).

Krasner, Stephen D., ed., *International Regimes*, (Ithaca: Cornell University Press, 1983).

Leckie, Scott and Huggins, Chris, *Conflict and Housing, Land and Property Rights: a Handbook on Issues, Frameworks and Solutions* (New York: Cambridge University Press, 2011).

Nakashima, Douglas J. et al., *Weathering Uncertainty: Traditional Knowledge for Climate Change Adaptation and Assessment* (Paris, Darwin: UNESCO/UNU, 2012).

Ponting, Clive, *The New Green History of the World: The Environment and the Collapse of Great Civilizations* (New York: Penguin Books, 2007).

Richard, Guy, *Ailleurs, l'herbe est plus verte: Histoire des migrations dans le monde* (Paris: Arléa-Cordet, 1996).

Sarooshi, Daniel, *The Art and Craft of International Environmental Agreements* (Cambridge, Mass.: Harvard University Press, 2011).

Singh Juss, Satvinder, *International Migration and Global Justice* (Farnham, UK and Burlington, Vt.: Ashgate Publishing Company, 2006).

Soltau, Friedrich, *Fairness in International Climate Change Law and Policy* (Cambridge: Cambridge University Press, 2009).

Yamin, Farhana and Depledge, Joanna, *The International Regime on Climate Change: A Guide to Rules, Procedures and Institutions* (Cambridge: Cambridge University Press, 2005).

Official Documents and Reports

European Union, Council Directive 21/55/EC, of 20 July 2001, *on minimum standards for giving temporary protection in the event of mass influx of displaced persons and on measures promoting a balance of efforts between Member States in receiving such persons and bearing the consequences thereof*, Official Journal of the European Communities, 7 August 2001, L 212/12–23.

European Union, European Parliament, *Report on the proposal for a directive of the European Parliament and of the Council on minimum standards for the qualification and status of third-country nationals or stateless persons as beneficiaries of international protection and the content of the protection granted*, 2002, document reference: COM (2001) 510-C5-0573/2001–2001/0207 (CNS), Committee on Civil Liberties, Justice and Home Affairs.

European Union, European Parliament, *European Declaration on the Status of Environmental Refugees*, 6 April 2004, nº 9/2004.

European Union, European Council, Directive 2004/83/CE, 29 April 2004, *on minimum standards for the qualification and status of third-country nationals or stateless persons as refugees or as persons who otherwise need international protection and the content of the protection granted*, Official Journal of the European Union, 30 September 2004, L 304/12ff.

European Union, European Commission, *Communication on Strengthening the Global Approach to Migration: Increasing Co-ordination, Coherence and Synergies*, 8 October 2008, COM (2008) 611/3. On the recent increase in European concern about the issue of 'climate refugees', see N. Itano, 'Europe Fears "Climate Migrants"' (2008), *Christian Science Monitor*, vol. 100, issue 76, p. 6.

Global Migration Group, 'Statement on the Impacts of Climate Change on Migration, adopted in Paris on 11 November 2011 by the Principals of the GMG', available at: http://www.un.org/esa/population/migration/GMG%20statement_Paris_english.pdf (5 May 2015).

Inter-American Court of Human Rights, 'Petition to the Inter-American Court of Human Rights Seeking Relief from Violations Resulting from Global Warming Caused by Acts or Omissions of the United States', Submitted by Sheila Watt-Cloutier with the support of the Circumpolar Inuit Conference, on Behalf of All Inuit of the Arctic Regions of the United States and Canada, 7 December 2005.

International Federation of the Red Cross and Red Crescent Societies, *Preparedness for Climate Change: a Study to Assess the Future Impact of Climatic Changes upon the Frequency and Severity of Disasters and the Implications for Humanitarian Responses and Preparedness*, 2–6 December 2003 (Report prepared for the 28th International Conference of the Red Cross and Red Crescent).

Law Association for Asia and the Pacific, Human Rights Committee, 'Draft Pacific Charter on Human Rights and Explanatory Memoranda' (Kensington, NSW, 1989).

Oxfam International, *The Right to Survive: the Humanitarian Challenge for the 21st Century* (Oxford: Oxfam Publishing, 2009).

United Nations, Commission on Economic, Social and Cultural Rights, General Comment nº 4, 'The Right to Adequate Housing (Article 11(1) of the Covenant)', 1991, E/1992/23.

United Nations, Commission on Economic, Social and Cultural Rights, General Comment nº 12, 'The Right to Adequate Food', 1999, E/C.12/1999/5.

United Nations, Commission on Economic Social and Cultural Rights, General Comment nº 14, 'The Right to the Highest Attainable Standard of Health', 2000, E/C.12/2000/4.

United Nations, General Assembly, Resolution A/RES/48/134, 'National Institutions for the Promotion and Protection of Human Rights' (with Paris Principles relating to the Status of National Institutions for the Promotion of Human Rights Annexed), 48th session, 4 March 1994.

United Nations, Un-Habitat, D. Fitzpatrick, 'Scoping Report: Addressing Land Issues after Natural Disasters' (UN-HABITAT; United Nations Settlement Program, 2007).

United Nations, High Commissioner on Human Rights, Regional Office of the Pacific, *Protection of Human Rights of Internally Displaced People: Challenges of the Pacific*, Discussion Paper (Fiji: Davui Printery, 2011).

United Nations, High Commissioner on Refugees, 'Climate Change, the Environment, Natural Disasters and Human Displacement: a UNHCR Perspective', (Policy and Evaluation Service), 29 January 2008.

United Nations, High Commissioner for Refugees, 'Forced Displacement in the Context of Climate Change: Challenges for States under International Law', Paper submitted by the Office of the UNHCR in co-operation with the Norwegian Refugee Council, the Representative of the Secretary-General on the Human Rights of Internally Displaced Persons, and the United Nations University, 6th session of the Ad-hoc Working Group on Long-Term Co-operative Action under the UNFCCC Convention (AWG-LCA 6), held in Bonn (Germany), 1–12 June 2009.

United Nations, Human Rights Committee, General Comment nº 31, 'Nature of the General Legal Obligation of State Parties to the Covenant', 2004, U.N. Doc. CCPR/C/21/Rev.1/Add.13.

United Nations, Human Rights Commission, Resolution 2005/74, 'National Institutions for the Promotion and Protection of Human Rights', 59th meeting, 20 April 2005.

United Nations, Inter-Agency Standing Committee, *Operational Guidelines for the Protection of Human Rights in Natural Disasters* (Bern: Brookings-Bern Project on Internal Displacement, 2006).

United Nations, Inter-Agency Standing Committee, *Protecting Persons Affected by Natural Disasters: Operational Guidelines on Human Rights and Natural Disasters*, Brookings-Bern Project (Washington, D.C.: IASC, June 2006).

United Nations, Inter-Agency Standing Committee, *Human Rights and Natural Disasters, Operational Guidelines and Field Manual on Human Rights Protection in Situations of Natural Disasters* (pilot version, March 2008).

United Nations, Inter-Agency Standing Committee, *Operational Guidelines and Field Manual on Human Rights Protection in Situations of Natural Disaster*, Brookings-Bern Project (Washington, D.C.: IASC, Pilot version March 2008).

United Nations, Office of the High Commissioner on Human Rights (Regional Office of the Pacific), *Ratification of International Human Rights Treaties: Added Value for the Pacific*, Discussion Paper (Suva, Fiji: OHCHRs Regional Office of the Pacific, July 2009).

United Nations, Office for the Co-ordination of Humanitarian Affairs, *Guiding Principles of Internal Displacement*, 2004, full text and further information available at: http://www.idpguidingprinciples.org/. (5 May 2015).

United Nations, Office of the High Commissioner for Human Rights (OHCHR), *Assessing the Effectiveness of National Human Rights Institutions* (Versoix: International Council for Human Rights Policy and OHCHR, 2005).

United Nations Environment Programme, *Environmental Refugees* (Nairobi: United Nations Environment Programme, 1988).

Chapter 3: The Challenge of Government Failure

Articles and Book Chapters

Anguie, Anthony, 'The "Heart of my Home": Colonialism, Environmental Damage and the Nauru Case', *Harvard International Law Journal*, vol. 34 (1993): 2, pp. 445–473.

Ayers, Alison J., 'The Illusion of an Epoch: Critiquing the Ideology of Failed States', *International Politics*, vol. 49 (2012): 5, pp. 568–590.

Beck, Ulrich, 'Vivir en la Sociedad del Riesgo Mundial', *Documentos CIBOB-Dinámicas Interculturales*, vol. 8 (2007), pp. 5–32.

Bedjaoui, Mohammed, 'Responsibility of States: Fault and Strict Liability', *Encyclopedia of Public International Law*, vol. 4 (1987), pp. 358–362.

Bilgin, Pinar and Morton, Adam D., 'Historicising Representations of "Failed States": beyond Cold War Annexation of the Social Sciences?', *Third World Quarterly*, vol. 23 (2002): 1, pp. 23–50.

Brock, Lothar, 'Peace through Parks: The Environment on the Peace Research Agenda', *Journal of Peace Research*, vol. 28 (November 1991): 4, pp. 407–423.

Brownlie, Ian, 'The Rights of Peoples in Modern International Law', in Crawford, James, ed., *The Rights of Peoples* (Oxford: Clarendon Press, 1988), pp. 1–16.

Cashman, Peter and Abbs, Ross, 'Liability in Tort for Damage Arising from Human-Induced Climate Change', in Lyster, Rosemary, ed., *In the Wilds of Climate Law* (Brisbane: Australian Academic Press, 2010), pp. 235–271.

Chandler, David, 'Great Power Responsibility and Failed States: Strengthening Sovereingty?', in Raue, Julia and Sutter, Patrick, eds., *Facets and Practices of State-Building* (Leiden: Martinus Nijhoff, 2009), pp. 15–30.

Christakis, Théodore, 'The State as a "Primary Fact": Some Thought on the Principle of Effectiveness', in Kohen, Marcelo G., ed., *Secession: International Law Perspectives* (Cambridge: Cambridge University Press, 2006), pp. 138–170.

Clapham, Christopher, 'Degrees of Statehood', *Review of International Studies*, vol. 24 (1998), pp. 143–157.

Classen, Claus Dieter, ' "Failed States" and the Prohibition to Use Force,' in *Les nouvelles menaces contre la paix et la sécurité internationales: Journée franco-allemande*, Société Française pour le Droit international (Paris: Pedone, 2004), pp. 129–140.

Combacau, Jean and Allan, Dennis, 'Primary and Secondary Rules in the Law of State Responsibility: Categorising Legal Obligations', *Netherlands Yearbook of International Law*, vol. 16 (1985), pp. 81–109.

Crawford, James, 'The Rights of Peoples: "Peoples" or "Governments"?', in Crawford, James, ed., *The Rights of Peoples* (Oxford: Clarendon Press, 1988), pp. 55–67.

Crisp, R., 'Ethics and International Environmental Law', in Besson, Samantha and Tasioulas, John, *The Philosophy of International Law* (Oxford: Oxford University Press, 2010), pp. 473–490.

Dalby, Simon, 'Geopolitics and Ecology: Rethinking the Contexts of Environmental Security', in Lowi, Miriam R., ed., *Environment and Security: Discourses and Practice* (New York: Palgrave McMillan, 2000), pp. 84–100.

Davidson, Marc D., 'Wrongful Harm to Future Generations: The Case of Climate Change', *Environmental Values*, vol. 17 (2008), pp. 471–488.

Delcourt, Barbara, 'L'introduction de la notion de responsabilité de protéger dans les autorisations données par le Conseil de sécurité : Enjeux politiques et paradoxes', in Mollard-Bannelier, Karine and Pison, Cyrille, eds., *Le recours à la force autorisé par le Conseil de sécurité : Droit et responsabilité* (Paris: Pedone, 2014), pp. 53–66.

Deudney, Daniel, 'The Case against Linking Environmental Degradation and National Security', *Millennium*, vol. 19 (winter 1991): 3, pp. 461–476.

Di John, J., 'Conceptualising the Causes and Consequences of Failed States: A Critical Review of the Literature', *Crisis States Working Paper Series*, vol. 2 (January 2008): paper nº 25, pp. 1–54.

Dorff, Robert H., 'State Failure and Responding to It', paper presented at the 43rd *Annual Convention of the International Studies Association*, held in New Orleans (USA), 24–27 March 2002, pp. 1–15.

Dugard, John and Raic, David, 'The Role of Recognition in the Law and Practice of Secession', in Kohen, Marcelo G., ed., *Secession: International Law Perspectives* (Cambridge: Cambridge University Press, 2006), pp. 94–137.

Dupuy, Pierre-Marie, 'International Liability for Transfrontier Pollution', in Bothe, M., *Trends in International Environmental Policy and Law* (Gland: International Union for the Conservation of Natural Resources, 1980), pp. 369–379.

Espósito, Carlos, 'Soberanía e igualdad en el Derecho internacional', *Estudios internacionales* (*Universidad de Chile*), vol. 165 (2010), pp. 171–196.

Franck, Thomas, 'Epistemology at a Time of Perplexity', *European Journal of International Law*, vol. 13 (2002): 4, pp. 1025–1030.

Fyodorov, Vladimir, 'The UN and a Comprehensive International Security System', *USSR Report: International Affairs* (September 1987), pp. 90–95.

Geiss, Robin, 'Failed States: Legal Aspects and Security Implications', *German Yearbook of International Law*, vol. 47 (2005), pp. 457–501.

Gleiditsch, Niels P., 'Armed Conflict and the Environment: A Critique on the Literature', *Journal of Peace Research*, vol. 35 (May 1998): 3, pp. 381–400.

Grubb, Michael, 'Seeking Fair Weather: Ethics and the International Debate on Climate Change', *International Affairs*, vol. 71 (1995): 3, pp. 463–496.

Grote Stoutenburg, Jenny, 'When Do States Disappear? Thresholds of Effective Statehood and the Continued Recognition of "Deterritorialised States"', in Gerrard, Michael and Wannier, Gregory, eds., *Threatened Island Nations: Legal Implications of Rising Seas and a Changing Climate* (Cambridge: Cambridge University Press, 2013), pp. 57–87.

Günter Brauch, Hans, 'Climate Change, Environmental Stress and Conflict', AFES-PRESS Report for the German Federal Ministry for the Environment, Nature Conservation and Nuclear Safety, reproduced in 'Climate Change and Conflict: Can Climate Change Impacts Increase Conflict Potentials? What is the Relevance of This Issue for the International Process on Climate Change?' *Environmental Policy*, Part II (November 2002), pp. 9–113.

Helman, Gerard and Ratner, Stephen, 'Saving Failed States', *Foreign Policy*, vol. 89 (1993), pp. 3–20.

Homer-Dixon, Thomas, 'Environmental Scarcities and Violent Conflict: Evidence from Cases', *International Security*, vol. 19 (winter 1994): 1, pp. 5–40.

Hurrell, Andrew, 'The State', in Dobson, Andrew and Eckersley, Robyn, eds., *Political Theory and the Ecological Challenge* (Cambridge, New York: Cambridge University Press, 2006), pp. 165–181.

Jackson, Robert H., 'Surrogate Sovereignty? Great Power's Responsibility and "Failed States"', *Institute of International Relations of the University of British Colombia*, Working Paper n° 25 (1998), pp. 1–17.

Jackson, Robert H., 'Quasi-States, Dual Regimes and Neoclassical Theory: International Jurisprudence and the Third World', *International Organization*, vol. 41 (1987): 4, pp. 519–549.

Kolbasov, Oleg, 'Modern Ecological Policy and the Utilization of a Global Environmental Protection Strategy', *Pace Environmental Law Review*, vol. 5 (spring 1988), pp. 2445–2454.

Kolbasov, Oleg, 'The Concept of Environmental Security: Legal Aspects', *Soviet State and Law*, vol. 12 (1988) [page numbers not available].

Korman, A. and Barcia, G., Rethinking Climate Change: Towards an International Court of Justice Advisory Opinion, *Yale Journal of International Law Online*, vol. 37 (2012), pp. 35–42.

Koskenniemi Martti, 'The Future of Statehood', *Harvard Journal of International Law*, vol. 32 (1991), pp. 397- 410.

Krasner, Stephen, 'Compromising Westphalia', *International Security*, vol. 20 (1995): 3, pp. 115–151.

Libiszewski, Stephan, 'What is an Environmental Conflict?', Occasional Paper nº 1, *Center for Security Studies and Conflict Research* (Bern: Swiss Peace Foundation, and Zurich, July 1992).

Malla, Katak, 'Climate Change Loss and Damage Compensation', *Indian Journal of International Law*, vol. 53 (2013), pp. 183–205.

Mickelson, Karin, 'Critical Approaches', in Dobansky, D., Brunée, J. and Hey, E., *The Oxford Handbook of International Environmental Law* (Oxford: Oxford University Press, 2007), pp. 262–290.

Millar, Ilona et al., 'Making Good the Loss: An Assessment of the Loss and Damage Mechanism under the UNFCCC Process', in Gerrard, Michael and Wannier, Gregory, eds., *Threatened Island Nations: Legal Implications of Rising Seas and a Changing Climate* (Cambridge: Cambridge University Press, 2013), pp. 433–472.

Minnerop, Petra, 'The Classification of States and the Creation of Status within the International Community', *Max Planck Yearbook of United Nations Law*, vol. 7 (2003), pp. 79–182.

Müllerson, Rein, 'Sources of International Law: New Tendencies in Soviet Thinking', *American Journal of International Law*, vol. 83 (1989), pp. 494–512.

Myers, Michael, 'Environment and Security', *Foreign Policy*, vol. 74 (spring 1989), pp. 23–41.

Najam, Adil, 'Why Environmental Politics Looks Different from the South', in Dauvergne, Peter, ed., *Handbook of Global Environmental Politics*, 1st ed. (Northampton: Edward Elgar Pub., 2005), pp. 111–126.

Nickel, J. and Magraw, D., 'Philosophical Issues in International Environmental Law', in Besson, Samantha and Tasioulas, John, *The Philosophy of International Law* (Oxford: Oxford University Press, 2010), pp. 453–471.

Nye, Joseph, 'Collective Economic Security', *International Affairs*, vol. 50 (1974): 4, pp. 584–598.

Prott, Lyndel V., 'Cultural Rights as Peoples' Rights in International Law', in Crawford, James, ed., *The Rights of Peoples* (Oxford: Clarendon Press, 1988), pp. 93–106.

Reisman, W. Michael and Arsanjani, Manhoush H., 'The Quest for an International Liability Regime for the Protection of the Global Commons', in Wellens, Karel C., ed., *International Law: Theory and Practice; Essays in Honour of Eric Suy* (The Hague: Martinus Nijhoff Publishers, 1998), pp. 469–492.

Robinson, Nicholas A., 'Perestroika and Priroda: Environmental Protection in the USSR', *Pace Environmental Law Review*, vol. 5 (spring 1988): 2, pp. 351–423.

Rotberg, Robert I., 'The New Nature of Nation-State Failure', *Washington Quarterly*, vol. 25 (2002): 3, pp. 85–96.

Rotberg, Robert I., 'Failed States, Collapsed States, Weak States: Causes and Indicators', in Rotberg, Robert I., ed. *State Failure and State Weakness in a Time of Terror* (Washington, DC: Brookings Institution, 2003).

Salmon, Jean-Michel, 'Quelle place pour l'État dans le Droit international d'aujourd'hui?', *Recueil des Cours de l'Académie de Droit international de La Haye*, vol. 347 (2010), pp. 9–78.

Schreuer, Christoph, 'The Waning of the Sovereign State: Toward a New Paradigm of International Law', *European Journal of International Law*, vol. 4 (1993): 1, pp. 447–471.

Shue, Henry, 'Global Environment and International Inequality', *International Affairs*, vol. 75 (1999): 3, pp. 531–545.

Spanger, Hans-Joachim, 'Failed State or Failed Concept? Objections and Suggestions', *Peace Research Institute of Frankfurt*, Paper presented at the Conference on Failed States III: Globalization and Failed States, held in Florence (Italy) 7–10 April 2000.

Stallworthy, Mark, 'Environmental Justice Imperatives for an Era of Climate Change', *Journal of Law and Society*, vol. 36 (March 2009): 1, pp. 55–74.

Thomson, Janice E., 'State Sovereignty and International Relations: Bridging the Gap between Theory and Empirical Research', *International Studies Quarterly*, vol. 39 (1995), pp. 213–233.

Timonshenko, Alexandre S., 'The Global Environmental Security: International Legal Aspects', *Soviet State and Law*, vol. 1 (1989), [page numbers not available].

Timoshenko, Alexandre S., 'Ecological Security: The International Aspect', *Pace Environmental Law Review*, vol. 7 (autumn 1989): 1, pp. 151–160.

Tomuschat, Christian, 'International Law: Ensuring the Survival of Mankind on the Eve of the New Century', General Course on Public International Law, *Recueil des Cours de l'Académie de Droit international de La Haye*, vol. 281 (1999), pp. 9–438.

Tol, Richard S.J. and Verheyen, Roda, 'State Responsibility and Compensation for Climate Change Damages: A Legal and Economic Assessment', *Energy Policy*, vol. 32 (2004), pp. 1109–1130.

Tuchman Mathews, Jessica, 'Redefining security', *Foreign Affairs*, vol. 68 (spring 1989): 2, pp. 162–177.

Vladimirsky, Piotr, 'Comprehensive Security Equal for All', *USSR Report: International Affairs* (October 1989), pp. 9–19.

Voigt, Christina, 'State Responsibility for Climate Change Damages', *Nordic Journal of International Law*, vol. 77 (2008): 1–2, pp. 1–22.

Waldron, Jeremy, 'Two Conceptions of Self-Determination', in Besson, Samantha and Tasioulas, John, *The Philosophy of International Law* (Oxford: Oxford University Press, 2010), pp. 397–413.

Wallerstein, Michael B., 'Whither the Role of Private Foundations in Support of International Security Policy?' *The Nonproliferation Review*, vol. 9 (spring 2002): 1, pp. 83–92.

Weisbrod, David and Collins, Clay, 'The Human Rights of Stateless Persons', Human Rights Quarterly, vol. 28 (2006), pp. 245–276.

Werksman, Jacob David, 'Could a Small Island Successfully Sue a Big Emitter?', in in Gerrard, Michael and Wannier, Gregory, eds., *Threatened Island Nations: Legal Implications of Rising Seas and a Changing Climate* (Cambridge: Cambridge University Press, 2013), pp. 409–431.

Wilde, Ralph, 'The Skewed Responsibility Narrative of the "Failed States" Concept', *ILSA Journal of International and Comparative Law*, vol. 9 (2003): 2, pp. 425–429.

Williams, Paul D., 'State Failure in Africa: Causes, Consequences and Responses', in *Africa South of the Sahara* (Oxford: Taylor & Francis: Europa World Yearbook, 39th ed. 2010), pp. 21–28.

Zemanek, Karl, 'Causes and Forms of International Liability', in Cheng, Bin and Duncan Brown, Eduard, eds., *Contemporary Problems in International Law: Essays in Honour of Georg Schwarzenberger on His Eightieth Birthday* (London: Stevens and Sons, 1988).

Zhang, Yongjin, 'A Regional Power by Default', Wesley-Smith, Terence and Porter, Edgar A., eds., *China in Oceania: Reshaping the Pacific?* (Oxford and New York: Bergham Books, 2010), pp. 49–66.

Books

Akpinarli, Neyire, *The Fragility of the 'Failed State' Paradigm: a Different International Law Perception of the Absence of Effective Government* (The Hague and Leiden: Martinus Nijhoff Publishers, 2010).

Besson, Samantha and Tasioulas, John, *The Philosophy of International Law* (Oxford: Oxford University Press, 2010).

Brown Weiss, Edith, *In Fairness to Future Generations: International Law, Common Patrimony and Intergenerational Equity* (Tokyo: United Nations University, 1989).

Chen, Ti-Chiang, *The International Law of Recognition – with Special Reference to Practice in Great Britain and the United States* (London: Stevens, 1951).

Crawford, James, ed., *The Rights of Peoples* (Oxford: Clarendon Press, 1988).

Dryzec, John, *The Politics of the Earth: Environmental Discourses* (New York: Oxford University Press, 1999).

Durgard, John, *Recognition and the United Nations* (Cambridge: Grotius, 1987).

Feshback, Murray and Friednly Jr., Alfred, *Ecocide in the USSR: Health and Nature under Siege* (New York: Basic Books, 1992).

Franck, Thomas, *The Power of Legitimacy among Nations* (New York and Oxford: Oxford University Press, 1990).

Goldman, Marshall I., *The Spoils of Progress: Environmental Pollution in the Soviet Union* (Cambridge, Mass.: MIT Press, 1972).

Gorbachev, Mikhail, *Perestroika: A New Thinking for Our Country and the World*, (New York: Harper and Row, 1987).

Higgins, Rosalyn, *Problems and Process: International Law and How We Use It* (Oxford: Clarendon Press, 1994).

Hoffman, Stanley, *Chaos and Violence: What Globalization, Failed States and Terrorism Mean for US Foreign Policy* (Lanham, Md.: Rowman and Littlefield, 2006).

Jackson, Robert H., *Quasi-States: Sovereignty, International Relations and the Third World* (Cambridge: Cambridge University Press, 1990).

Khan, Ali, *The Extinction of Nation-States: A World without Borders* (Michigan: Kluwer Law International, 1996).

Kooijmans, Pieter H., *The Doctrine of Legal Equality of States: An Enquiry into the Foundations of International Law* (Leiden: Sythoff, 1964).

Koskenniemi, Martti, *The Gentle Civiliser: the Rise and Fall of International Law: 1870–1960* (Cambridge: Cambridge University Press, 2006).

Lauterpacht, Hersch, *Recognition in International Law* (Cambridge: Cambridge University Press, 1947).

Oppenheim, Lassa, *The Future of International Law* (Oxford: Clarendon Press, 1921).

Peterson, Martha J., *Recognition of Governments: Legal Doctrine and State Practice, 1815–1995* (New York: Macmillan, 1997).

Scanlon, Helen L., *European Governments in Exile* (Washington: Carnegie Endowment for International Peace, 1943).

Scelle, Georges, *Manuel de Droit international public* (Paris: Domat Montchrestien, 1948).

Schmitt, Carl, *Political Theology: Four Chapters on the Concept of Sovereignty* (Chicago: University of Chicago Press, 1985).

Shain, Yossi, *Governments-in-Exile in Contemporary World Politics* (London: Routledge, 1991).

Simpson, Gerry, *Great Powers and Outlaw States: Unequal Sovereigns in the International Legal Order* (Cambridge: Cambridge University Press, 2004).

Talmon, Stephan, *Recognition of Governments in International Law – with Particular Reference to Governments in Exile* (Oxford: Clarendon Press, 1998).

Verheyen, Roda, *Climate Change Damage and International Law: Prevention Duties and State Responsibility* (The Hague and Leiden: Martinus Nijhoff, 2005).

Zartman, William, *Collapsed States* (Boulder, Colo.: Lynne Rienner, 1995).

Official Documents and Reports

Adelphi Research, Carius, Alexander, Maas, Achim and Barkemeyer, Janina, *Climate Change and Security: Two Scenarios for the Indian–Pacific Ocean Island States* (Berlin: Adelphi Research, 2009).

Adelphi Research, Carius, Alexander and Maas, Achim, *Climate Change and Security: Three Scenarios for Middle America* (Berlin: Adelphi Research, 2009).

Adelphi Research, Carius, Alexander, Maas, Achim and Fritzsche, Kerstin, *Climate Change and Security: Three Scenarios for South West Asia* (Berlin: Adelphi Research, 2009).

Adelphi Research, Carius, Alexander, Maas, Achim and Fritzsche, Kerstin, *Climate Change and Security: Three Scenarios for South West Asia* (Berlin: Adelphi Research, 2009).

Adelphi Research, Carius, Alexander, Maas, Achim and Fritzsche, Kerstin, *Climate Change and Security: Scenarios for South East Asia* (Berlin: Adelphi Research, 2009).

European Union, Council of the European Union, 'Presidency Conclusions by M. Hans Pöttering', 21–22 June 2007, document reference: 11177/1/07–Rev.1–Conc. 2.

European Union, Council of the European Union, 'Climate Change and International Security: Paper from the High European Union, Representative and the European Commission to the European Council', 14 March 2008, document reference: S/113/08.

European Union, Council of the European Union, 'Council Conclusions on Climate Change and Security', 2985th Foreign Affairs Council Meeting, 8 December 2009.

European Union, High Representative of the EU Javier Solana, 'Joint Process Report and Follow-up Recommendations on Climate Change and International Security', 18 December 2008, document reference: CL08–241EN.

German Advisory Council on Climate Change (WBGU), *World in Transition: Climate Change as a Security Risk*', *Summary for Policy-Makers* (Berlin: WBGU, 29 May 2007), pp. 1–13.

International Institute for Sustainable Development, Brown, Oli and Crawford, Alec, *Assessing the Security Implications of Climate Change for West Africa: Country Case Studies of Ghana and Burkina Faso* (Winnipeg: IISD, 2008).

International Institute for Sustainable Development, Brown, Oli and Crawford, Alec, *Rising Temperatures, Rising Tensions: Climate Change and the Risk of Violent Conflict in the Middle East* (Winnipeg: IISD, 2009).

International Institute for Sustainable Development, Brown, Oli and Crawford, Alec, *Climate Change and Security in Africa: a Study for the Nordic–African Foreign Ministers Meeting* (Winnipeg: IISD, March 2009).

International Law Commission, 'First Report on Prevention of Transboundary Damage from Hazardous Activities', Special Rapporteur: Pemmaraju Sreenivasa Rao, 18 March 1998, document reference: A/CN.4/487 and Add.1.

International Law Commission, 'Third Report on International Liability for Injurious Consequences Arising out of Acts not Prohibited by International Law (Prevention of Transboundary Damage for Hazardous Activities)', Special Rapporteur: Pemmaraju Sreenivasa Rao, 9 June 2000, document reference: A/CN.4/510.

Royal United Services Institute, Fetzek, Shiloh, *Climate-Related Impacts on National Security in Mexico and Central America* (London: Royal United Services Institute, 2009).

United Nations, General Assembly, Resolution 41/92, 'Establishment of a Comprehensive System of International Peace and Security', adopted on 4 December 1986, 96th plenary meeting, UN doc.: A/RES/41/92.

United Nations, Statement by Mikhail Gorbachev, 'Reality and Safeguards for a Secure World', 18 September 1987, and attachment of the Letter of the Deputy Head of the Delegation of the Union of Soviet Socialist Republics to the 42nd session addressed to the Secretary-General, UN doc.: A/42/574.

United Nations, General Assembly, Verbatim Record of the 21st meeting of the Second Committee of the General Assembly, 23 October 1987, UN doc.: A/C.2/42/SR.21.

United Nations, General Assembly, Draft Resolution presented to the Second Committee by Czechoslovakia and Ukrainian Soviet Socialist Republic, 'International Ecological Security', 30 October 1987, UN doc.: A/C.2/42/L.34.

United Nations, General Assembly, Resolution 42/184, 'International Co-operation in the Field of Environment', adopted during the 96th plenary meeting, on 11 December 1987, UN doc.: A/RES/42/184.

United Nations, General Assembly, Verbatim Record of the 47th to 51st meetings of the First Committee of the General Assembly, 23–30 November 1988, UN doc.: A/C.1/43/PV.47–PV.54.

United Nations, General Assembly, Speech of Mikhail Gorbachev reproduced in Verbatim Record, 8 December 1988, UN doc.: A/43/PV.72.

United Nations, General Assembly, Speech of Eduard Shevardnadze, reproduced in Verbatim Record, 1989, UN doc.: A/44/PV.6, pp. 26–50.

United Nations, General Assembly, Resolution 56/82, 'Report of the International Law Commission on the Work of its Fifty-third Session', 12 December 2001, UN doc.: A/RES/56/82.

United Nations, General Assembly, Resolution 61/36, 'Allocation of Loss in the Case of Transboundary Harm Arising out of Hazardous Activities', 4 December 2006, UN doc.: A/RES/61/36.

United Nations, General Assembly, Resolution 62/68, 'Consideration of Prevention of Transboundary Harm from Hazardous Activities and Allocation of Loss in the Case of Such Harm', 6 December 2007, UN doc.: A/RES/62/68.

United Nations, General Assembly, Draft Resolution on Security and Climate Change, 27 October 2008, document reference: A/63/L.8.

United Nations, General Assembly Amendments to Draft Resolution A/63/L.8, 17 November 2008, UN doc.: A/63/L.30.

United Nations, General Assembly, Resolution 63/281 on Climate Change and Its Possible Security Implications, 3 June 2009, 85th plenary meeting, UN doc.: A/RES/63/281.

United Nations, General Assembly, Verbatim Record, document reference: A/63/PV.85, 3 June 2009, 10 a.m., 85th plenary meeting.

United Nations, General Assembly, 'Climate Change and its Possible Security Implications: Report of the Secretary-General', 11 September 2009, UN doc.: A/64/150.

United Nations, General Assembly, Resolution 65/28, 'Consideration of Prevention of Transboundary Harm from Hazardous Activities and Allocation of Loss in the Case of Such Harm, 6 December 2010, UN doc.: A/RES/65/28.

United Nations, General Assembly, Statement of Johnson Toribiong, President of the Republic of Palau, to the 66th Regular Session of the United Nations General Assembly (22 September 2011), available at the official website of Palau's Permanent Mission to the United Nations, at: http://gadebate.un.org/66/palau (5 May 2015).

United Nations, Security Council, Letter dated 12 April 2007 from the Chargé d'affaires of the Permanent Mission of Cuba to the United Nations addressed to the President of the Security Council, UN doc.: S/2007/203.

United Nations, Security Council, Letter dated 16 April 2007 from the Permanent Representative of Pakistan to the United Nations addressed to the President of the Security Council, UN doc.: S/2007/211.

United Nations, Security Council, Letter of the German Presidency of the Security Council to the UN Secretary-General, 'Maintenance of International Peace and Security: Impacts of Climate Change', Concept Note, 1 July 2011, UN doc.: S/2011/408.

United Nations, Security Council, Verbatim Record S/PV.6587, of the 6587th meeting of the Security Council, 20 July 2011, 10 a.m.

United Nations, Security Council, Presidential Statement on Maintenance of International Peace and Security and Impacts of Climate Change, 20 July 2011, UN doc.: S/PRST/2011/15.

United Nations, Submission of Vanuatu on behalf of the Association of Small Island States, 'Elements for a Framework Convention on Climate Change', 18 June 1991, UN doc.: UN Doc. A/AC.237/Misc.1/Add.3, 22–32.

Index

Printed in the United States
By Bookmasters